AF361020

Bioceramic Composites

Bioceramic Composites

Editors

Corrado Piconi
Simone Sprio

MDPI • Basel • Beijing • Wuhan • Barcelona • Belgrade • Manchester • Tokyo • Cluj • Tianjin

Editors
Corrado Piconi
Italian National Research
Council, Institute for Science and
Technology of Ceramics
(CNR-ISTEC)
Italy

Simone Sprio
Italian National Research
Council, Institute for Science and
Technology of Ceramics
(CNR-ISTEC)
Italy

Editorial Office
MDPI
St. Alban-Anlage 66
4052 Basel, Switzerland

This is a reprint of articles from the Special Issue published online in the open access journal *Journal of Composites Science* (ISSN 2504-477X) (available at: http://www.mdpi.com).

For citation purposes, cite each article independently as indicated on the article page online and as indicated below:

LastName, A.A.; LastName, B.B.; LastName, C.C. Article Title. *Journal Name* **Year**, *Volume Number*, Page Range.

ISBN 978-3-0365-3633-0 (Hbk)
ISBN 978-3-0365-3634-7 (PDF)

Cover image courtesy of ISTEC-CNR
micrograph of foamed calcium phosphate-titania scaffolds: detail of pore edge.

Contents

About the Editors

Corrado Piconi, Senior Associate at the Institute of Science and Technology for Ceramics of the Italian National Research Council (ISTEC-CNR). He received his M.Sc. in biomaterials and has more than 40 years of experience with ceramic materials, especially those used in oxides for joint replacement. Since 1989, when he was involved in the New Materials Division of the Italian Authority for New Technologies, Energy and the Environment (ENEA), his main research interest has been devoted to zirconia ceramics with an emphasis on precursors that are suitable for the production of implantable medical devices. Previously, he acted as the R&D Project Manager in Tecnobiomedica S.p.A (Pomezia, Italy) and worked on the development of medical devices for orthopedic and cardiovascular applications, and he was then the Scientific Director of the MeLab R&D facility at GHIMAS S.p.A. (Brindisi, Italy), developing scaffolds for bone tissue engineering and dental implants using advanced ceramic composites and via the 3D laser sintering of titanium powders. From 1997 to 2017, he was a biomaterials lectrurer, Orthopedics Dept., School of Medicine, Catholic University "Sacro Cuore", Roma, Italy, and from 2006 to 2012, he was a Scientific Consultant, Dental Materials Dept., School of Dentistry, University "Tor Vergata", Roma, Italy. He is the author of more than 130 articles and book chapters have been read about 29.000 times and that have received 3,550 citations (ResearchGate, March 2022). He has been the Scientific Cordinator and the WP Leader in several national and EC-funded projects. Currently, he is a member of the Scientific Advisory Board of the European Society for Ceramic Implantology (ESCI) and an active member of the Italian Society for Biomaterials (SIB), of the European Society for Biomaterials (ESB), of the International Society for Ceramics in Medicine (ISCM), and of the Italian Society for Research in Orthopedics (IORS).

Simone Sprio, Senior Researcher M.Sc. in Physics, Ph.D. in Chemistry, and 25 years of experience in materials science, particularly in the development of ceramic materials and devices for biomedical applications. Chief Scientist for the research on Bioceramics for Regenerative Medicine at ISTEC-CNR. His main fields of investigation are biomimetic ceramic phases and 3D scaffolds with cell-instructive ability; intrinsic antibacterial biomaterials; bioactive injectable pastes and bone cements; biomorphic transformation processes by 3-D heterogeneous chemistry; and bone scaffolds as drug delivery systems. He is the author of more than 120 peer-reviewed papers (H-index: 32) and book chapters and has edited a book and various Special Issues in international journals. He is the inventor of nine funded international patents that are related to new processes, materials, and devices for regenerative medicine. He has delivered more than 100 communications at international conferences, including 25 invited lectures. He has been the Chief Scientist or WP Leader on several national and EC-funded projects. He has organized several international congresses, symposia, and workshops.He has supervised various M.Sc. and Ph.D students and young fellow researchers.

Journal of
Composites Science

Editorial

Editorial for the Special Issue on Bioceramic Composites

Corrado Piconi * and Simone Sprio

National Research Council of Italy, Institute of Science and Technology for Ceramics (CNR-ISTEC),
48018 Faenza, Italy; simone.sprio@istec.cnr.it
* Correspondence: corpico@libero.it

Citation: Piconi, C.; Sprio, S. Editorial for the Special Issue on Bioceramic Composites. *J. Compos. Sci.* **2022**, *6*, 65. https://doi.org/10.3390/jcs6030065

Received: 16 February 2022
Accepted: 18 February 2022
Published: 22 February 2022

Publisher's Note: MDPI stays neutral with regard to jurisdictional claims in published maps and institutional affiliations.

This Special Issue on bioceramic composites and its published papers, addressing a number of current topics from industry and academia, are intended to be a reference for students and scholars in the field of biomaterials science, giving an insight into challenges and research topics in the field bioceramic composites. Moreover, its aim is to stimulate the interest of young and experienced researchers introducing novel research topics as inspiration for future work.

At present, bioceramic composites have two wide areas of application in the biomedical field. The first is in load-bearing devices, such as the joints of hip, knee, and shoulder joint replacements, as well as in dental implants. In this field, zirconia–alumina composites have attracted a great deal of attention because of their superior mechanical behavior, and new compositions are under development. The other field where bioceramic composites are intensively investigated is bone regeneration. Particular emphasis is given to calcium phosphates and silicates, as well as to doping with bioactive ions, aiming to enhance osteogenic ability and bioresorbability. On the other hand, hybrid biopolymer/ceramic materials mimicking the complex composition and multiscale structure of bone tissue are a new class of biomimetic materials that are very promising in regenerative medicine.

The paper by Piconi and Sprio provides the reader with an overview of the state of the art of bioceramic composites in orthopedics [1]. Their development is placed in a historical perspective, and the characteristics of the different materials now on the market are outlined. These themes are further developed by Burger and Kiefer [2], who offer an exhaustive presentation on the issues that led to the production of BIOLOXdelta®, the ceramic composite that is today's golden standard in the bearings for total hip replacements. Furthermore, they demonstrate improved mechanical properties using ceramic composites made of zirconia stabilized by an yttria and ceria matrix with dispersed alumina and hexagonal platelets, and highlight their potential for use in dental applications. Moreover, they underline the need for special raw materials to achieve the expected behavior, and the relevance of the raw material processing and of the feedstock for injection molding, today representing the standard for production of dental implants.

The relevance of the physical–chemical characteristics of the stabilizing and toughening additives to zirconia, as well as of their concentrations, was investigated by Magnani et al., with special attention to their influence on the mechanics of toughening and, hence, on the mechanical properties [3]. The study of the in situ toughening mechanism induced by the tetragonal–monoclinic (t–m) transformation of zirconia allowed modeling the opposite effect played by the grain size and the tetragonality of the zirconia lattice on the mechanical properties. In this way, the design of materials with customized properties such as fracture toughness and bending strength is feasible, opening new perspectives for the development of high-performance zirconia composites for orthopedic implants with high hydrothermal resistance.

The hydrothermal resistance of zirconia is one of the key issues for its use as a biomaterial. The spontaneous transition into the monoclinic toughening tetragonal phase, enhanced in a wet environment, can lead to the degradation of the mechanical properties of the ceramic. This is a relevant issue for load-bearing zirconia–alumina composites, where a higher monoclinic content than expected from in vitro simulations has been observed. Notwithstanding the number of explanations formulated thus far, none have been fully satisfactory. The preferred method to quantify the amount of transformed monoclinic zirconia is Raman spectroscopy. Nevertheless, many sources of errors may negatively affect Raman results, leading to errors in their interpretation. This issue is addressed by Porporati et al. [4], who put in evidence the critical aspects associated with the used equation for the calculation of the volume fraction of the monoclinic phase and with the definition of the related calibration coefficients. Raman spectroscopy is a delicate procedure that does not offer consistent and unique results for the analysis of surface degradation, and the need for a standard is clearly put in evidence by this paper. Solarino et al. [5] highlight the clinical relevance of the improvement in the outcomes of total hip replacements thanks to the use of bioceramic composite components. Implants with composite components are especially suited for high-demand patients, with no material-related side effects, no ceramic fracture, and no mechanical loosening of the implant components. Clinical cases that support this conclusion are described, as well as a comprehensive discussion of the literature.

Tavoni et al. provide a comprehensive overview of the state of the art of calcium phosphate (CaP) biomaterials, widely accepted today to promote the regeneration of bone tissue [6]. They discuss the reported strategies to develop and optimize bioceramics while also highlighting future perspectives in the development of bioactive ceramic composites for bone tissue regeneration. The co-existence of various factors such as the bioactive chemical composition, nanostructure, and bone-like mechanical performance is a major problem with ceramics due to the need for sintering and the difficulty of achieving complex bone-mimicking 3D structures. This paper puts in evidence how several technologies for the manufacturing of a highly porous bioceramic-based scaffold from traditional methods (partial sintering, replica method, sacrificial templates, and direct foaming, as well as various 3D printing technologies) do not result in the expected outcomes. The future of bioactive and effective bone scaffolds is strongly related to the development of new approaches that can generate bone scaffolds endowed with bone-mimicking features yielding an effective regenerative ability. Besides the biological aspects, a relevant issue of bone scaffolds is their intrinsic brittleness. This limits their applications, especially in the case of large bone defects in load-bearing sites. The work conducted in the past to develop processes enhancing both the strength and toughness of bioceramics is reviewed by Abbas et al. [7]. To this aim, fiber reinforcement is a promising approach, although further work is necessary to improve the fiber–matrix interface and control thermal fiber decomposition.

Ahlhelm et al. [8] report the manufacturing of load-bearing scaffolds with improved mechanical behavior. The process set up to obtain this result consists in a combination of additive manufacturing and freeze foaming. In this way, they obtained complex-shaped structural composites that not only unite the structural features of a real bone (dense and porous sections) but also reach similar and improved compressive strengths. This process has the potential for the production of bone replacements suitable for a number of bone defects, especially long-bone and load-bearing positions.

Hybrid nanostructured materials obtained by biomineralization are a special class of composites for bone regeneration, consisting in organic (e.g., polymer) and inorganic (e.g., hydroxyapatite) components. The process of biomineralization is described in detail in the paper by Campodoni et al. [9], who devoted special attention to calcium biomineralization, a process that can lead to highly biomimetic and biocompatible materials resembling natural hard tissues such as bones and teeth. Hints are offered on the numerous applications of biomimetic materials, whose behavior can be finely tuned by changing the environmental conditions (e.g., pH), doping ions, and organic network. The technologies to obtain hybrid scaffolds combining the tunable macro/microporosity and osteoinductive properties of ceramic materials with the chemical/physical properties of biodegradable polymers are the subject of the comprehensive review by Ozcan et al. [10]. The porosity of the scaffolds can be tuned for optimum results using conventional and additive manufacturing techniques and, more recently, 3D and 4D printing. The authors put in evidence that, facing the growing demand, only a limited number of biodegradable materials are currently available for the manufacture of materials and composites, particularly by 3D printing techniques. The conclusion highlights the need for research to develop new biomaterials for hybrid composites with adjustable properties that can restore functionality at the application site, providing optimum printability, mechanical stability, and better integration with the host.

The need for specific feedstocks for 3D printing of ceramics is also addressed by Magnani et al. [3]. The feedstock they prepared allowed the production of an alumina-toughened zirconia (ATZ) dental implant starting from a blend of selected ceramic precursors and a photopolymeric resin. This demonstrates the effectiveness of additive manufacturing in the small-batch production of medical devices with complex shapes using bioceramic composites.

Conflicts of Interest: The authors declare no conflict of interest.

References

1. Piconi, C.; Sprio, S. Oxide Bioceramic Composites in Orthopedics and Dentistry. *J. Compos. Sci.* **2021**, *5*, 206. [CrossRef]
2. Burger, W.; Kiefer, G. Alumina, Zirconia and Their Composite Ceramics with Properties Tailored for Medical Applications. *J. Compos. Sci.* **2021**, *5*, 306. [CrossRef]
3. Magnani, G.; Fabbri, P.; Leoni, E.; Salernitano, E.; Mazzanti, F. New Perspectives on Zirconia Composites as Biomaterials. *J. Compos. Sci.* **2021**, *5*, 244. [CrossRef]
4. Porporati, A.A.; Gremillard, L.; Chevalier, J.; Pitto, R.; Deluca, M. Is Surface Metastability of Today's Ceramic Bearings a Clinical Issue? *J. Compos. Sci.* **2021**, *5*, 273. [CrossRef]
5. Solarino, G.; Spinarelli, A.; Virgilio, A.; Simone, F.; Baglioni, M.; Moretti, B. Outcomes of Ceramic Composite in Total Hip Replacement Bearings: A Single-Center Series. *J. Compos. Sci.* **2021**, *5*, 320. [CrossRef]
6. Tavoni, M.; Dapporto, M.; Tampieri, A.; Sprio, S. Bioactive Calcium Phosphate-Based Composites for Bone Regeneration. *J. Compos. Sci.* **2021**, *5*, 227. [CrossRef]
7. Abbas, Z.; Dapporto, M.; Tampieri, A.; Sprio, S. Toughening of Bioceramic Composites for Bone Regeneration. *J. Compos. Sci.* **2021**, *5*, 259. [CrossRef]
8. Ahlhelm, M.; Latorre, S.H.; Mayr, H.O.; Storch, C.; Freytag, C.; Werner, D.; Schwarzer-Fischer, E.; Seidenstücker, M. Mechanically Stable _-TCP Structural Hybrid Scaffolds for Potential Bone Replacement. *J. Compos. Sci.* **2021**, *5*, 281. [CrossRef]
9. Campodoni, E.; Montanari, M.; Artusi, C.; Bassi, G.; Furlani, F.; Montesi, M.; Panseri, S.; Sandri, M.; Tampieri, A. Calcium-Based Biomineralization: A smart Approach for the Design of Novel Multifunctional Hybrid Materials. *J. Compos. Sci.* **2021**, *5*, 278. [CrossRef]
10. Özcan, M.; Hotza, D.; Fredel, M.C.; Cruz, A.; Volpato, C.A.M. Materials and Manufacturing Techniques for Polymeric and Ceramic Scaffolds Used in implant Dentistry. *J. Compos. Sci.* **2021**, *5*, 78. [CrossRef]

Journal of
Composites Science

Review

Oxide Bioceramic Composites in Orthopedics and Dentistry

Corrado Piconi * and Simone Sprio

National Research Council of Italy, Institute of Science and Technology for Ceramics (CNR-ISTEC),
48018 Faenza, Italy; simone.sprio@istec.cnr.it
* Correspondence: corpico@libero.it

Abstract: Ceramic composites based on alumina and zirconia have found a wide field of application in the present century in orthopedic joint replacements, and their use in dentistry is spreading. The development of this class of bioceramic composites was started in the 1980s, but the first clinical applications of the total hip replacement joint were introduced in the market only in the early 2000s. Since then, several composite systems were introduced in joint replacements. These materials are classified as Zirconia-Toughened Alumina if alumina is the main component or as Alumina-Toughened Zirconia when zirconia is the main component. In addition, some of them may contain a third phase based on strontium exa-aluminate. The flexibility in device design due to the excellent mechanical behavior of this class of bioceramics results in a number of innovative devices for joint replacements in the hip, the knee, and the shoulder, as well in dental implants. This paper gives an overview of the different materials available and on orthopedic and dental devices made out of oxide bioceramic composites today on the market or under development.

Keywords: alumina; zirconia; Alumina-Toughened Zirconia; Zirconia-Toughened Alumina; hip arthroplasty; dental implants

Citation: Piconi, C.; Sprio, S. Oxide Bioceramic Composites in Orthopedics and Dentistry. *J. Compos. Sci.* **2021**, *5*, 206. https://doi.org/10.3390/jcs5080206

Academic Editor: Francesco Tornabene

Received: 6 July 2021
Accepted: 30 July 2021
Published: 3 August 2021

Publisher's Note: MDPI stays neutral with regard to jurisdictional claims in published maps and institutional affiliations.

1. Introduction

Oxides are among the most stable inorganic materials since no further oxidative processes (e.g., corrosion, ion release) can take place. This is a major reason for the use of oxides as ideal bioceramic materials since the 1960s because their chemical inertness was considered as the basis for biocompatibility. The first ceramic oxide used in orthopedics was alumina (Al_2O_3), while ceramic composite oxides, prevalently made of alumina-zirconia (Al_2O_3-ZrO_2) were subsequently developed, seeking improved mechanical performance [1].

The first use of alumina as a biomaterial is due to Dr. Sami Sandhaus, a Swiss dentist, who in 1962 developed a screw-shaped dental implant named Crystalline Bone Screw—CBS® and used it in a significant number of cases [2]. In 1963, L.W. Smith and J.F. Estes (Haeger Potteries, Dundee, IL, USA) developed Cerosium™ as a bone substitute in case of large bone defects, i.e., a silica aluminate matrix where pores (about 50% in volume) were filled with epoxy resin [3].

At that time, hip arthroplasty was taking its first steps. Although keenly interested by the potential of such a procedure, orthopedic surgeons were very concerned about the failures of implants due to the wear of bearings. The cooperation between Dr. Boutin—an orthopedic surgeon working in Pau, a town in Southern France—and one of his patients, the manager of a factory sited nearby manufacturing high alumina electric insulator, led to the first total hip replacement (THR) with an alumina-on-bearing in 1970 [4]. Such an implant had a stainless-steel stem and a Ultra-High Molecular Weight Polyethylene (UHMWPE) socket—soon replaced by a socket made of alumina—both cemented into bone.

In the same years, several scientists in Germany (i.e., G. Langer in Keramed, G. Heimke in Friedrichfeld, H. Dörre in Feldmüle, M. Saltzer in Rosenthal) gave a decisive contribution to the development of alumina for orthopedic components and overall for alumina as a biomaterial. This resulted in the development of a number of ceramic orthopedic devices, among which it is worthwhile to mention the BIOLOX® alumina developed by H. Dörre in Feldmüle (now CeramTec GmbH), which became *"the ceramic"* in orthopedics until this attribute was overtaken by the higher-performing BIOLOX® delta alumina–zirconia ceramic composite [5].

Indeed, alumina showed critical issues particularly related to failure of THR implants [6], leading some manufacturers to withdraw from the market of implantable ceramics. Alumina exhibits low fracture strength and toughness and is very sensitive to microstructural flaws that lead to a poor resistance to stress concentration or mechanical impact. As the presence of intergranular pores and large grain size are the main microstructural features that affect the mechanical strength of alumina, the efforts of ceramists were focused on decreasing the porosity and the grain size in alumina ceramics. This was obtained by the selection of proper precursors (e.g., alkoxide-derived powders) and by the optimization of the overall manufacturing process, from batch preparation to final densification by hot isostatic pressing (HIP). Such improvements resulted in the so-called "third generation alumina" based on high-purity precursors and characterized by finer grain size and density near the theoretical one, as illustrated in Table 1.

Table 1. Selected properties of alumina, evidencing the development of the material [5].

Property	Units	First-Generation Alumina (1970s)	Second-Generation (BIOLOX®, Since 1974)	Third-Generation (BIOLOX® Forte Since 1995)
Al_2O_3 content	vol.%	99.1–99.6	99.7	>99.8
Density	$g\,cm^{-3}$	3.90–3.95	3.95	3.97
Av. grain size	μm	≤4.5	4	1.75
Flexural strength	MPa	>300	400	630
Young's modulus	GPa	380	410	407
Hardness	HV	1800	1900	2000

Nevertheless, the feasibility to obtain alumina components with specific design was limited by its typical brittle fracture behavior [7]. In response to these issues, the company Desmarquest (now Saint Gobain Céramiques Avancées Desmarquest—SGCAD, Evreux, France) followed a different approach: they focused their attention on a different, intrinsically tough ceramic, Yttria-stabilized Zirconia Polycrystal (Y-TZP). Zirconia (zirconium dioxide, ZrO_2) is characterized by the polymorphism of its crystal lattice; therefore, it exists in three thermodynamically stable crystalline phases: monoclinic (up to 1170 °C), tetragonal (1170–2370 °C), and cubic (2370–2680 °C). Effective applications of zirconia ceramics in medicine were made possible since the discovery of the stabilization of the tetragonal phase at room temperature based on the introduction of small amounts of oxide phases as stabilizers. Such a discovery led to the development of Partially Stabilized Zirconia (PSZ) [8] because the low concentration of the stabilizing oxide did not allow the full stabilization of the cubic phase. PSZ was firstly obtained by using calcium oxide as stabilizer, but successively either magnesium oxide (magnesia, MgO) or yttrium oxide (yttria, Y_2O_3) were used for this purpose. The real breakthrough in the development of zirconia ceramics occurred in 1975 with the publication of the research paper "Ceramic Steel?" by Garvie, Hannink, and Pascoe [9]. They reported the increase in toughness in MgO-stabilized PSZ (Mg-PSZ) due to the transformation of the tetragonal phase into monoclinic. Such a transformation, taking place in a "martensitic" way as in some steels, results in an effective dissipative mechanism for fracture energy and, finally, in a self-toughening effect.

More recently, Gupta et al. [10] reported that a tetragonal zirconia ceramic showing grain size ≈0.3 to 0.5 μm could be obtained by using 2–3 mol% of Yttrium Oxide (Yttria— Y_2O_3) as a stabilizer, thus resulting in minimal residual cubic and monoclinic zirconia. Since then, although many studies had been dedicated to materials stabilized by CaO, MgO, and CeO, the main zirconia ceramic that was developed industrially for the production of medical devices was the one stabilized by Y_2O_3. The tetragonal grains in Y-TZP—being metastable—can shift to the monoclinic natural form at the expense of an external source of energy, i.e., the elastic stress field that yields an advancing crack. Thus, phase transformation results as an efficient dissipative mechanism for the energy that otherwise would lead to fracture. Indeed, it was evaluated that the tetragonal–monoclinic phase transformation implies (for a free grain) a volume expansion 4–5 vol %. As each grain is constrained by its neighbors (the matrix), the constrained phase transformation generates a compressive stress field that increases the energy threshold that a crack has to overcome to develop further. These concurrent, energy-dissipative, microscopic-scale phenomena above outlined result, at a macroscopic level, in the remarkable bending strength and toughness of Y-TZP (see Table 2).

Table 2. Selected properties of Y-TZP. Alumina data reported for comparison.

Property	Units	Alumina (1970s)	Y-TZP
Al_2O_3 Content	vol %	99.1–99.6	–
ZrO_2 Content	vol %	–	>99
Av. Grain Size Al_2O_3	μm	≤4.5	0.3
Density	g/cm^3	3.90–3.95	6.02
Thermal Conductivity	W/mK	30	2.5
Hardness	HV	2000	1200
Flexural Strength	MPa	>300	1000
Fracture Toughness	MPa m$^{\frac{1}{2}}$	3.5	4.5
Young Modulus	GPa	380	210

Several manufacturers worldwide started the production of Y-TZP ball heads. Among them, SGCAD—the main manufacturer of zirconia (Y-TZP) ball heads worldwide—and Kyocera (Kyoto, Japan) developed also zirconia knee condylar components for total knee replacements (TKR).

The metastability of Y-TZP is the key for its outstanding mechanical performances. However, the metastability of Y-TZP was a cause of concern since the beginning of the clinical use of Y-TZP components. In the biologic environment, Y-TZP may spontaneously transform from tetragonal to monoclinic, drastically decreasing its mechanical properties [11]. Such an undesired phenomenon is named aging or low-temperature degradation (LTD). In THR bearings, the onset of LTD at the surface of the component is related to an increase of the surface roughness, in turn leading to the wear increase of UHMWPE acetabular cups that are usually coupled to Y-TZP heads [3]. The studies on the physicochemical mechanisms giving rise to LTD are still running: several models have been proposed to explain such a behavior, but none have been fully satisfactory to date [12], although it is acknowledged that LTD kinetics is promoted by temperature (especially for T > 100–150 °C), by the presence of water in the environment, and by applied stresses. In addition, different LTD kinetics were observed in Y-TZP ball heads obtained from different manufacturers or from different batches produced by the same manufacturer, thus leaving open the main questions about the possible influence of the production process on LTD [13]. Finally, the unexpected high rates of failure in some batches of Y-TZP since 2000 [14] led to the abandon of its use in orthopedics.

2. Biocompatibility of Alumina and Zirconia Composites

The biological safety of alumina and zirconia and of alumina–zirconia composites has been established for a long time and was recently confirmed [15–18]. Tests on alumina, zirconia, and alumina–zirconia composites were performed using materials in the form of powders or dense ceramics, particularly addressing physicochemical features such as surface reactivity, chemical composition, impurity content, and type of stabilizer. The in vitro assays were performed using extracts in various media, in either direct or indirect contact, by using various cell lines such as macrophages, lymphocytes, fibroblasts, and osteoblasts. Similar considerations can be made on the in vivo tests, which had been performed in several implantation sites in different animal models, to analyze either adverse reactions in soft tissue and/or bone, as well as systemic toxicity. An absence of adverse reactions in cell culture either in tissues or organs after in vivo implants was observed whatever the culture conditions or the implants site.

3. Alumina Zirconia Composites: Early Studies

The abandon of zirconia in 2001 opened a technological gap, leaving unmet the urgent need of ceramic components for arthroplasty with effective design and increased reliability and longevity. Then, materials scientists turned their attention toward different zirconia-toughened ceramics (ZTCs), and promising results were obtained in the development of ceramic composites to be used as biomaterials in orthopedics. The work was focused on composites having alumina (in Zirconia-Toughened Alumina—ZTA) as the main component or zirconia (Alumina-Toughened Zirconia—ATZ).

Special attention was devoted to ZTAs. The basic concept of a ZTA material is to substantially increase the material fracture toughness and strength with respect to alumina, while maintaining relevant properties of alumina such as hardness, stiffness, and thermal conductivity, which are key factors for its successful clinical use in joint replacements. This is achieved by exploiting the tetragonal-to-monoclinic phase transformation of zirconia, which is introduced in ZTA as a reinforcing element. The key point for the excellent mechanical properties of ZTA ceramics is the transformability of the tetragonal zirconia. As a consequence, essential aspects are to retain a significant amount of the zirconia tetragonal phase at body temperature and its degree of stabilization in order to reach the desired toughening mechanism. The proper selection of the stabilizing oxide, the homogeneous and finely distribution of Y-TZP in the alumina matrix, and the control of the microstructure and grain size are key parameters to "tune" the stability of the tetragonal phase. In addition, the compressive residual stresses that develop on cooling, due to the mismatch in thermal expansion coefficients between the alumina matrix and the dispersed zirconia phase, increase the energy threshold for the T-M phase transformation, contributing to the strength of the composite [16].

This aspect is critical in the design of a ZTA composite: higher tetragonal zirconia stabilization as a consequence, for example, of a too high yttria concentration would lead to suppression of the zirconia phase transformation, then losing almost all the improvements of ZTA. On the other hand, poor zirconia stabilization—i.e., due to zirconia uncontrolled grain growth because of inappropriate sintering processes—would enhance the LTD of the material. Nevertheless, in the latter case, the mechanical properties might be outstanding, but the material could have unreliable performance, thus leading to catastrophic consequences due to the LTD of the zirconia phase.

Furthermore, it is perceived that the toughening mechanisms in monolithic Y-TZP and ZTA are significantly different. In monolithic zirconia, the stress induced by the single transformed ceramic grain makes the neighbor tetragonal zirconia transform as well, consequently spreading the transformation effect throughout the material. Such a transformation results in LTD and deterioration under long-term usage. In the case of alumina–zirconia composites, the zirconia phase is constrained in the stable alumina matrix, thereby preventing the transformation of the adjacent grains. Hence, the ZTA has a better retention of the tetragonal phase compared to the monolithic zirconia, when exposed to hydrothermal conditions in vitro [19].

The first studies on alumina–zirconia composites as biomaterials started during the mid-1980s by French researchers (INSA-Lyon, Ecole Centrale de Lyon) looking for a material strong and tough as zirconia but characterized by better resistance to LTD [20]. Among the tested material, a hot-pressed ZTA (Alumina-20 vol % Y-TZP) showed bending strength (four-point bending) higher than 1100 MPa and fracture toughness of about 10 MPa$\sqrt{m}$. Aging tests carried out using small bars implanted under the skin of Wistar rats—then in unloaded conditions –showed a limited decrease of the strength due to LTD. Laboratory wear tests (pin-on-disk, cylinder-on-flat) carried out against UHMWPE demonstrated an improved friction and wear behavior of hot-pressed ZTA in comparison with Y-TZP [21].

A further research project was carried out in the framework of the EUREKA programme (project EU 294) under the scientific coordination of the Italian Ceramic Centre (Bologna, Italy). The main goals attained were the production by slip casting of ZTA ball heads with several Z/A ratios [22,23] and the assessment in a hip simulator of the wear behavior of ZTA-UHMWPE bearings. Wear tests were carried out in a hip simulator using Alumina-UHMWPE bearings as the reference. The results did not show significant differences between the experimental and reference material [24]. Cytotoxicity assays confirmed the absence of harmful effects elicited by the composite materials [25].

4. Alumina–Zirconia Composites in Orthopedics

Zirconia-Toughened Alumina

The first ceramic composite introduced into the orthopedic market in 2002 was the alumina matrix composite (AMC) BIOLOX®delta, made by CeramTec GmbH, Plochingen, Germany [26,27]. As reported by Burger [28], such a material was made of a fine-grained, high-purity alumina matrix (approximately 80 vol %) combined with three different oxides, so as to retain the relevant properties typical of alumina such as stiffness, hardness, thermal conductivity, but improving fracture toughness and strength. To this aim, such an alumina matrix contained a tailored amount of zirconia phase (i.e., $\approx$17 vol %), on the basis of indications provided by Pecharromán et al. [29], to obtain the best compromise between appropriate mechanical performance and chemical stability.

Previous works described the addition of strontium oxide (SrO) to the base formulation, with the purpose of activating solid-state reaction with alumina during sintering and triggering the in situ growth of elongated strontium hexa-aluminate ($SrAl_{12}O_{19}$—SHA) crystals with a magneto-plumbite structure. These platelet-shaped SHA grains, homogenously dispersed in the ceramic composite matrix, increased the toughness of the material through a mechanism of crack deflection/bridging. The effectiveness of this approach was demonstrated by Cutler et al. [30], who investigated a three-phase system formed by a 12 wt % Ce-TZP matrix containing Al_2O_3 and SHA platelets nucleated in situ during sintering. The maximum platelets length is about 5 μm with an aspect ratio of 5–10. Figure 1 shows the microstructure of BIOLOX®delta, where the gray-colored grains represent the alumina matrix, while the white-colored grains represent the zirconia phase.

Then, two mechanisms are concurring to toughen and to reinforce such a composite: on one hand, the phase transformation of Y-TZP triggered by the tensile stresses in proximity of a crack tip is an effective energy-dissipative mechanism. On the other hand, the volume expansion associated to the T-M transformation is contrasted by the high stiffness of the alumina matrix, thus resulting in a compressive stresses field, which is highly effective in blocking the cracks propagation. Furthermore, the elongated zirconium aluminate crystals—formed upon solid-state reaction between alumina and zirconia during the sintering process—gave an additional contribution to the enhancement of the mechanical performance, acting indeed as short fibers capable of increasing the fracture strength and exerting an additional toughening effect. In addition to the reinforcing components, there are minor stabilizing oxides added to the material, giving additional effects such as specific coloring, such as the example of doping alumina with little chromium oxide amount (Cr_2O_3), which gives the material its characteristic pink color.

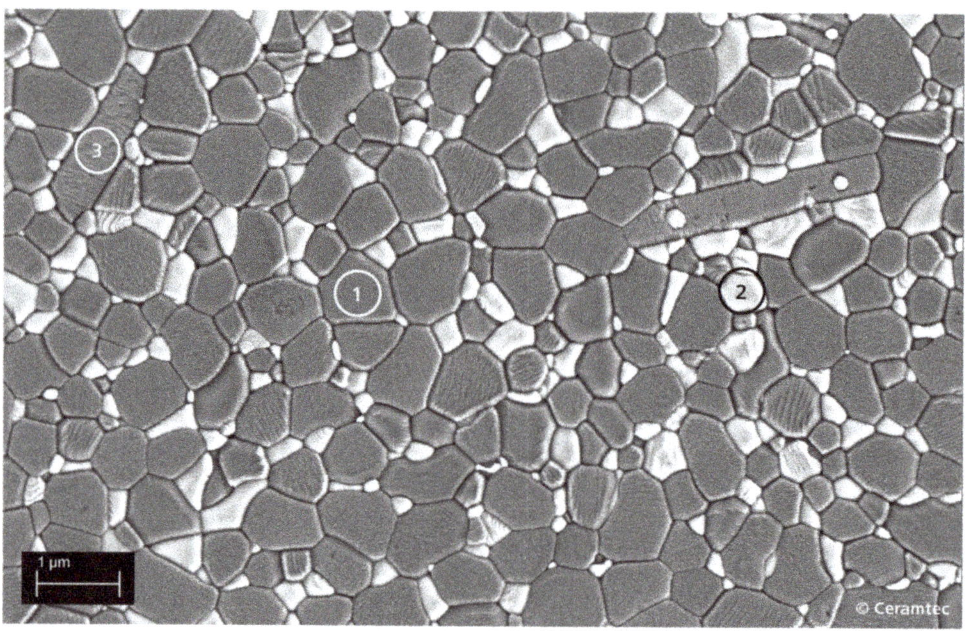

Figure 1. SEM image of the microstructure of the AMC composite ceramic BIOLOX®delta (Courtesy CeramTec GmbH, Plochingen, Germany). The gray grains (1) are the alumina matrix, the white grains (2) are the zirconia phase, and the elongated grains (3) consist of reinforcing strontium hexaaluminate (SHA) platelets.

CeramTec BIOLOX®delta is considered as the golden standard for ceramics in joint replacements bearings. BIOLOX®delta—which is characterized by a pink color, patented in the European market—is presently used for the production of ball heads that are manufactured in nine different diameters from 22 to 44 mm with four different neck lengths. The production of inserts is more differentiated, because it considers not only the inner but also the external design of the insert, to comply with the metallic shells of the different cup design now on the market.

A number of new ceramic devices were developed thanks to the behavior of AMC, as illustrated in Figure 2. This is including THR ball heads expressly designed for use during revision surgeries, ceramic components for knee replacements, humeral heads for shoulder replacements, and ceramic hip resurfacing implants [31].

Figure 2. AMC BIOLOX® delta products (Courtesy CeramTec GmbH, Plochingen, Germany).

Kyocera Medical Corp (Kyoto, Japan—formerly Japan Medical Materials) followed an original approach. The company is a historical manufacturer of fine ceramics for biomedical applications, the first implants of its alumina Bioceram® may be traced back to 1976. In 2010, Kyocera developed a ZTA (BioCeram® AZ209) where zirconia is stabilized in the tetragonal state by residual stresses only [32]. This is obtained by fine tuning of the grain size during cooling.

A more complex composite was released to the market in 2015 under the trademark of Bioceram®AZUL, which was likely chosen for its characteristic blue color [33,34]. AZUL contains approximately 19 wt % zirconia and 2 wt % oxide additives mixed with high-purity fine alumina. The addition of a mixture of oxides characterized by platelet-like grains, such as TiO_2, MgO, Co_3O_4, SiO_2, and SrO, aimed to reduce the grain size and induce the precipitation of platelet-like SHA grains. Additives used in the ZTA synthesis, except for silicon, were located in that platelet-like grain, substituting for strontium and/or aluminum [35]. Stress relaxation due to the zirconia particles and the crack-diverting effect of S6H crystals are relevant factors promoting the high strength and toughness of Bioceram®AZUL.

A further ZTA (25 vol % Y-TZP, alumina balance) was introduced on the orthopedic market under the trade name Symarec® on 2015 by Mathys Orthopedie (Bettlach, Switzerland), who incorporated the company Keramed (formerly Keramik Werk Hermsdorf—KWK Moesdorf, Germany). Keramed is among the pioneers in bioceramics, manufacturing alumina Bionit® since the early 1970s. Symarec® contains 25 vol % Y-TZP finely dispersed in the alumina matrix (75 vol %) [36].

5. Alumina-Toughened Zirconia

Similarly to the ZTA, the ATZ aimed to combine the advantages of the two monolithic materials. Although ATZ may appear of relatively simple design, the properties of these special materials depend significantly on the production steps and conditions, as already previously pointed out.

So far, Mathys Orthopedie (Bettlach, Switzerland) introduced the only ATZ for orthopedic applications in the market in 2007 under the trade name Ceramys® [37]. The mechanical behavior of the material is remarkable and, in some cases, it produced better results than the ZTA alternatives (see Table 3).

Table 3. Selected properties of alumina–zirconia composites for orthopedic applications. Adapted from [31]. (n.s.: not specified; *: measured by ring-on-ring bending).

Property	Units	CeramTec GmbH		Mathys AG		Kyocera Medical	
		Alumina BIOLOX forte	BIOLOX Delta	Symarec ZTA	Ceramys ATZ	AZ209	AZUL
Al_2O_3 Content	vol %	>99.8	79	75	20	84	79
ZrO_2 Content	vol %	–	17	25	80	14	19
Other Oxides	Vol %	–	1	n.s.	n.s	2	2
Density	g/cm^3	3.97	4.37	4.37	5.51	4.35	n.s.
Av. Grain Size Al_2O_3	μm	1.75	0.56	0.8	0.4	0.35	0.3
Vickers Hardness	GPa	20	19	20 (HV1)	15(HV20)	17	17.4
Flexural Strength (4-Point Bend)	MPa	631	1384	≥700 (*)	≥900 (*)	1200	1399
Fracture Toughness	MPa m$^{\frac{1}{2}}$	4.5	6.5	≥5	≥7	4.3	4.5

Ceramys® is presently the oxide ceramic composite with the higher toughness now on the market. The production of ball heads covers the most significant diameters now in clinical use (29, 32, 36 mm). Mathys is manufacturing also sleeved ball heads using the ATZ Ceramys®, for use in revision surgery or in primary implants thanks to extra-long sleeves.

The Ceramys® is sometimes described as a composite made of 80 wt % zirconia and 20 wt % alumina. More in detail, Schneider, et al. [38] describe the Ceramys® as formed by 61% tetragonal zirconia, 17% cubic zirconia, approximately 1% monoclinic zirconia, and alpha-alumina. The tetragonal zirconia phase is stabilized with 3 mol % yttria as for the standard monolithic zirconia (3Y-TZP). The alumina grains are finely dispersed in the zirconia matrix, and the average grain size approaches 0.4 μm, both for ZrO_2 and Al_2O_3.

According to Oberbach et al. [39] the biaxial flexural strength (acc. to standard ISO 6474) surpassed 1200 MPa. The pin-on-disk wear test with ATZ/ATZ couplings using serum as a fluid test medium was shown to be comparable to Al_2O_3/Al_2O_3 couplings (i.e., 0.152 mm^3 and 0.157 mm^3 of weight loss, for ATZ and Al_2O_3, respectively) [40].

Some properties of currently manufactured alumina–zirconia ceramic composites for orthopedic applications are summarized in Table 3.

The interest for zirconia-toughened ceramics (ZTCs) in dentistry is mainly due to the present evolution in the device design. Namely, metal-free dental implants are becoming more and more demanded by patients, and in the last 20 years, Y-TZP is used to produce crowns and bridges as a structural ceramic as well as in dental implants [41].

Several alumina–zirconia ceramic composites are already in use in dentistry. The one with the longer clinical record is known under the trade name NANOZR® (Matsushita/Panasonic, Osaka, Japan). Developed by Nawa et al. in the late 1990s, it is an ATZ composite ceramic constituted of 10 mol % Ce-TZP matrix with 30 vol % Al_2O_3 [42,43]. This material shows an intergranular type nanostructure, in which several 10–100 nm Al_2O_3 particles are trapped within the ZrO_2 grains and several 10 nm ZrO_2 particles are trapped within the Al_2O_3 grains. NANOZR® is now in use in CAD/CAM milling blanks for dental restorations (crown, bridges, etc.), and tests are now in progress in the view of its use in dental implants [44].

Dental implants made of Y-TZP are rather common today, and the large majority of the devices on the market are single-piece devices. As an alternative, a growing number of manufacturers are proposing two-piece dental implants made of Y-TZP. The design of the connection in two-piece implants is a challenging issue because of the small overall diameter of the device (3.5–4 mm) and the reduction in thickness of the walls in the zone of connection. The reliability of these devices can be improved by the use of composites. In addition, alumina–zirconia ceramic composites are expected to increase the reliability of small diameter implants (e.g., 3.5–3.25 mm in diameter) that are the most critical from the mechanical viewpoint.

The ATZ now in clinical use for dental implants is BioHip® (Metoxit, Tahingen, Switzerland) [45]. This material is a fine and homogeneous dispersion of Al_2O_3 grains (D ≈0.4 μm) in a submicron-size Y-TZP matrix, and it shows a marked improvement in the mechanical behavior with respect to Y-TZP (see Table 3). Zeralock and Zeramex Plus dental implants too (Dentalpoint, Zurich, Switzerland) are made out of ATZ BioHip®. The retrospective analysis of the company database demonstrates the excellent success rate of these two systems, 98.5% at >2 years follow up (Zeralock) and 99.4% at >1 year follow up (Zeramex Plus) [46].

6. Developments in Progress

The developments in progress in alumina–zirconia ceramic composite as biomaterials are presently based on the development of new medical devices using materials already established. This is true especially for applications in dental implants fostered by manufacturers developing two-piece and small diameter (<4 mm) dental implants. In addition, the present regulatory framework is making it difficult to introduce new materials in dental implants due to their class of risk (IIb).

As an illustrative example of this situation, it is noted that the small diameter implants now under development (2021) are based on a Ce-TZP-based triphasic composite already described six years ago [47,48]. Ceramic composites replacing Y-TZP by Ce-TZP are being studied for application in dentistry, i.e., by Apel et al. [49], who reported the processing of a ceria-doped tetragonal zirconia polycrystal-based composite (10Ce-TZP/16 vol % $MgAl_2O_4$), obtained by slip casting and die pressing of commercially available powders. Their work shows that the material is enhanced by the inter- or intra-granular dispersion of nano-scaled (max 200 nm) magnesia spinel, mostly at the grain boundaries of the Ce-TZP matrix but also within the zirconia grains. Moreover, this material contains very fine zirconia crystals within the spinel grains. Negligible LTD was observed after aging at 134 °C in saturated steam for 450 h [49].

Ce-TZP/Al$_2$O$_3$ composites were extensively studied in the framework of two EU-funded research projects, named LONGLIFE and SISCERA. The Ce-TZP/8 vol % Al$_2$O$_3$/8 vol % SHA composites (referred to as ZA8Sr8) were recently developed using as a precursor a powder obtained by the surface-coating route [50]. Reveron et al. reported that using coated powders allows homogeneous distribution of the second phases in the zirconia matrix, the tailoring of grain size and morphology, and a close control of the stoichiometry [51]. The mechanical properties of the triphasic composite ZA8Sr8 are summarized in Table 4. In addition, ZA8Sr8 shows optimum LTD resistance; the monoclinic fraction after 50 h of hydrothermal treatment (134 °C, sat. steam) is about 10 vol % in comparison with 70 vol % in Y-TZP treated in the same conditions [52].

Table 4. Selected characteristics of some ZTC of interest as materials for dental implants. Flexural strength measured on 4-p bending bars. * Biaxial measure on discs following ISO 6872.

Material		Ref	Composition	Flexural Strength (MPa)	Toughness (MPa·$\sqrt{m}$)	Hardness(GPa)
Reference	Y-TZP	16	ZrO$_2$/3 mol% Y$_2$O$_3$	900–1200	6–10	12–14
Commercial	NANOZR®	42	10 mol% Ce-TZP/30 vol.% Al$_2$O$_3$	1290 (*)	8.6	11.5
	ATZ BioHip®	38	Y-TZP/20 vol % Al$_2$O$_3$	1400	8	14
Under development	ZCA5P	54	Y-Ce-TZP/Al$_2$O$_3$/LaAl$_{11}$O$_{18}$	1250	8,5	13.5
	ZA8Sr8	51	8 vol %Al2O3/8 vol % SHA/Ce-TZP	1197(*)	10.2	–
	HTZ500	55	ZrO$_2$/2 mol% Y$_2$O$_3$/5 vol % SHA	1628	8.6	12.6

A further approach to increase the strength and toughness of ceramics is based on the introduction of elongated phases that increase the toughness of the material by crack deflection/bridging as discussed previously in the case of SHA platelets in an alumina matrix. Similar platelet-like grains made out of lanthanum hexaluminate (LaAl$_{11}$O$_{18}$—LHA) nucleated in Ce-TZP during sintering were reported by Miura et al. [53].

Similar structures reinforced by LHA can be obtained in a Y-TZP matrix, leading to materials with variable behavior depending on the volume fractions of the different phases, i.e., the ZCA10P composite (Oximatec GmbH, Hochsdorf, Germany) containing 10 vol % platelets described by Burger [54], illustrated in Figure 3.

Further materials that are expected to originate innovative devices in dental implantology are the ones described by Gottwik et al. [55]. They disclosed a new material with remarkable mechanical behavior. Identified as HTZ500, it consists of a matrix of 2 mol % Y-TZP (grain size: 0.27 μm) containing a dispersion of 5 vol % SHA platelets [55]. The bending strength is ≈1.6 times the Y-TZP one. This material is especially of interest for its strength after scratching or after indentation, which is defined by the author as "damage tolerance" [56]. Namely, the abutment of one-piece dental implants is finished by grinding, which may originate a network of subsurface cracks that may grow until fracture, which is driven by the stresses applied during clinical use [57].

Other approaches to increase the mechanical behavior of ZTA were followed by Duntu et al. [58]. They observed a relevant increase in hardness and toughness after the addition of graphene to the material. In hot pressed ceramics, fracture toughness was increased up to 115%, while the fracture toughness (KIC) of alumina—10 wt % ZrO$_2$–was increased up to 164% thanks to the refinement in grain size of the alumina matrix and to the toughening effects of intergranular graphene and ZrO$_2$ grains.

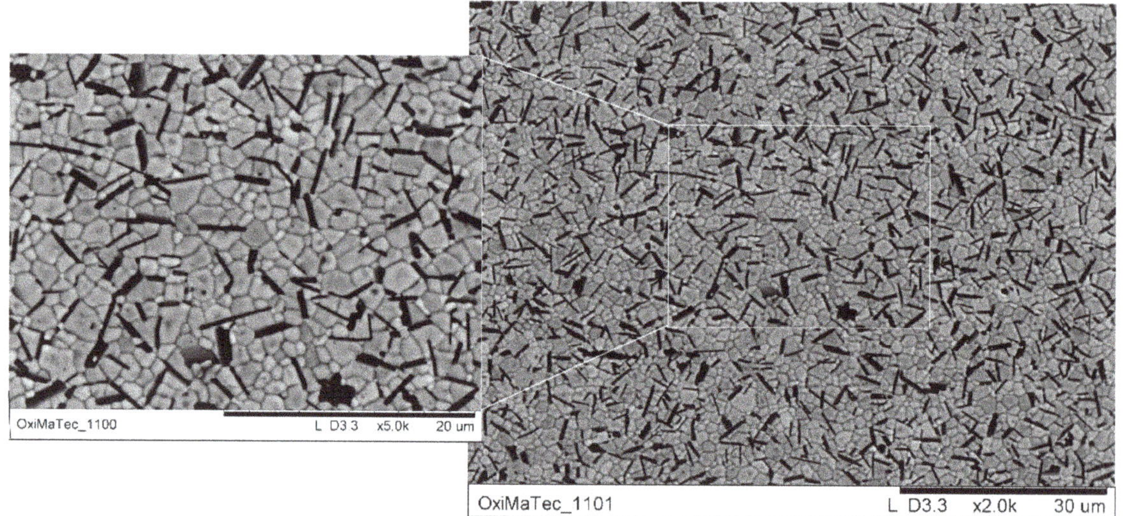

Figure 3. Microstructure of Ce-TZP/LaAl$_{11}$O$_{18}$ composite (ZCA10P) (courtesy Dr. W. Burger, Oxymatec Gmbh, Hochdorf, Germany).

A further approach to improve the mechanical behavior of ZTA consists in adding TiO$_2$ to the composite, thus increasing the composite density [59,60].

Khaskhoussi et al. [61] reported the results of the biological characterization in vitro of experimental ternary alumina/12 mol % ceria-stabilized zirconia/titania ceramic composites. Adding TiO$_2$ in ZTA is known as an effective way to increase alumina and zirconia density [60]. The tests demonstrated that these composites may be able to promote cell adhesion and bonding in the tissue and implant interface, but these promising results must be confirmed in vivo. In addition, genotoxic effects were observed in the composites with higher concentration of TiO$_2$ tested (10 wt %), indicating the need to establish a threshold for this component in the ceramic.

7. Conclusions

Alumina—zirconia ceramic composites have wide acceptance in orthopedics, and their relevance is growing in dentistry. This due to the biological safety of the materials now in production, their stability, and their outstanding mechanical properties.

In orthopedics, the large majority of orthopedic companies rely on BIOLOX®delta for ceramic components for THR bearings. This ceramic covers more than 90% of the worldwide market, while the ones made by competitors rely on niche markets only. More than 5 million patients so far are taking advantage of ceramic components made out of this material.

BIOLOX®delta is today "the ceramic" in hip replacement bearings worldwide, and it is becoming the material of choice for ceramic bearings in the growing field of shoulder arthroplasty. Attempts to use ZTA components in knee arthroplasty are limited to niche markets mainly due to the high pricing of bioceramic knee replacements because of their design, which is much more complex that the spherical joint of the hip or of the glenoid component for shoulder replacements.

In dentistry, especially in implantology, ceramic devices had to cope with the clinical success of titanium implants. Although ceramic dental implants have had a clinical outcome analogous to the titanium ones, the share of ceramic implants—although steadily growing—is today no more than 10% of the market.

The diffusion of ceramic implant in dentistry is depending on the formation of practitioners, who could hence add these devices to the array of solutions that they propose to their patients.

The standard material for ceramic dental implants is Y-TZP, but some implant manufacturers already use ATZ. The growing interest for two-piece implants and for small diameter may widen the application of oxide ceramic composites, especially the ternary ones making use of ceria-stabilized zirconia and SHA platelets, because of their mechanical properties.

Although several experimental materials are under study, their future appears uncertain due to the strict rules of the regulatory systems for medical devices that are making the approval of new materials more and more difficult.

We note the growing attention on ceramic implants by the global players of the dental market, which will lead to significant changes in the dental implantology field during the next few years. The major companies recently made agreements with, or acquisition of, companies involved in material development and/or in the manufacture of ceramic dental implants. A new momentum to metal-free implantology and to oxide ceramic composites biomaterials in this field can be expected from these operations.

Author Contributions: writing—original draft preparation, P.C.; writing—review and editing, S.S. All authors have read and agreed to the published version of the manuscript.

Funding: This research received no external funding.

Conflicts of Interest: The authors declare no conflict of interest.

Abbreviations

AMC	Alumina Matrix Composite
ATZ	Alumina-Toughened Zirconia
HIP	Hot Isostatic Pressing
LTD	Low-Temperature Degradation
SHA	Strontium Exaluminate
TKR	Total Knee Replacement
THR	Total Hip Replacement
UHMWPE	Ultra-High Molecular Weight Polyethylene
Y-TZP	Yttria-stabilized Tetragonal Zirconia Polycrystal
ZTA	Zirconia-Toughened Alumina
ZTC	Zirconia-Toughened Ceramics

References

1. Piconi, C.; Maccauro, G.; Muratori, E.; del Brach Prever, E. Alumina and zirconia ceramics in joint replacements: A review. *J. Appl. Biomat. Biomech.* **2003**, *1*, 19–32.
2. Sandhaus, S. *Noveaux Aspects de l'Implantologie. L'Implant CBS*; Gessler SA: Sion, Switzerland, 1969.
3. Smith, L. Ceramic-Plastic Material as a Bone Substitute. *Arch. Surg.* **1963**, *87*, 653–661. [CrossRef]
4. Boutin, P. Arthroplastie totale de la hanche par prosthéses en alumine fritté. *Rev. Chir. Orthop.* **1972**, *58*, 230–246.
5. Piconi, C. Alumina. In *Comprehensive Biomaterials II*; Ducheyne, P., Grainger, D.W., Healy, K.E., Hutmacher, D.W., Kirkpatrick, C.J., Eds.; Elsevier: Oxford, UK, 2017; Volume 1, pp. 92–121.
6. Piconi, C.; Labanti, M.; Magnani, G.; Caporale, M.; Maccauro, G.; Magliocchetti, G. Analysis of a failed alumina THR ball head. *Biomaterials* **1999**, *20*, 1637–1646. [CrossRef]
7. Traina, F.; De Fine, M.; Di Martino, A.; Faldini, C. Fracture of Ceramic Bearing Surfaces following Total Hip Replacement: A Systematic Review. *BioMed Res. Int.* **2013**, *2013*, 1–8. [CrossRef]
8. Garvie, R.C.; Nicholson, P.S. Structure and thermodynamical properties of Partially Stabilized Zirconia in the CaO-ZrO$_2$ System. *J. Am. Ceram. Soc.* **1972**, *55*, 152–157. [CrossRef]
9. Garvie, R.C.; Hannink, R.H.J.; Pascoe, R.T. Ceramic Steel? *Nature* **1975**, *258*, 703–704. [CrossRef]
10. Gupta, T.K.; Bechtold, J.H.; Kuznicki, R.C.; Cadoff, L.H.; Rossing, B.R. Stabilization of tetragonal phase in polycrystalline zirconia. *J. Mater. Sci.* **1977**, *12*, 2421–2426. [CrossRef]
11. Piconi, C.; Maccauro, G. Zirconia as a ceramic biomaterial. *Biomaterials* **1999**, *20*, 1–25. [CrossRef]
12. Lughi, V.; Sergo, V. Low temperature degradation -aging- of zirconia: A critical review of the relevant aspects in dentistry. *Dent. Mater.* **2010**, *26*, 807–820. [CrossRef]

13. Piconi, C.; Maccauro, G.; Angeloni, M.; Rossi, B.; Learmonh, I.D. Zirconia heads in perspective: A survey of zirconia outcomes in total hip replacements. *Hip. Intl.* **2007**, *17*, 119–130. [CrossRef]

14. Piconi, C.; Maccauro, G.; Pilloni, L.; Burger, W.; Muratori, F.; Richter, H.G. On the fracture of a zirconia ball head. *J. Mater. Sci. Mater. Med.* **2006**, *17*, 289–300. [CrossRef]

15. Kretzer, J.P.; Mueller, U.; Streit, M.R.; Kiefer, H.; Sonntag, R.; Streicher, R.M.; Reinders, J. Ion release in ceramic bearings for total hip replacement: Results from an in vitro and an in vivo study. *Int.Orthop.* **2018**, *42*, 65–70. [CrossRef]

16. Piconi, C.; Porporati, A. Bioinert ceramics: Zirconia and alumina. In *Handbook of Bioceramics and Biocomposites*; Antoniac, I., Ed.; Springer: Berlin, Germany, 2016; Volume 1, pp. 59–90.

17. Alasvand, N.; Banijamali, S.; Milan, P.B.; Mozafari, M. Cellular response to alumina. In *Handbook of Biomaterials Biocompatibility*; Elsevier BV: Amsterdam, The Netherlands, 2020; pp. 335–352.

18. Manicone, P.F.; Ziranu, A.; Perna, A.; Maccauro, G. Cellular response to zirconia. In *Handbook of Biomaterials Biocompatibility*; Mozafari, M., Ed.; Woodhead: Cambridge, UK, 2020; pp. 335–352.

19. Fabbri, P.; Piconi, C.; Burresi, E.; Magnani, G.; Mazzanti, F.; Mingazzini, C. Lifetime estimation of a zirconia-alumina composite for biomedical applications. *Dent. Mater.* **2014**, *30*, 138–142. [CrossRef]

20. Mandrino, A.; Moyen, B.; Ben Abdallah, A.; Treheux, D.; Orange, D. Aluminas with dispersoids. *Tribol. Prop. In Vivo Aging. Biomater.* **1990**, *11*, 88–91.

21. Ben Abdallah, A.; Treheux, D. Frictional and wear of ultrahigh molecular weight polyethylene against various new ceramics. *Wear* **1991**, *142*, 43–56. [CrossRef]

22. Agathopoulos, S.; Nikolopoulos, P.; Salomoni, A.; Tucci, A.; Stamenkovich, I. Preparation and properties of binary oxide bioceramics. *J. Mater. Sci. Mater. Med.* **1996**, *7*, 629–636. [CrossRef]

23. Salomoni, A.; Tucci, A.; Esposito, L.; Stamenkovich, I. Forming and sintering of multiphase bioceramics. *J. Mater. Sci. Mater. Med.* **1994**, *5*, 651–653. [CrossRef]

24. Affatato, S.; Testoni, M.; Cacciari, G.L.; Toni, A. Mixed-oxide prosthetic ceramic ball heads. Part 1: Effect of the ZrO2 fraction on the wear of ceramic on polyethylene joints. *Biomaterials* **1999**, *20*, 971–975. [CrossRef]

25. Ciapetti, G.; Verri, E.; Savioli, F.; Pizzoferrato, A. *Vitro Cytotoxicity Testing of Bioceramics for Hip Prosthesis. Proc. 9th SIMCER*; Palmonari, C., Stamenkovic, I., Eds.; Centro Ceramico: Bologna, Italy, 2000; pp. 39–42.

26. Burger, W.; Richter, H.G. High strength and toughness alumina matrix composites by transformation toughening and "in situ" platelet reinforcement (ZPTA). *Key Eng. Mater.* **2001**, *192–195*, 545–548.

27. Heimke, G.; Leyen, S.; Willmann, G. Knee arthoplasty: Recently developed ceramics offer new solutions. *Biomaterials* **2002**, *23*, 1539–1551. [CrossRef]

28. Burger, W. Oxidkeramik wieder im Trend-neue Werkstoffe für die Medizintechnik und industrielle Anwendungen. *Keram. Z.* **2012**, *2*, 134–137.

29. Pecharromán, C.; Bartolome, J.F.; Requena, J.; Moya, J.; Deville, S.; Chevalier, J.; Fantozzi, G.; Torrecillas, R. Percolative Mechanism of Aging in Zirconia-Containing Ceramics for Medical Applications. *Adv. Mater.* **2003**, *15*, 507–511. [CrossRef]

30. Cutler, R.A.; Lindemann, J.M.; Ulvensøen, J.H.; Lange, H.I. Damage-resistant SrO doped Ce-TZP/Al2O3 composites. *Mater. Des.* **1994**, *15*, 123–133. [CrossRef]

31. Piconi, C. Ceramics for joint replacement: Design and application of commercial bearings. In *Advances in Ceramic Biomaterials*; Palmero, P., Cambier, F., De Barra, E., Eds.; Woodhead: Duxford, UK, 2017; pp. 129–179.

32. Ikeda, J.; Mabuchi, M.; Nakanishi, T.; Miyaji, F.; Ueno, M.; Pezzotti, G. Phase Stability and Fracture Toughness of Zirconia Toughened Alumina for Joint Replacement. *Damage Assess. Struct. VII* **2007**, *361*, 767–770. [CrossRef]

33. Ikeda, J.; Murakami, T.; Shimozono, T.; Watanabe, R.; Iwamoto, M.; Nakanishi, T. Characteristics of Low Temperature Degradation Free ZTA for Artificial Joint. *Key Eng. Mater.* **2014**, *631*, 18–22. [CrossRef]

34. Available online: https://global.kyocera.com/prdct/medical/technology/bioceram/material.html (accessed on 30 June 2021).

35. Ikeda, J.; Murakami, T.; Sasaki, T.; Shimozono, T.; Shouyama, Y.; Iwamoto, M. Wear amd corrosion resistance of low temperature degradation free ZTA for artificial joints. *Key Eng. Mater.* **2017**, *720*, 296–300. [CrossRef]

36. Al-Hajjar, M.; Jennings, L.M.; Begand, S.; Oberbach, T.; Delfosse, D.; Fisher, J. Wear of novel ceramic-on-ceramic bearings under adverse and clinically relevant hip simulator conditions. *J. Biomed. Mater. Res. Part B Appl. Biomater.* **2013**, *101*, 1456–1462. [CrossRef]

37. Begand, S.; Glien, W.; Oberbach, T. ATZ–A New Material with a High Potential in Joint Replacement. *Key Eng. Mater.* **2005**, *284-286*, 983–986. [CrossRef]

38. Schneider, J.; Begand, S.; Kriegel, R.; Kaps, C.; Glien, W.; Oberbach, T. Low Temperature aging of Alumina Toughened Zirconia. *J. Am. Ceram. Soc.* **2008**, *91*, 3613–3618. [CrossRef]

39. Oberbach, T.; Begand, S.; Glien,W. In-vitro wear of different ceramic couplings. *Key Eng. Mater.* **2007**, *330–332*, 1231–1234. [CrossRef]

40. Oberbach, T.; Ortmann, C.; Begand, S.; Glien, W. Investigations of an alumina ceramic with zirconia gradient for the application as load bearing implant for joint prostheses. *Key Eng. Mater* **2009**, *309-311*, 1247–1250. [CrossRef]

41. Piconi, C.; Rimondini, L.; Cerroni, L. *La Zirconia in Odontoiatria*; Masson Elsevier: Milano, Italy, 2008.

42. Nawa, M.; Nakamoto, S.; Sekino, T.; Niihara, K. Tough and strong Ce-TZP/Alumina nanocomposites doped with titania. *Ceram. Int.* **1998**, *24*, 497–506. [CrossRef]

43. Ban, S.; Sato, H.; Suehiro, Y.; Nakanishi, H.; Nawa, M. Biaxial Flexure Strength and Low Temperature Degradation of Ce-TZP/Al2O3 Nanocompsite and Y-TZP as Dental Restoratives. *J. Biomed. Mater. Res. Appl. Biomater.* **2008**, *87*, 492–498. [CrossRef] [PubMed]
44. Oshima, Y.; Iwasa, F.; Tachi, K.; Baba, K. Effect of Nanofeatured Topography on Ceria-Stabilized Zirconia/Alumina Nanocomposite on Osteogenesis and Osseointegration. *Int. J. Oral Maxillofac. Implant.* **2017**, *32*, 81–91. [CrossRef]
45. Rieger, W.; Leyen, S.; Weber, W. The use of bioceramics in dental and medical applications. *Digit. Dent. News* **2009**, *3*, 6–13.
46. Spies, B.C.; Balmer, M.; Patzelt, S.B.M.; Vach, K.; Kohal, R.J. Clinical and patient-reported outcomes of a zirconia oral implant: Three-year results of a prospective cohort investigation. *J. Dent. Res.* **2015**, *94*, 1385–1391. [CrossRef]
47. Palmero, P.; Fornabaio, M.; Montanaro, L.; Reveron, H.; Esnouf, C.; Chevalier, J. Towards long lasting zirconia-based composites for dental implants. Part I: Innovative synthesis, microstructural characterization and in vitro stability. *Biomaterials* **2015**, *50*, 38–46. [CrossRef]
48. Burkhardt, F.; Harlass, M.; Adolfsson, E.; Vach, K.; Spies, B.; Kohal, R.-J. A Novel Zirconia-Based Composite Presents an Aging Resistant Material for Narrow-Diameter Ceramic Implants. *Materials* **2021**, *14*, 2151. [CrossRef]
49. Apel, E.; Ritzberger, C.; Courtois, N.; Reveron, H.; Chevalier, J.; Schweiger, M.; Rothbrust, F.; Rheinberger, V.M.; Höland, W. Introduction to a tough, strong and stable Ce-TZP/MgAl2O4 composite for biomedical applications. *J. Eur. Ceram. Soc.* **2012**, *32*, 2697–2703. [CrossRef]
50. Fornabaio, M.; Reveron, H.; Adolfsson, E.; Montanaro, L.; Chevalier, J.; Palmero, P. Design and development of dental ceramics: Ecxamples of current innovations and future concepts. In *Advances in Ceramic Biomaterials*; Palmero, P., Cambier, F., De Barra, E., Eds.; Woodhead: Duxford, UK, 2017; pp. 355–390.
51. Reveron, H.; Fornabaio, M.; Palmero, P.; Fürderer, T.; Adolfsson, E.; Lughi, V.; Bonifacio, A.; Sergo, V.; Montanaro, L.; Chevalier, J. Towards long lasting zirconia-based composites for dental implants: Transformation induced plasticity and its consequence on ceramic reliability. *Acta. Biomater.* **2017**, *48*, 423–432. [CrossRef] [PubMed]
52. Altmann, B.; Karygianni, L.; Al-Ahmad, A.; Butz, F.; Bächle, M.; Adolfsson, E.; Fürderer, T.; Courtois, N.; Palmero, P.; Follo, M.; et al. Assessment of novel long-lasting ceria-stabilized zirconia-based ceramics with different surface topographies as implant materials. *Adv. Funct. Mater.* **2017**, *27*, 1702512. [CrossRef]
53. Miura, M.; Hongoh, H.; Yogo, T.; Hirano, S. Formation of plate-like lanthanum-β-aluminate crystal in Ce-TZP matrix. *J. Mater. Sci.* **1994**, *29*, 262–268. [CrossRef]
54. Burger, W.; Kiefer, G. Ceramic composites for tailored material properties. *J. Compos. Sci.* **2021**, submitted.
55. Gottwik, L.; Wippermann, A.; Kuntz, M.; Denkena, B. Effect of strontium hexaluminate on the damage tolerance of yttria-stabilized zirconia. *Ceram. Intl.* **2017**, *43*, 15891–15898. [CrossRef]
56. Denkena, B.; Wippermann, A.; Busemann, S.; Kuntz, M.; Gottwik, L. Comparison of residual strength behavior after indentation, scratching and grinding of zirconia-based ceramics for medical-technical applications. *J. Eur. Ceram. Soc.* **2018**, *38*, 1760–1768. [CrossRef]
57. Bethke, A.; Pieralli, S.; Kohal, R.-J.; Burkhardt, F.; Von Stein-Lausnitz, M.; Vach, K.; Spies, B.C. Fracture Resistance of Zirconia Oral Implants In Vitro: A Systematic Review and Meta-Analysis. *Materials* **2020**, *13*, 562. [CrossRef] [PubMed]
58. Duntu, S.H.; Tetteh, F.; Ahmad, I.; Islam, M.; Boakye-Yiadom, S. Characterization of the structure and properties of processed alumina-graphene and alumina-zirconia composites. *Ceram. Int.* **2021**, *47*, 367–380. [CrossRef]
59. Manshor, H.; Aris, S.M.; Azhar, A.Z.A.; Abdullah, E.C.; Ahmad, Z.A. Effect of Titania (TiO$_2$) Addition to Zirconia Toughened Alumina (ZTA) Phases, Hardness and Microstructure. *Adv. Mater. Res.* **2015**, *1087*, 293–298. [CrossRef]
60. Wahsh, M.M.S.; Khattab, R.M.; Zawrah, M.F. Sintering and technological properties of alumina/zirconia/nano-TiO$_2$ ceramic composites. *Mater. Res. Bull.* **2013**, *48*, 1411–1414. [CrossRef]
61. Khaskhoussi, A.; Calabrese, L.; Currò, M.; Ientile, R.; Bouaziz, J.; Proverbio, E. Effect of the Compositions on the Biocompatibility of New Alumina–Zirconia–Titania Dental Ceramic Composites. *Materials* **2020**, *13*, 1374. [CrossRef] [PubMed]

Journal of
Composites Science

Review

Alumina, Zirconia and Their Composite Ceramics with Properties Tailored for Medical Applications

Wolfgang Burger * and Gundula Kiefer

OxiMaTec GmbH, D-73269 Hochdorf, Germany; gundi.kiefer@gmail.com
* Correspondence: wolfbur@web.de

Abstract: Although in 1977 the first ceramic composite material had been introduced into the market, it was a long time before composite materials were qualified for medical applications. For a long period high purity alumina ceramics have been used as ball-heads and cups. Because of their brittleness, in 1986 yttria stabilized zirconia has been introduced into this application, because of higher strength and fracture toughness. However, due to its hydrothermal instability this material disappeared in orthopaedic applications in 2000. Meanwhile a composite materials based on an alumina matrix with dispersed metastable tetragonal zirconia particles and in-situ formed hexagonal platelets became the standard material for ceramic ball-heads, because of their excellent mechanical strength, hardness and improved fracture toughness. Especially fracture toughness can be improved further by special material formulations and tailored microstructure. It has been shown that a mixed stabilisation of zirconia by yttria and ceria with dispersed alumina and hexagonal platelets overcomes the hydrothermal instability and excellent materials properties can be achieved. Such materials do have big potential to be used in dental applications. Furthermore, these materials also can be seen as a new generation for ball-heads, because of their enhanced fracture toughness. All materials are described within these articles. In order to achieve the required properties of the materials, special raw materials are required. Therefore, it is quite important to understand and know the raw material manufacturing procedures.

Keywords: alumina; zirconia; transformation toughening; platelet reinforcement

Citation: Burger, W.; Kiefer, G. Alumina, Zirconia and Their Composite Ceramics with Properties Tailored for Medical Applications. *J. Compos. Sci.* **2021**, *5*, 306. https://doi.org/10.3390/jcs5110306

Academic Editor: Francesco Tornabene

Received: 4 October 2021
Accepted: 12 November 2021
Published: 22 November 2021

Publisher's Note: MDPI stays neutral with regard to jurisdictional claims in published maps and institutional affiliations.

1. Introduction

Ceramic materials have played an important role for many years. Their main use has been in bricks and in pottery. Because of its inertness and corrosion resistance, table wear is a quite popular, centuries-old application of ceramics. All goods named are made out of natural raw materials. About 90 years ago, the first publications showed alumina based ceramic materials based on synthetic alumina (Al_2O_3) powders [1].

In the early 50s of the last century, alumina ceramics became popular for some wear applications. Especially in the textile industry, it is still used as the most wear resistant material against yarns. Additional developments related to several applications, like cutting tools, lead to further improvements of the materials [2].

Due to its reversible phase transformation, zirconia (ZrO_2) didn't play a significant role for 80 years, because of the need of stabilizing oxides in order to keep a stable cubic crystal structure. In refractories, such zirconia became popular, because of their low thermal conductivity. However, mechanical strength properties of such kinds are quite limited. In mid 70s of the last century, it has been found that by a partial stabilization of the zirconia, special properties by microstructural design have been achieved [3]. Finally, in the early 80s it has been found that by stabilization of zirconia by yttria (Y_2O_3), the tetragonal modification can be stabilized. Such materials show a very high mechanical strength, and since that time, many new developments have been made.

Mixtures of alumina and zirconia were introduced in 1977 for ceramic cutting tools [4]. By incorporation of very fine grained zirconia particles into an alumina matrix, mechanical

19

strength properties and fracture toughness can be enhanced. Finally, in 1991 it was published that reinforcement of alumina with zirconia and hexagonal platelets led to a further improvement of the materials.

One of the major uses of yttria stabilized zirconia started in 1986 when ball-heads for total hip prosthesis have been introduced into the market [5]. Today ball-heads on the bases of yttria stabilized zirconia do not play a role in this field. They have been substituted by composite ceramics. However, in the dental industry yttria stabilized zirconia ceramics are quite popular in prosthetics and as implants [6].

In the following, alumina, zirconia and its composites will be described more in detail.

2. High Purity Alumina

The most popular processing technology for making alumina powder is a disintegration of bauxite with caustic soda. By this process, called the Bayer-process, the alumina is dissolved at temperatures of about 120–140 °C and a pressure of about 2–3 bar. The $Al(OH)_4{}^-$-ions are then precipitated by the addition of seeds. These $Al(OH)_3$ (aluminium hydroxide) precipitates normally contain impurities like magnesium, calcium, silica, iron and sodium ions. After a thermal treatment, the hydroxide changes to several intermediate oxides, before finally the thermodynamically stable α-phase of alumina is reached. Usually, this powder then is used for making synthetic ceramic materials [7].

One of the first applications for an alumina ceramic material has been the isolating component of spark plugs. Since densification during the sintering process of such alumina powders is very difficult, additional components are added. Therefore, most of the alumina ceramic materials contain silica (SiO_2), magnesia (MgO) and/or calcium oxide (CaO) [8,9]. By addition of these components, glassy phases are formed and such glassy phases support the densification behaviour significantly. Besides spark plugs, which are usually based on 97 wt-% of alumina and additional silica and calcium oxide or magnesia, seal-discs and substrates for electronic applications are based on similar formulations. Unfortunately, these glassy-phase containing alumina ceramics tend to corrode under long term treatment in humid atmospheres.

Alumina ceramics with a purity of 99.7% based on Bayer-alumina still contain a certain small amount of calcium and silica impurities and do not have a very high sintering activity. Therefore, in order to achieve a density of at least 97.5% it is mandatory to apply sintering temperatures up to 1700 °C. By application of such high sintering temperatures a significant grain-growth occurs. Furthermore, grain-growth control is very difficult. Even the addition of magnesium oxide, which is well known as grain-growth inhibitor, does not help any more to control the growth of the grains at these high temperatures [10]. As a result, discontinuous grain growth occurs. Grain-size of such a kind varies significantly, and single grains with a size of 20 μm or more are quite often found in the microstructures. The typical mechanical strength of these materials is about 250–300 MPa. Figure 1 shows the microstructure of such a ceramic.

Looking back to the 1960s, only alumina powders based on the Bayer process have been available on the market. In parallel, already electro corundum has been applied for grinding applications. Electro corundum are alumina single crystals with a size of about 15–300 μm. Due to the melting process, which occurs in order to achieve the single crystals, the purity of these grains is much higher compared to alumina powders derived from the Bayer process. The only remaining impurities are small amounts of silica (SiO_2) and sodium oxide (Na_2O).

Erhard Dörre, a pioneer in the development of high purity alumina ceramics, recognized the advantage of the higher purity of electro corundum, milled the single crystals down to a size of about 0.6 μm, cleaned the milled powder with hydrofluoric acid in order to get rid of the silica impurities, followed by a second cleaning step with hydrochloric acid in order to get rid of sodium oxide. By this approach, he realized a high purity powder for making high purity ceramics. In addition, this ceramic material could be sintered at temperatures of less than 1600 °C. By addition of magnesia as grain-growth inhibitor a very uniform microstructure with a mean grain-size of about 5 µm has been realized (Figure 2) [11]. Furthermore, the density achieved has been at 99%. As a result of this, the mechanical strength could be increased to 420 MPa.

Figure 1. Microstructure of alumina with a purity of 99.7% sintered at 1650 °C.

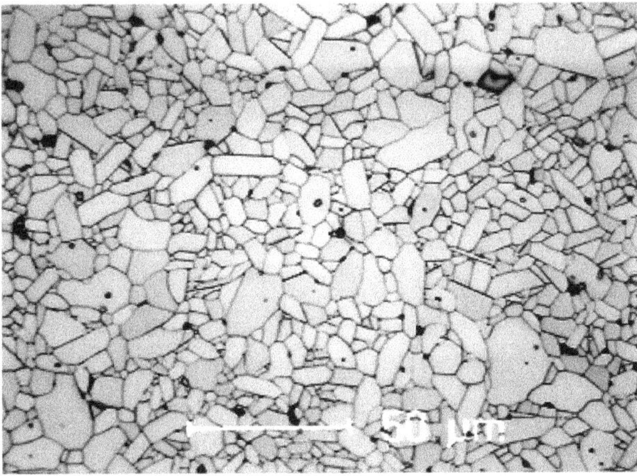

Figure 2. Microstructure of 99.7% alumina based on purified and milled electro corundum.

The pioneering work of Dörre has been underestimated for a long period. Furthermore, the costs of Dörre's material have been significantly higher. Therefore, managing people fought against this material, because they only have seen the higher costs and didn't want to see the unique performance of this material.

Dörre's approach shows the importance of the raw material in relation to the mechanical properties. Furthermore, by his approach he could avoid impurities, which are sensitive to corrosion. While in the Bayer alumina raw materials always small amounts of calcium oxide and silica are present, the high purity alumina ceramics made by Dörre no longer had any impurities. As a consequence of this, besides the improved mechanical properties, this material also shows a significant higher corrosion resistance.

Finally, Dörre achieved the break-through with this material in a publicly funded project, which has been related to bioceramic ball-heads. Mechanical strength and corrosion resistance are mandatory for a long-term stability. Since Dörre's material fulfilled these requirements, finally his material has been qualified for bioceramic applications as ball-heads in total HIP replacement systems, while approaches on the basis of Bayer alumina only had limited success.

In 1970s, new powder processing routes for high purity alumina have been developed. All of these processes start from defined chemicals, which are isolated and afterwards transferred to high purity alumina. Typical precursor salts are Ammoniumaluminiumsulfate (Alaun) $(NH_4)Al(SO_4)2 \cdot 12H_2O$, Aluminiumchloride $[Al(H_2O)_6]Cl_3$, Ammoniumaluminiumcarbonate $(NH_4)Al(CO_3)_2$ or Aluminiumalkoxide $Al(OR)_3$. These salts are easily dissolved and precipitation can be controlled, which means that the primary crystallites formed are influenced by the precipitation method. Especially for the Alkoxides by precipitation either hexagonal or ball-like precipitates can be tailored.

At this point it has to be stated that chemically derived alumina powders are significantly higher in costs compared to Bayer alumina. However, these powders can be sintered already below 1500 °C to a final density of 99.7% of the theoretical density with a very homogeneous fine grained microstructure of about 2.5–3 µm in oxidizing atmosphere (Figure 3, left) or between 1–2 µm after HIPing (Figure 3, right) in mean grain-size. In case these ceramics are only pre-fired to a density of about 97–98% and afterwards are hot isostatic pressed, the theoretical density of 3.98 g/cm^3 is achieved. By this approach the homogeneous microstructure with a mean grain-size of about 1.5 µm or even less can be realized, and mechanical properties can be enhanced to about 620–650 MPa [12].

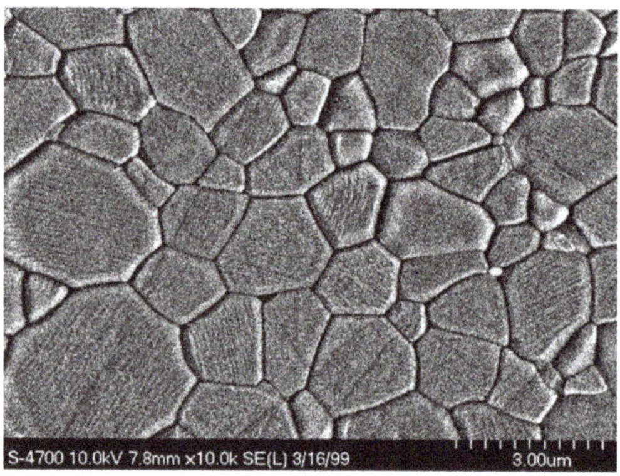

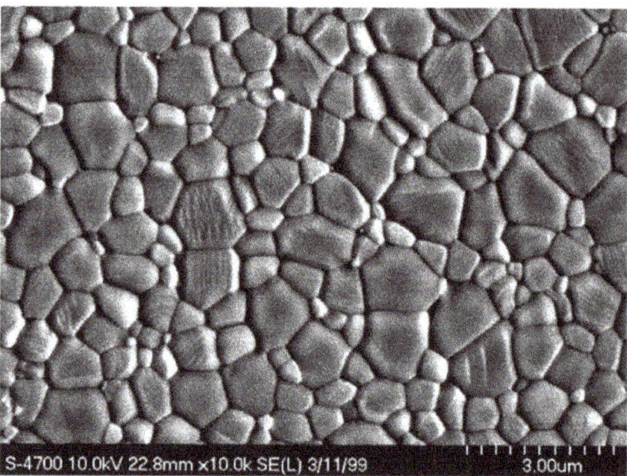

Figure 3. Microstructure of alumina ceramics; **left**: starting powder with a mean grain-size of 0.33 µm; **right**: starting powder with a mean grain-size of 0.22 µm.

Because of the higher mechanical strength properties of high purity alumina ceramics, chemically derived powders have substituted Dörre's alumina in 1987, followed by introduction of the HIP process in 1994. However, all of these materials with improved mechanical strength properties are extremely brittle and very stiff. It is well known that a brittle material with failures on the surface has a catastrophic breakage. This means that a failed part, i.e., a fractured ball-head, generates many fine particles. These have to be removed before a new ball-head can be replaced in the hip.

Taking into account that within a period of 20 years the mechanical strength properties have been increased by about 50% compared to the original material, it can be concluded that a continuous process improvement including new alumina powders, has been quite successful in order to enhance the mechanical strength properties and therefore the safety of ball-heads made out of alumina ceramics [13].

3. Zirconia

Zirconia did not play a significant role in engineering ceramic applications for a long period because it has a reversible phase transformation. While at room temperature the monoclinic phase is stable, it transfers at 1174 °C diffusion-less into a tetragonal modification. By cooling down, re-transformation into the monoclinic phase takes place. Figure 4 shows the hysteresis, which occurs during phase transformation in pure zirconia in comparison to doped zirconia with a different amount of stabilizing calcium oxide (CaO) [14]. Only after the addition of 19.5 Mol-% of CaO the expansion behaviour becomes reversible, while at lower concentrations, i.e., 5 Mol-%, does the hysteresis effect still occur. The phase transformation from monoclinic to tetragonal is combined with a re-orientation of the ions within the lattice. Due to the higher symmetry of the tetragonal modification, the density of it is 6.1 g/cm^3, while in the monoclinic phase it is only 5.85 g/cm^3. This means that for a non-stabilized sample, sintering takes place in the tetragonal modification and by cooling down, it re-transforms to monoclinic combined with a volume increase of about 4%. As a consequence of this, cracks are induced within the ceramic body and it is not stable. Figure 5 shows the symmetries in both modifications of the [111]-direction of the cationic lattice.

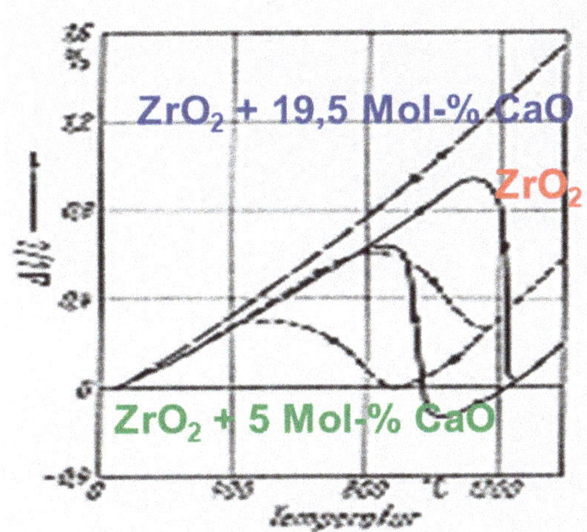

Figure 4. Hysteresis curve of the reversible phase transformation in zirconia.

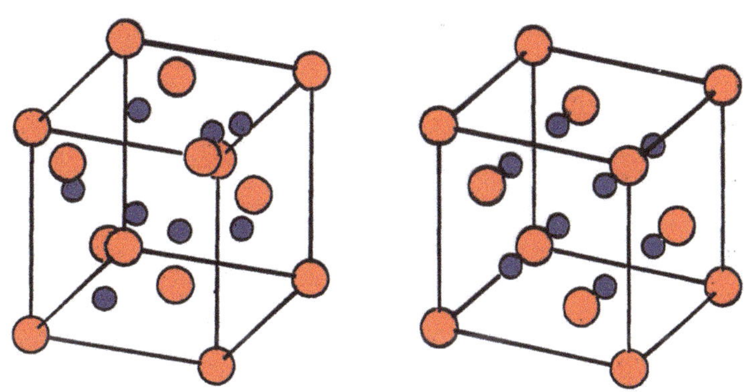

Figure 5. Orientation of the Zr-ions in the monoclinic and tetragonal lattice.

3.1. PSZ Ceramics

In order to overcome the re-transformation from the higher symmetric modification to its monoclinic phase, MgO and/or CaO are added in a relatively high amount. As it can be seen in the zirconia-rich side of the phase diagram of the system ZrO_2-MgO (Figure 6), the concentration of 12–13 Mol-% the cubic zirconia phase remains stable because of formation of a solid solution. Within this solid solution Mg^{2+}-ions replace Zr^{4+}-ions within the cationic lattice. In order to achieve neutrality, oxygen vacancies remain in the anionic lattice [15].

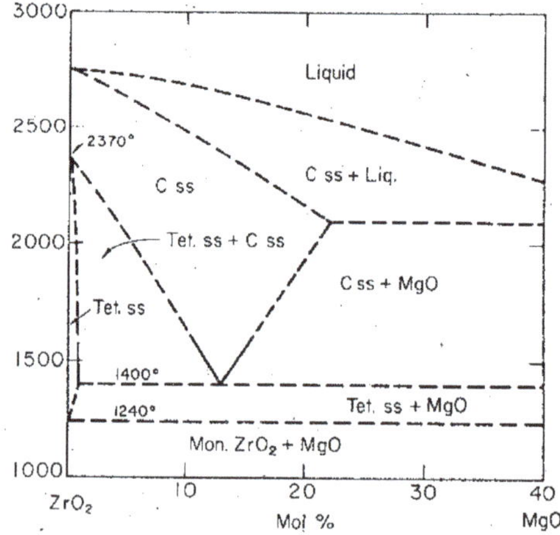

Figure 6. Phase diagram of the system ZrO_2-MgO.

Cubic fully stabilized zirconia doesn't play an important role in high performance applications. Its major application has been in refractories as isolating material, because zirconia itself has a very low thermal conductivity. Besides, the reversible phase transformation zirconia powders are much more expensive compared to many alumina powders. From an economic point of view, this has been an additional argument against a broad application of cubic zirconia ceramics.

The raw material sources itself are mainly zircon sand, a zirconiumsilicate ($ZrSiO_4$), which contains rare earth impurities, including uranium oxide (UO_2). In South Africa the mineral Baddeleyite, a monoclinic zirconia material is found. Again, this mineral is usually accompanied by rare earth oxides and actinides [16].

For making the zirconia powders, there are different processing methods available. The simplest method is the thermal decomposition of zircon sand, $ZrSiO_4$. As can be seen in the phase diagram (Figure 7), it decomposes at temperatures of 1685 °C into a solid solution of ZrO_2 and SiO_2 [16]. However, by this approach for making zirconia, a relatively high amount of impurities remains within the zirconia. The major impurity is remaining silica. Furthermore, all rare earth and actinides remain; i.e., the accompanying uranium- and thorium oxide remains within the zirconia grains as solid solution. Due to these impurities, it is obvious that such a raw material cannot be used in biomedical applications.

A more suitable approach is the alkaline disintegration. By this approach, zircon sand is dissolved in caustic soda and the soluble sodiumzirconate Na_2ZrO_3 then is transformed into a defined chemical composition, i.e., with sulfuric acid into $Zr_5O_8(SO_4)_2 \cdot H_2O$ or with hydrochloric acid into zirconylchloride ($ZrOCl_2$). The first salt can be thermally decomposed into ZrO_2. $ZrOCl_2$ can be cleaned and finally by shifting of the pH with ammonia into alkaline region, zirconiumoxyhydroxide ($ZrO(OH)_2$) is precipitated. By calcination of the hydroxide finally a pure zirconia is achieved. It has to be remarked that in this processing route radioactive impurities can be reduced to a very low level, even below the detection limit [17].

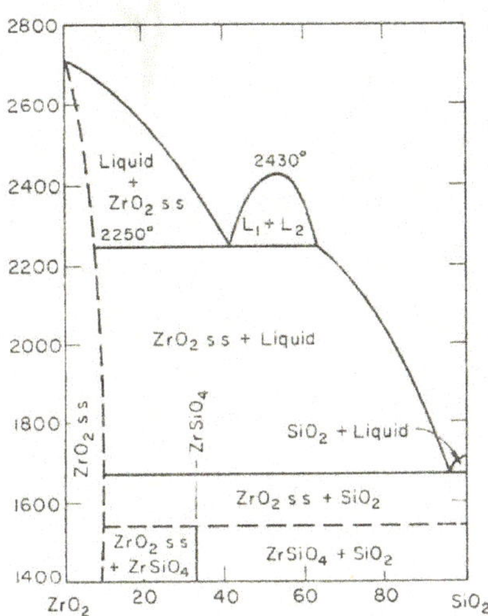

Figure 7. Phase diagram of $ZrSiO_4$.

It has to be remarked that neither by thermal separation nor by alkaline disintegration the hafnia (HfO_2) can be separated from zirconia. This means that normally zirconia always contains an amount of about 1.5–2.0 wt-% of hafnia.

A third alternative approach for making zirconia powder is the carbo-thermic disintegration. By this method, zircon sand is mixed with carbon and chlorine and heat-treated at 900–1300 °C:

$$ZrSiO_4{:}Hf,U,Th + C + Cl_2 \rightarrow ZrCl_4 + HfCl_4 + SiCl_4 + UCl_4 + ThCl_4 + CO_2$$

The formation of the chlorides is the only processing route in order to separate Zr^{4+}- and Hf^{4+}-ions. Furthermore, it is a very effective method to get rid of the radioactive impurities [18,19]. The separation of the Chlorides is made in a condensation column. Finally, the zirconium tetrachloride is directly transferred at about 250 °C in water steam into zirconia:

$$ZrCl_4 + 2H_2O \rightarrow ZrO_2 + 4HCl$$

By this method an extremely high purity zirconia powder with a nano-scaled particle size can be made. This powder behaves completely differently from all other known high purity zirconia powders. It is the most promising raw material in order to achieve excellent yttria stabilized zirconia ceramics. Furthermore, it is the best raw material to be composed with alumina in order to make composite materials. However, before being used in the ceramic body, it is mandatory to make a chemical pre-treatment of it. Figure 8 shows an image of the particles in comparison with a standard grade zirconia material. Table 1 summarizes the chemical impurity levels of the zirconia powders derived by different processing routes [17].

As already mentioned, in order to have stable zirconia ceramics, additional oxides have to be added in order to retain either the cubic phase, or in the case of a partial stabilisation either to directly keep the tetragonal phase metastable until room temperature, or to have cubic matrix grains and within these grains tetragonal precipitates. The last named ceramics are well known under the name "PSZ" (partially stabilized zirconia). The most convenient stabilizing oxide for these kinds of ceramic materials is MgO [15,20].

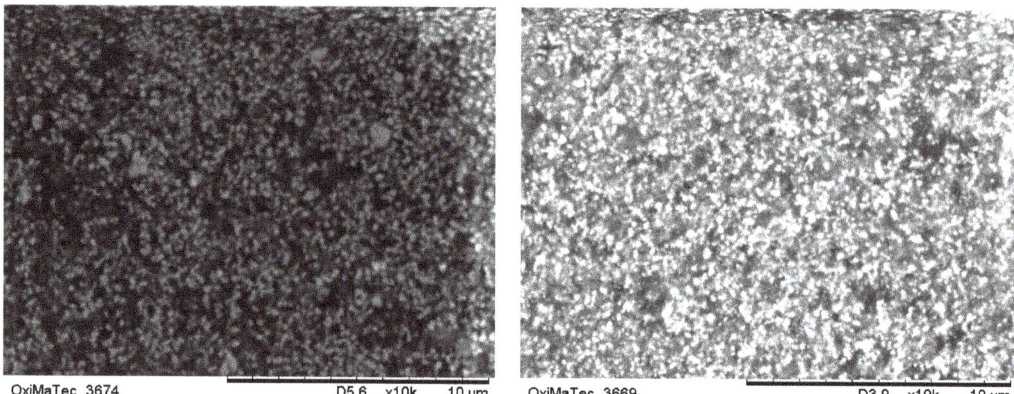

Figure 8. Powder morphology of standard grade zirconia powder (**left**) and the nano-scaled powder (**right**) made by synthesis without calcination.

Table 1. Overview table of zirconia powders derived by different processing routes.

Impurity Material	Unit	Baddeleyite	Thermal Decomposition of $ZrSiO_4$	Alkaline Disintegrartion of $ZrSiO_4$ without Cleaning	Cleaned $ZrO(OH)_2$ by $Zr_5O_8(SO_4)_2$	Cleaned $ZrO(OH)_2$ by $Zr_5O_8(SO_4)_2$ and $ZrOCl_2$	Carbothermal Chlorination and Cleaning
Na_2O	[ppm]	<100	<100	<200	<100	<100	<100
MgO	[ppm]	<100	<100	<100	<100	<100	<100
CaO	[ppm]	50	200	350	200	<100	<100
Al_2O_3	[ppm]	100	800	500	<100	<100	<100
SiO_2	[ppm]	2600	100	500	100	<100	<100
TiO_2	[ppm]	1200	100	1100	200	<100	<100
Fe_2O_3	[ppm]	300	150	200	<100	<100	<100
HfO_2	[ppm]	20,000	20,000	20,000	20,000	20,000	<100
UO_2	[ppm]	800	300	35	35	<1	<1
ThO_2	[ppm]	200	150	10	10	<1	<1

Figure 6 shows the zirconia riche side of the phase diagram ZrO_2-MgO. As can be seen in the phase diagram, the region, where the tetragonal phase is existing, is quite limited. It is only exisiting at high temperatures up to a concentration of about 0.5 Mol-% and disappears already below 1200 °C. Such a low concentration cannot stabilize the tetragonal zirconia modification.

Garvie et al. have shown a very interesting approach by a partial stabilization of zirconia with MgO. They reduced the amount of stabilizing MgO to about 9.2 Mol-% (3.2 wt-%), sintered at a temperature of 1750 °C and cooled the system quickly until about 800 °C [21]. Sintering at 1750 °C means that this process takes place in the region, where only a cubic solid solution is existing. During cooling, the material has to pass a region where tetragonal and cubic solid solution exists in parallel, i.e., during cooling tetragonal precipitates are formed within the cubic matrix grains. Either by optimizing the cooling rate or by heat treatment after sintering, these tetragonal precipitates can be developed within the cubic matrix grains. These cubic grains are in the size of 30–70 μm, while the tetragonal particles within the cubic grains are limited up to 0.2 μm in order to retain their tetragonal modification. A bigger size of these tetragonal precipitates lead to an immediate phase transformation to the monoclinic phase at room-temperature. So, the closer the tetragonal precipitates come to the critical coherence length, the higher the mechanical strength and fracture toughness. Figure 9 shows the typical microstructure of Mg-PSZ and its tetragonal precipitates.

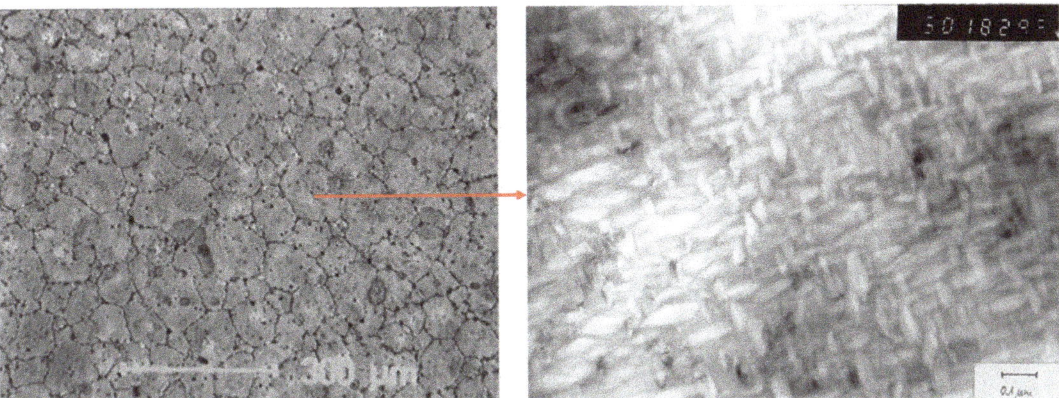

Figure 9. Microstructure of Mg-PSZ (**left**) and its tetragonal precipitates within one grain (**right**).

Compared to cubic stabilized zirconia, which only has a mechanical strength of about 200–250 MPa and is very brittle, the Mg-PSZ grades show strength levels of 500–750 MPa. However, in the case of tetragonal precipitates being close to the critical coherence size, the material can only be used at room temperatures. In the event elevated temperatures are required in the application, the size of the precipitates has to be reduced and therefore the mechanical strength goes down to about 500 MPa. However, in this case the materials can be used up to about 800 °C. Due to its unique microstructure with its coarse grain-size, it fractures trans-granular at a high Weibull modulus of up to 25 (see Figure 10) [22].

Figure 10. Fractured surface of Mg-PSZ.

Usually, Mg-PSZ ceramics show a hardness of $HV_{10} \approx 1150$–1200. Its fracture toughness is about 5 MPa$\sqrt{m}$. While pure Mg-PSZ is quite critical in its behaviour related to phase transformation, a system based on co-stabilisation of MgO and Y_2O_3 is thermodynamically much more stable. Figure 11 shows the strength decrease of Mg-PSZ and Y-Mg-PSZ after treatment at 1100 °C. It is evident that mixed stabilisation causes significant benefits compared to stabilisation with only one single oxide [22,23].

For making PSZ ceramics, raw materials are used, which come from the thermal decomposition process of zircon sand. Their chemical purity is limited (see Table 1) and most of these materials have a yellow colour, which goes back to the trace impurities of radioactive elements. Small amounts of silica, up to about 0.2 wt-% are not very critical for these kind of ceramic materials. Silica remains in the grain-boundaries and forms forsterite ($MgSiO_4$). As a consequence of this, the zirconia matrix is destabilized. In order to avoid the formation of forsterite, it has been proposed to add small amounts of strontium oxide in order to make $SrSiO_4$ and therefore retain the MgO within the cubic matrix grains [24].

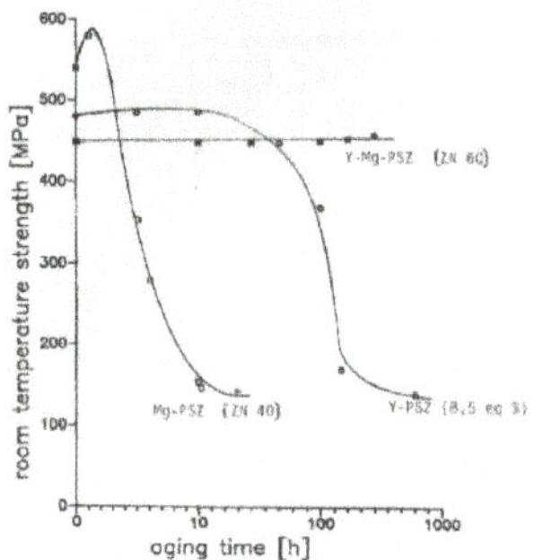

Figure 11. Thermal stability of Y-Mg-PSZ vs. Mg-PSZ.

In the past there have been attempts to qualify Mg-PSZ–based on high purity zirconia powders–ceramics for total hip replacement systems. Compared to alumina, Mg-PSZ has a higher fracture toughness and lower hardness. Mechanical strength properties are comparable. However, there was no breakthrough with these materials, although these kinds of materials do not undergo a hydrothermal decomposition reaction like Y-TZP ceramics.

3.2. TZP Ceramics

Yttria stabilized zirconia (Y-TZP) materials are quite popular, because within the system ZrO_2-Y_2O_3 there is a broader range of tetragonal solid solution (see Figure 12). The tetragonal solid solution region exists up to an yttria content of 3 Mol-%. In practice, this means that by having powders with a high sintering activity, the tetragonal phase can be retained stable as long as the sintering temperature does not exceed 1450 °C.

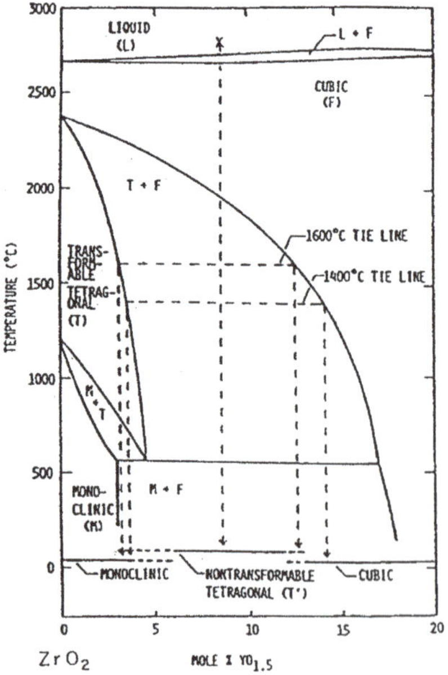

Figure 12. Phase diagram of the system ZrO_2-Y_2O_3.

Normally, high sintering activity of a powder is given when they are fine grained and have a high specific surface area. However, handling of these powders is much more difficult in pressing compared to more coarse powders, which are used for PSZ ceramics. Suitable zirconia powders are derived from purified $ZrOCl_2$- and $Zr_5O_8(SO_4)_2$-precursor salts. During precipitation the formation of the crystallites is controlled severely, and very fine-grained particles of $ZrO(OH)_2$ are precipitated. These particles can be calcined at different temperatures. Finally, the most popular powders—relatively easy to handle for making ceramics—have a specific surface area of about 8 m^2/g. For high-end materials with a very high sintering activity, the specific surface area is about 15 m^2/g.

For many years the distribution of yttria within the zirconia grains has already been made in the precursor salt solution through addition of a solution of YCl_3. In the following precipitation process, the yttria is extremely homogeneously distributed within the zirconia grains. During the following calcination step a solid solution of yttria dissolved in the zirconia matrix takes place. Its particle size is about 0.2–0.3 µm. This approach is well known as the so-called "coprecipitation" process.

As it has been already described in detail, the carbo-clorination process also leads the very fine-grained zirconia powder (see Figure 8 right image). Although in principle yttria could be dispersed in the precursor salt, in practice this has never been done. In the late 1980s it was found that the surface properties of this zirconia raw material can be tailored by chemical treatment in order to adopt yttria onto the surface of the nano-scaled zirconia powder. This approach is the so-called "coating" process. Figure 13 shows the principle differences between these two different approaches for stabilisation of the tetragonal phase in the system ZrO_2-Y_2O_3.

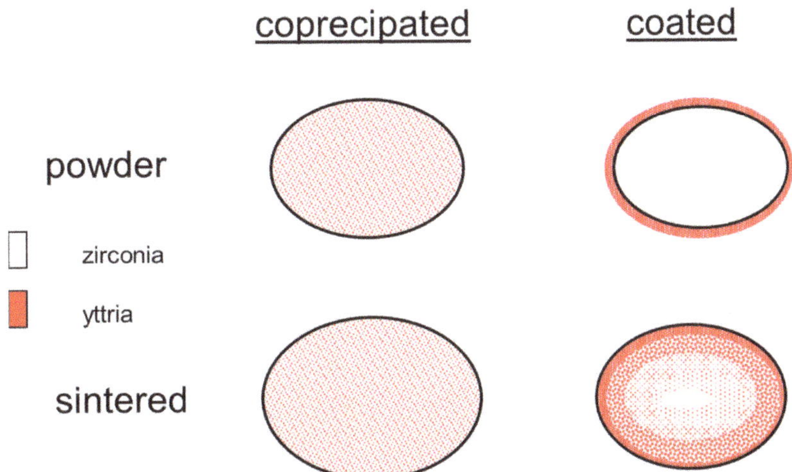

Figure 13. Comparison of "coprecipitation" and "coating" technology (principle!).

Figure 13 illustrates the difference between "coprecipiatated" and "coated" powder. While in the starting grains of the coprecipitated material, there is a very homogeneous distribution of yttria, in the coated material the stabilizing yttria is on the surface of the grains. During sintering there is diffusion of yttria into zirconia with the formation of the solid solution. This diffusion reaction prevents and limits grain-growth. However, it has to be taken into account that the sintering temperature is kept at reasonably low conditions in order to realize a gradient of yttria within the zirconia grains. In case of a high sintering temperature, the gradient disappears. Since in the coprecipitated powder almost 75% of the particles are in its tetragonal modification, during sintering slight grain-growth occurs.

A quite interesting investigation was conducted in the early 1990s through application of high temperature X-ray investigations on the two different powders: while the coprecipitated powder already has about 75% of tetragonal phase, the remaining monoclinic phase transforms already at a temperature of about 1000 °C into tetragonal. By cooling to room temperature, the tetragonal phase remains stable [25]. Opposite to this behaviour, at the transition temperature of 1174 °C a first phase transformation from monoclinic is observed. Only by a higher temperature and time is there the transformation to the tetragonal phase. Both X-ray diffraction patterns can be seen in Figure 14. In the DTA-analysis there is also observed a small endothermal effect in the yttria coated powder at the transition temperature of 1174 °C (Figure 15). This grants proof of a slightly diffusion controlled phase transformation by yttria.

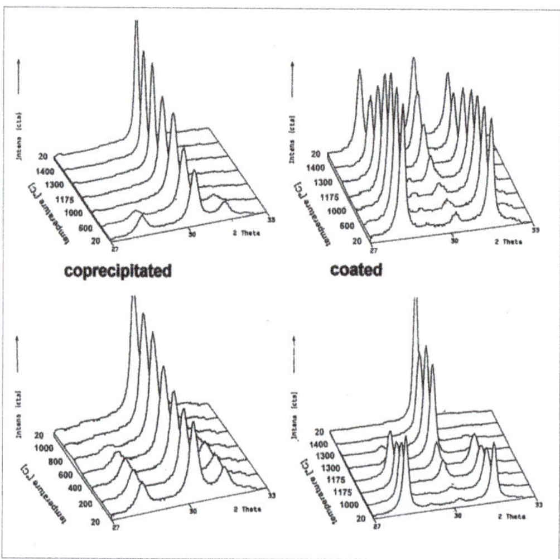

Figure 14. High temperature X-ray diffraction pattern of coprecipitated and coated Y_2O_3-ZrO_2 composition.

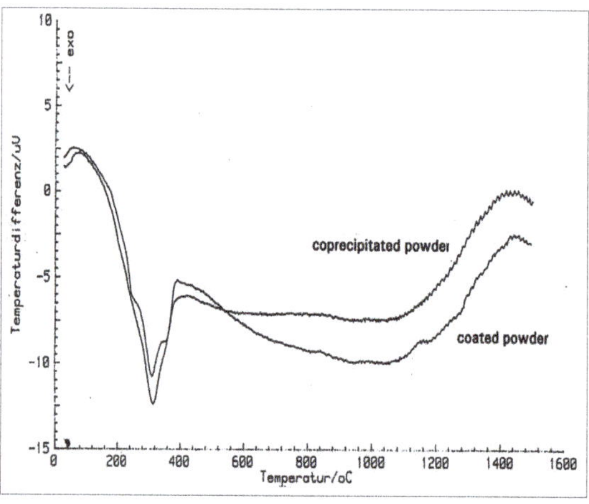

Figure 15. DTA-analysis of coprecipitated and coated yttria-zirconia composites.

The sintering characteristics of the two Y-TZP materials are different: while the coprecipitated material starts earlier in the shrinkage, it requires higher sintering temperatures in order to reach the final density. Opposite to this, densification of coated powder starts later, but the shrinkage rate is higher and lower sintering temperatures are required. Dilatation experiments of the shrinkage behaviour are shown in Figure 16 [26]. For the coated material only 1390 °C are required in order to reach a density of 99.7%; the coprecipitated powder requires 1450 °C. Figure 17 shows the microstructure of the two materials. As it can be taken out of the picture, the microstructure of the coated material is finer.

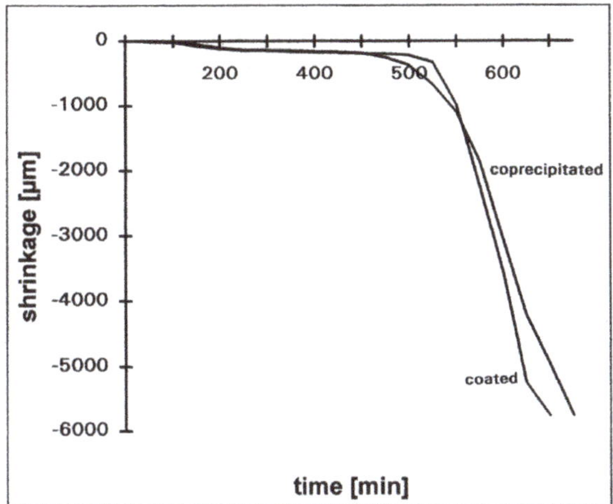

Figure 16. Shrinkage behaviour of different Y-TZP materials.

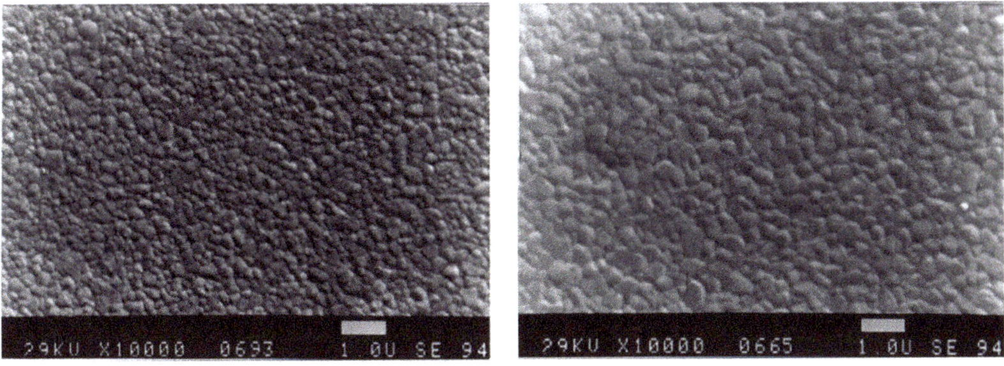

Figure 17. Microstructure of Y-TZP derived by coating (**left**) and coprecipitation (**right**) method (magnification 10,000×).

Mechanical strength properties of the materials are comparable; its mechanical strength after sintering in oxidizing atmosphere is about $\sigma \approx 1000$–1100 MPa. Due to the yttria gradient in the coated material, its fracture toughness is higher. The real difference between the two materials is in their hydrothermal stability. Aging experiments at 135 °C with a vapor pressure of 2 bar have been made in comparison with bio-grade alumina [27]. In both cases a decrease of strength is measured; however, it is more drastic for the coprecipitated material [28,29]. The relative strength decrease is shown in Figure 18; Figure 19 contains the development of the monoclinic phase with increasing aging time, and Figure 20 shows the thickness of the corroded layer. In Figure 21 the corroded layer of the two materials is shown after 48 h treatment.

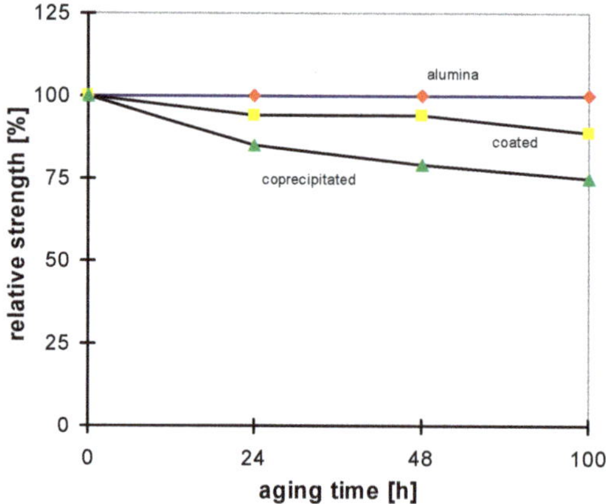

Figure 18. Strength decrease of Y-TZP after hydrothermal treatment.

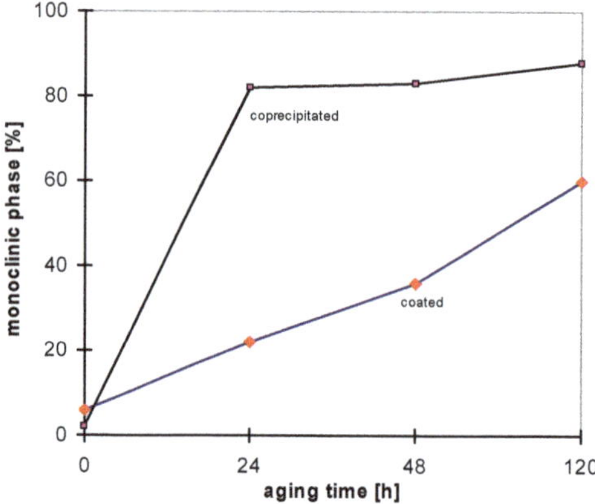

Figure 19. Monoclinic phase of Y-TZP after hydrothermal ageing.

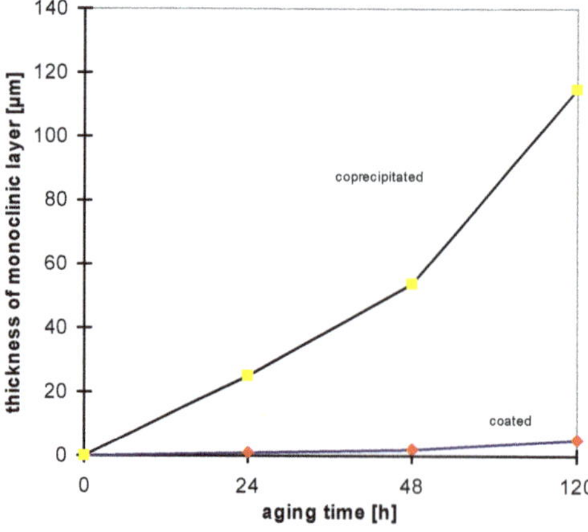

Figure 20. Thickness of the corroded layer after hydrothermal treatment.

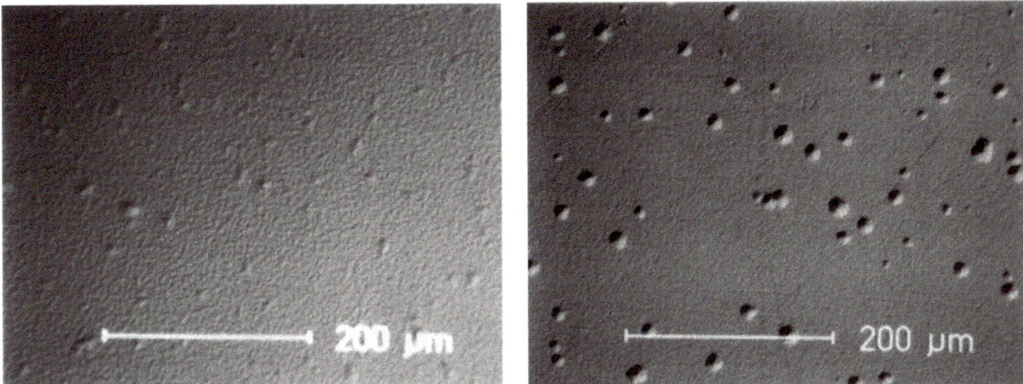

Figure 21. Cross-sectioned polished surface after hydrothermal treatment for 48 h (**left**: coating technology, **right**: coprecipitation).

As it is shown in the above mentioned experiments, yttria stabilized zirconia based on the coating process has an enhanced resistance against hydrothermal decomposition compared to coprecipitated materials. Y-TZP ceramics have been very popular for being used in orthopaedic applications as ball-heads mated against UHMWPE (ultra high molecular weight polyethylene). Due to a production problem, many ball-heads fractured and therefore Y-TZP ceramics disappeared for this application.

Although Y-TZP based on coated zirconia has a higher stability, when the coprecipitated materials are made in a proper way, they also fulfill the requirements for being used as dental implants [30]. Because of the higher fracture toughness and its improved hydrothermal stability, coated ceramics are preferred. They are already on the market in the premium segment applications. Recent experiments related to the hydrothermal stability (treatment at 134 °C up to 100 h in an autoclave system) have shown its superior behaviour related to aging (Figure 22). Aging according to the linear drawing shows a parabolic behaviour (Figure 23). By Mehl–Avrami–Johnson drawing the linearity is demonstrated (Figure 24). The gradient is n = 0.273 ± 0.016 and lnk = −1.39 ± 0.045, which corresponds to a velocity constant of k = 0.25 h^{-1} [31]. Compared to Chevalier, aging of ZY is significantly slower than in coprecipitated Y-TZP [32].

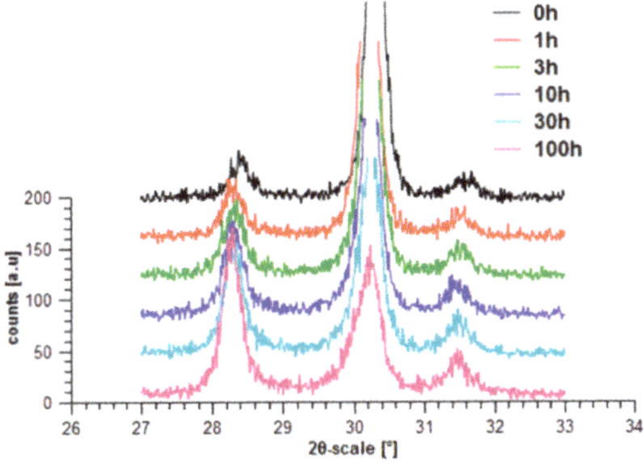

Figure 22. Hydrothermal treatment of ZY (Y-TZP based on coating technology).

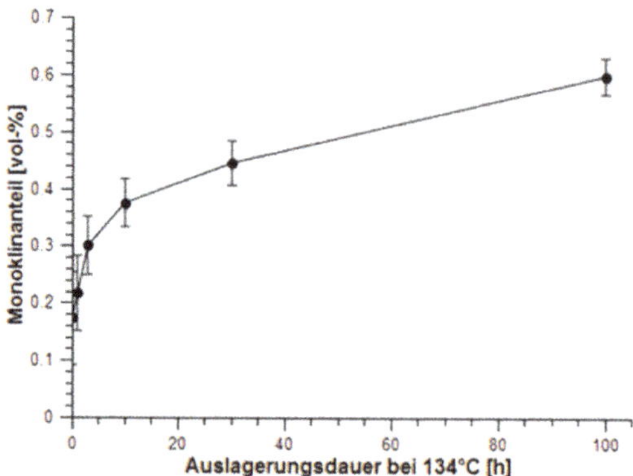

Figure 23. Increase of monoclinic phase–linearly.

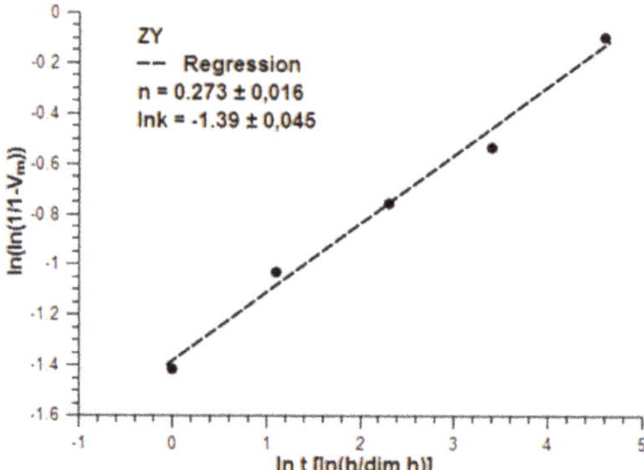

Figure 24. Increase of monoclinic phase according to Mehl-Avrami-Johnson-kinetics.

Due to their biocompatibility and aesthetics, Y-TZP ceramics are of great interest in dental applications. It sounds that even in coprecipitated materials the longterm stability is good enough for these materials to be used in this application. For sure, the higher hydrothermal stability of Y-TZP made by the coating method is higher, but the more important argument is their higher fracture toughness.

4. Alumina Matrix Composites

In a certain sense, alumina with an amount of 92–96 wt-% contain additional oxides. But usually, only alumina, spinel and/or mullite can be detected in X-ray diffraction. In many cases X-ray diffraction is reduced to alumina, because the ingredients form a glassy phase, which reduces sintering temperatures. Opposite to these materials, composites are based on different materials. One of the oldest composites is the dispersion of Titaniumcarbide in an alumina matrix. Such materials have been used as ceramic cutting tools and are made by hot-pressing [33].

Dispersing of zirconia in an alumina matrix was discovered in the early 1970s. From the phase diagram it is well known that alumina and zirconia do not have a chemical reaction. The only interesting thing is that at a temperature of 1660 °C and a composition of 42.6 wt-% (47.2 Mol-%) alumina and 57.4 wt-% (52.8 Mol-%) zirconia there is a eutectic point (Figure 25) [34]. Due to the fact that there is no chemical reaction between alumina and zirconia, such a composite material has become of great interest. While first attempts at dispersing monoclinic zirconia particles within the alumina matrix increased the fracture toughness by inducing microcracks, no breakthrough was achieved because of its limited mechanical strength properties [35].

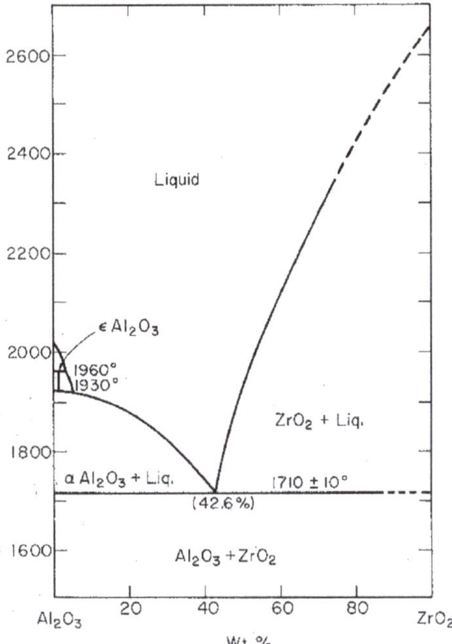

Figure 25. Phase diagram of the system alumina-zirconia.

In 1977 it was recognized by Dworak and Olapinski that by dispersion of nanoscaled zirconia grains within the alumina matrix, they can be retained as metastable in its tetragonal modification at room temperature; i.e., alumina works as a stabilizing matrix. The zirconia particles, which have been based on the carbothermal disintergration process, may not be agglomerated, but rather homogeneously distributed within the alumina matrix [2]. In theory, each zirconia particle should be surrounded by alumina grains. Such a distribution guarantees a constraint of the tetragonal zirconia grains and keeps the tetragonal modification metastable until room-temperature. Later, it was found that the optimal size of the zirconia grains should be between 0.2–0.6 µm [36]. In case the particles are coarser, there is a spontaneous phase transformation to monoclinic, while the finer grains do not have a tendency to re-transform into their monoclinic phase. Without any additional stabilizing oxide, a zirconia concentration of 5–10 wt-% is relatively easy to handle [37]. It becomes very difficult at a concentration of up to 15-wt-% (10 Vol-%). Higher amounts of zirconia dispersed within the alumina matrix cannot be retained in its tetragonal modification. Figure 26 shows a typical microstructure of the system 90 wt-% Al_2O_3 and 10 wt-% (6.9 Vol-%) ZrO_2.

Within a dispersion of 75 wt-% Al_2O_3 and 25 wt-% (18 Vol-%) ZrO_2 it is impossible to retain the tetragonal modification of zirconia without any additional stabilizing oxide. The most common stabilizing oxide for zirconia is yttria. The very conventional approach is to use a coprecipitated yttria containing zirconia powder and disperse it in alumina. Normally stabilisation of zirconia is made with 3 Mol-% of yttria. However, due to the fact that we still have the stabilizing effect of alumina, such a stabilisation is too high. In order to reduce the stabilizing oxide, attempts have been made by using a mixture of 3 Mol-% yttria containing zirconia with a non-stabilized zirconia. Unfortunately, this approach has had only limited success.

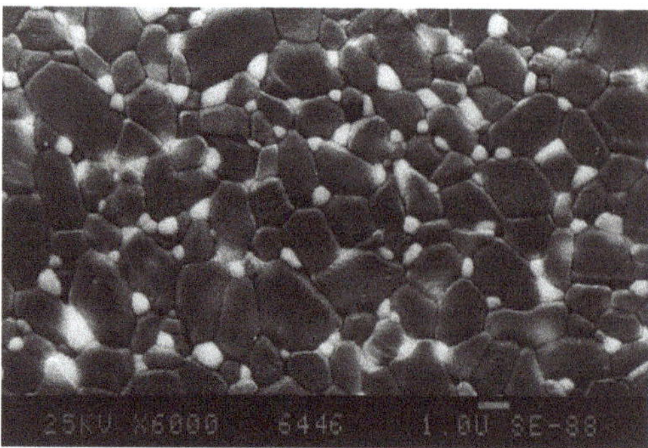

Figure 26. Microstructure of ZTA-ceramics containing 10 wt-% of ZrO_2.

Experimental work has shown that working with a 3 Mol-% yttria containing zirconia powder increases the mechanical strength, but decreases the fracture toughness compared to the approach with zirconia coming from the carbo-chlorination process in addition with the coating technology. However, because of the solubility of yttria in water suspension, it has been very difficult to retain the slurry stable during the milling process. During milling the solubility of yttria is enhanced further and it has been quite difficult to control the rheology. In laboratory experiments it has been shown that the higher amount of zirconia within the alumina matrix leads to an enhanced fracture toughness [38,39]. Corresponding to the higher fracture toughness, the hardness decreased. This is shown in Figure 27.

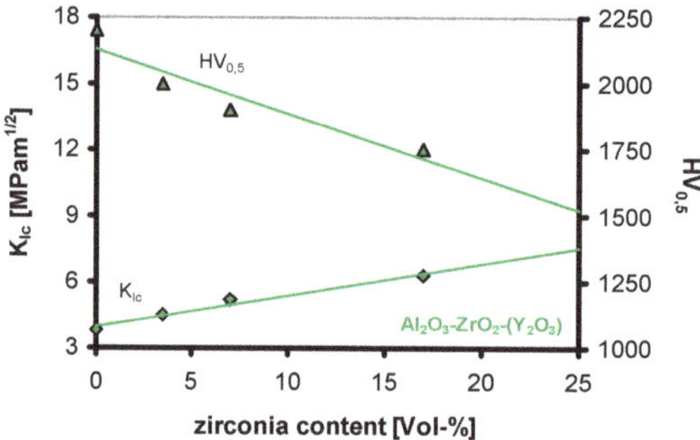

Figure 27. Fracture toughness and hardness in the system Al_2O_3-ZrO_2(Y).

While in the two-phase system based on alumina and zirconia, the rheology of the slurry is still not difficult, the situation immediately changes through the addition of yttria as mentioned above. This makes it extremely difficult to have a stable suspension during the body preparation process. During milling the formation of hydroxides influences the stability of the slurry significantly.

In the late 80th of the last century, the know-how related to rheology with defined chemicals has not been established. It has been quite popular to use commercially available dispersing agents. Unfortunately, the suppliers of these dispersing agents were not prepared to disclose any functional groups within their systems. So, it has been very difficult to understand the colloidal chemistry within a slurry. Since at this time the focus has been dedicated to new inorganic material formulations, processing technology, especially understanding colloidal chemistry, has been very limited. As a consequence of this, finally, in order to have a stable suspension, the following idea has been created: synthesis of a stable ternary chemical composition with yttria.

Due to the fact that the solubility of chromia in alumina is well known and in the literature has been described that only amounts of chromia of at least 1 wt-% shall lead to any influence in the alumina, it has been decided to synthesise the ternary material yttriumchromite, $YCrO_3$. Furthermore, it has been assumed that during the sintering process the following chemical reaction may take place:

$$Al_2O_3 + ZrO_2 + YCrO_3 \rightarrow Al_2O_3{:}Cr + ZrO_2{:}Y$$

In detail this means that during the sintering process the ternary component is destroyed and the formation of solid solutions takes place.

Body preparation of the system containing alumina, zirconia and yttriumchromite ($YCrO_3$) worked pretty well and the rheology of the slurry remained completely stable. Pressed and sintered parts are pink-coloured. In X-ray diffraction analysis, besides alumina, only tetragonal zirconia has been detected. At this time, it has been very surprising that the addition of only a small amount of chromia already caused a significant hardness increase combined with brittleness of the ceramic material. Even at very high amounts of zirconia, the material still stayed at high hardness and remained very brittle compared to the chromia-free system (Figure 28).

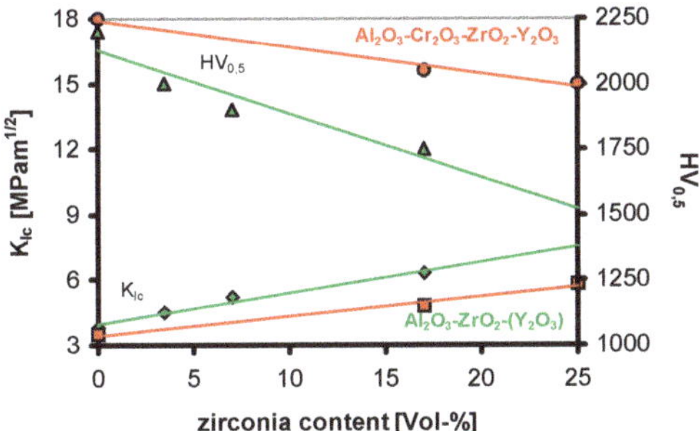

Figure 28. Fracture toughness and hardness within the system Al_2O_3-ZrO_2:Y-Cr_2O_3.

During the period that this development work was made, a group in Ceramatec (Salt Lake City, UT, USA), worked in the system CeO_2-ZrO_2-Al_2O_3-SrO. They have found, besides formation of tetragonal zirconia, hexagonal platelets with the chemical composition $SrAl_{12}O_{19}$ in the sintered bodies. At a certain concentration of strontium aluminate, these platelets enhanced fracture toughness significantly [40].

Following the approach of Cutler et al.,the addition of small amounts of strontium oxide within the system Al_2O_3-Cr_2O_3-ZrO_2-Y_2O_3 finally showed a significantly higher fracture toughness at a high hardness compared to the four component system (Figure 29). In addition, this composition also had a very high mechanical strength of about 1000 MPa after sintering in oxidizing atmosphere. Application of HIPing leads to a strength of about 1200 MPa.

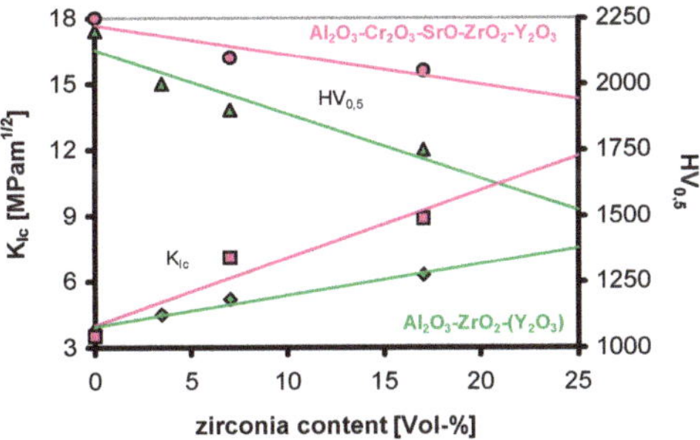

Figure 29. Fracture toughness and hardness in the system Al_2O_3-Cr_2O_3-ZrO_2-Y_2O_3-SrO.

Since strontium oxide is also not very stable in water suspension, again the approach of making a ternary oxide has been made: synthesis of strontiumzirconate ($SrZrO_3$). Commercially available $SrZrO_3$-powders could not be used because of their radioactive impurities. Therefore, the ternary oxide was made by a solid state chemical reaction using strontium carbonate and high purity zirconia.

The material system of the composite material has therefore been based on Al_2O_3-ZrO_2-$YCrO_3$-$SrZrO_3$. During the sintering process therefore the following solid state chemical reactions take place:

$$Al_2O_3 + ZrO_2 + YCrO_3 \rightarrow Al_2O_3{:}Cr + ZrO_2{:}Y$$

$$Al_2O_3{:}Cr + ZrO_2{:}Y + SrZrO_3 \rightarrow Al_2O_3{:}Cr + SrAl_{12-x}Cr_xO_{19} + ZrO_2{:}Y$$

In order to really have a reproducible product, the sintering schedule has to be kept under controlled conditions. During the thermal treatment, the above-mentioned solid state chemical reactions take place. At this point it has to be clearly stated that it is very important to use a pre-treated zirconia powder coming from the carbo-chlorine process, because of the formation of the gradient of yttria within the zirconia grains.

When only the mechanical strength is taken into account, a different zirconia powder can be used. However, only zirconia coming from the carbo-chlorination process finally forms a gradient with yttria, while conventional calcined zirconia powders don't show this effect; they behave like coprecipitated zirconia powders. Certainly, there is no negative effect on the mechanical strength, but its fracture toughness decreases compared to the materials based on the yttria gradient in zirconia of the system. Furthermore, its behaviour related to aging can be compared to Y-TZP based on coprecipitated and coated zirconia powders.

As already described above, at higher zirconia concentrations it is mandatory to stabilize the zirconia grains at least in parts, due to the fact that even at higher zirconia additions, there is a constraint of the metastable tetragonal zirconia particles. However, in this case they have to be partly stabilized by addition of a stabilizing oxide, i.e., yttria. The amount of required yttria for stabilization depends on the amount and also on the size of the particles. For the commercially available material "Biolox delta", which is based on the above described details, the yttria concentration required is 1.5 Mol-% related to zirconia—optimized for zirconia from carbo-clorination process—and its nanoscaled powder size. Figure 30 shows the typical microstructure of the material with zirconia coming from carbo-chlorine disintegration. In Figure 31 a fractured surface is shown. Table 2 summarizes the typical mechanical properties of this material.

Figure 30. Microstructure of the original "Biolox delta", containing zirconia from carbo-chlorine processing route.

Figure 31. Fractured surface of the original "Biolox delta".

Table 2. Material properties of original "Biolox delta".

	Mechanical Properties
Density	4.37g/cm^3
Young's Modulus	350 GPa
Strength	1200 MPa
Fracture Toughness	$7 \text{ MPam}^{1/2}$
Vickers Hardness	2000 ($HV_{0.5}$)

The above-mentioned composite material are three phase ceramics: alumina, metastable tetragonal zirconia and hexagonal platelets. Without platelets, the system is brittle. During the development work made for "Biolox delta", we have also investigated the system Al_2O_3-Cr_2O_3-ZrO_2-CeO_2-SrO [38,39]. Within this system fracture toughness increases significantly, while the hardness remains reasonably high (Figure 32). However, in its mechanical strength properties it remains limited, because it is only about 550 MPa. Sintering only can take place in oxidizing atmosphere. By HIP treatment ceria is reduced from Ce^{4+} to Ce^{3+}, which causes a complete de-stabilisation of the zirconia to its monoclinic modification.

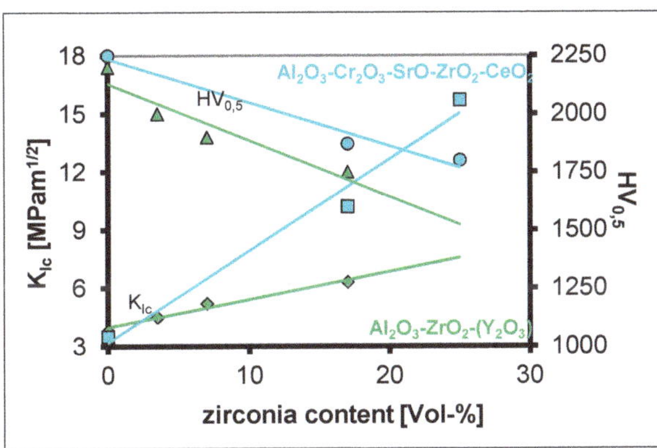

Figure 32. Hardness and fracture toughness of the system Al_2O_3-Cr_2O_3-ZrO_2-CeO_2-SrO.

Although the mechanical strength level is limited, the material has a very high fracture toughness at a reasonable hardness. The mechanical results obtained are summarized in Table 3. Figure 33 shows the corresponding microstructure.

Table 3. Material properties of an experimental grade composite material, using cerium oxide as stabilizing oxide for zirconia.

Mechanical Properties	
Density	$4.66\ \mathrm{g/cm^3}$
Young's Modulus	300 GPa
Strength	550 MPa
Fracture Toughness	$14.9\ \mathrm{MPam^{1/2}}$
Vickers Hardness	1500 $(HV_{0.5})$

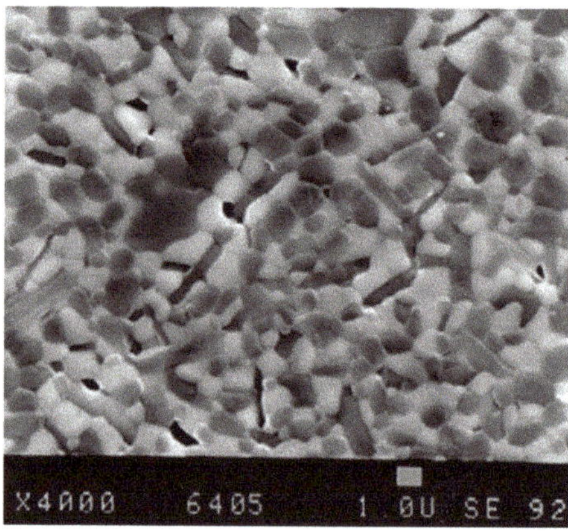

Figure 33. Microstructure of a ceria stabilized alumina matrix composite.

In principle, the combination of strength properties coming from stabilisation of zirconia by yttria and fracture toughness coming from stabilisation of zirconia by ceria in combination with the platelet reinforcement might be an optimum material. So, taking into account the results obtained for the described system, as a logical consequence zirconia might be stabilised by yttria and ceria.

Another aspect which has to be taken into account are the results obtained in the system Al_2O_3-La_2O_3. Yasuoka et al. published a significant increase of strength and fracture toughness in an alumina matrix composite containing 20 Vol-% of $LaAl_{11}O_{18}$ [41]. Own experiments within the system Al_2O_3-Cr_2O_3-ZrO_2-Y_2O_3-La_2O_3 have shown that platelet formation of $LaAl_{11}O_{18}$ is more extended compared to SrO. Furthermore, it is obvious that a higher amount of partially stabilized zirconia gives an additional benefit with respect to fracture toughness [42].

It is well known that ceria stabilized zirconia leads to ceramic materials with high fracture toughness, but limited strength. Dispersing of zirconia in an alumina matrix with yttria and ceria as stabilizing agents for zirconia has been studied intensively in a publicly funded European project (GRD1-199-10585). Within these investigations the following chemical composition has been found to be very promising:

55.3 wt-% Al_2O_3
50.7 wt-% Cr_2O_3
54.0 wt-% La_2O_3
536.45 wt-% ZrO_2
51.05 wt-% Y_2O_3
52.5 wt-% CeO_2

This chemical composition combines relatively high hardness, high mechanical strength and high fracture toughness in the sintered body. Its typical properties after HIPing are summarized in Table 4. Detailed stress analysis investigations have shown high compressive stresses at very low monoclinic content. This material also has been investigated very detailed with respect to its biocompatibility. Cytotoxicity and in-vitro cancerogeniticity tests have been made. No cytotoxic effect has been found. Finally, in-vivo experiments (implantation of ceramic parts into the bones of NewZealand white rabbits) confirmed the good biocompatibility of the material [43].

Table 4. Mechanical properties of a composition based on an alumina matrix, containing platelets and about 40 wt-% of zirconia, partially stabilized by yttria and ceria.

Mechanical Properties	
Density	$4.685 \, \text{g/cm}^3$
Young's Modulus	305 GPa
Strength	1000 MPa
Weibull Modulus	11
Fracture Toughness	$6.8 \, \text{MPam}^{1/2}$
Vickers Hardness	1600 ($\text{HV}_{0.5}$)

Although these promising results already had been established until 2005, such a material formulation has not been regarded, when the ISO standard 6474/partII has been established. Its maximum amount of zirconia dispersed within the alumina matrix in this standard is limited to 30 wt-%. Since the results described above have been ignored for the new ISO-standard 6474/partII, such a promising material cannot be exploited in the bioceramic field. In order to realize a material with higher fracture toughness, it is mandatory to reduce the zirconia content to less than 30 wt-% and therefore, results obtained in the former development project GRD1-199-10585 cannot be exploited.

Meanwhile, the concept of zirconia stabilisation with a mixture of yttria and ceria in an alumina matrix was transferred to a material system with only about 25 wt-% of zirconia. During the development phase the following chemical composition sorted out, which combines high hardness, fracture toughness and mechanical strength:
572.5 wt-% Al_2O_3
52.5 wt-% La_2O_3
523.4 wt-% ZrO_2
50.4 wt-% Y_2O_3
51.0 wt-% CeO_2
50.2 wt-% Pr_6O_{11}
The typical material properties of this chemical composition are summarized in Table 5. The chemical material formulation contains in addition a small amount of praseodymium oxide. This oxide works as a "bridging" component between alumina and zirconia [44].

Table 5. Material properties of an alumina matrix composite containing plateles and about 25 wt-% of zirconia, partially stabilized by yttria and ceria.

Mechanical Properties	
Density	$4.40 \, \text{g/cm}^3$
Young's Modulus	330 GPa
Strength	1100 MPa
Weibull Modulus	11
Fracture Toughness	$\approx 7 \, \text{MPam}^{1/2}$
Vickers Hardness	1700 (HV_{10})

Again, it is confirmed that the stabilisation of zirconia by yttria and ceria in addition with platelets based on lanthanum aluminium oxide leads to good materials, especially taking into account the fracture toughness. So, the higher the fracture toughness, the better the reliability of the material; it becomes more resistant to slow crack growth behaviour.

Unfortunately, none of the new approaches in order to enhance the quality of these composite materials further was exploited to-date. For 20 years, there have been no material innovations for ceramic ball-heads. Meanwhile, the zirconia within "Biolox delta" has been substituted by a conventional zirconia. It appears as though nobody wants to take the risk to introduce a new, more innovative material into the field of orthopaedic applications because of long-term qualification procedures.

The above described material innovations show that it is possible to enhance the material properties further, and as a consequence of this, make the materials even more safe than they are today. It sounds like the hurdles, coming from regulatory affairs, have limited the amount of zirconia addition due to the existing ISO-standard and the high costs for qualification for bioceramic applications compared to the state-of-the-art, are too high and therefore the limiting factors.

While alumina ceramics have been improved step by step, it sounds like the existing composite material qualified for bioceramics does not have such a steady state improvement, but is optimized related to cost savings because of the use of a cheaper zirconia material.

5. Zirconia Matrix Composites

Alumina matrix composites, especially when doped with chromia, became quite famous in bioceramic applications. As it has been described above, there are made different developments in order to further enhance the fracture toughness of such kind of materials in order to further enhance the safety aspects related to potential fractures of ball-heads, which are quite popular in THR surgeries.

For a couple of years, it has been quite popular to use hard-hard pairings in hip surgeries. Meanwhile, mating of ceramic ball-heads with UHMWPE cups again achieved a higher priority, because of more or less no wear of the cup. Such a system also was popular in the late 80th and early 90th by using zirconia ball-heads in combination with UHMWPE. Unfortunately, by the limited hydrothermal stability, finally this solution disappeared, although Y-TZP based on the coating technology didn't show such a big disadvantage compared to ceramics based on co-precipitated powders.

On the other hand, yttria stabilized zirconia has become quite popular in the dental industry. Nowadays, crowns based on this ceramic material are quite popular. Abutments and implants have also started to become interesting in dental restorations. Especially Y-TZP based on yttria coated zirconia powders have extremely good success in dental implantology [44]. Its major benefit is the better hydrothermal stability compared to the corresponding ceramics based on co-precipitated powders. As an alternative material for dental implants, yttria stabilized zirconia containing 20 wt-% of alumina is under discussion because of its high strength, but limited fracture toughness.

Since dental implants are embedded into natural bone, mechanical strength is less important. It is much more important to have a material available which is resistant against shear forces. Resistance against shear forces can be realized in ceramic materials with a high fracture toughness. Pure Ce-TZP ceramics have a very high fracture toughness of up to 20 MPa$\sqrt{m}$, but their mechanical strength of about 400–500 MPa is limited [45]. By addition of about 10 wt-% of alumina, the mechanical strength can be enhanced, but fracture toughness decreases to 8.62 MPa$\sqrt{m}$ [46]. Based on these results and taking into account the experience of the alumina matrix composites, it makes sense to combine the yttria/ceria stabilisation and platelet reinforcement in a zirconia matrix.

In a very first approach, yttria stabilized zirconia derived by the coating method has been combined with platelets on the basis of lanthanum aluminium oxide. In practice, a mixture of 90 wt-% of yttria coated zirconia and 10 wt-% of Al_2O_3/La_2O_3 has been made (ZYA10P) [47]. The ratio of alumina and lanthanum oxide has been made in order to form about 60% of hexagonal platelets within the zirconia matrix during sintering. The sintering process itself is combined with a solid state chemical reaction and therefore has to be made in a very accurate manner; i.e., within the process, nuclides for platelet formation have to be made, before a homogeneous distribution of platelets within the microstructure can be obtained. Its principle mechanical properties are summarized in Table 6. Figure 34 shows the typical microstructure.

Table 6. Typical mechanical properties of ZYA10P.

Property		
ρ_E [g/cm^3]		≈ 5.80
HV$_{0.5}$	Sintered	1419 ± 12
	HIP	1384 ± 27
σ_{3B} [1)] [MPa]	Sintered	1247 ± 136
	HIP	1557 ± 138
K_{Ic} [2)] [MPa$\sqrt{m}$]	Sintered	$9.0 \pm 1.0^*$
	HIP	$7.3 \pm 0.2^*$

[1)] 3-Pkt. (ISO 6872); [2)] HV$_{10}$-indent.

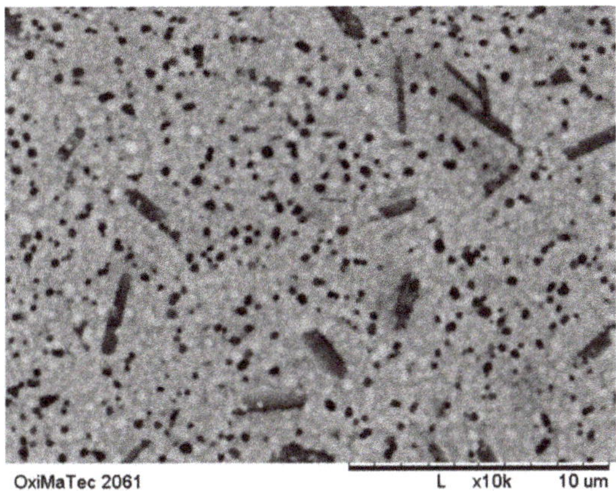

Figure 34. Microstructure of ZYA10P, a material based on Y-TZP and LaAl$_{11}$O$_{18}$.

As can be seen within the table, mechanical strength properties correspond to Y-TZP. This material shows a higher fracture toughness compared to the pure Y-TZP material after HIP treatment. A small remaining porosity in the only sintered material shows a very high fracture toughness. This effect has to be related to the remaining micro-porosity within the microstructure. Its aging behaviour under hydrothermal conditions corresponds to Y-TZP made by coating technology.

In order to enhance the fracture toughness further, we have analysed the stress-strain behaviour of Y-TZP and Ce-TZP (both based on coating technology). Figure 35 shows the results obtained. While in Y-TZP we have linear-elastic behaviour, in Ce-TZP there is a plastic deformation.

First investigations in the system Ce-TZP/SrO/Al$_2$O$_3$ have been made by Cutler et al. [37]. They have found a significant increase in mechanical strength in composites based on a Ce-TZP matrix and containing 15 Vol-% and 30 Vol-% of alumina, where about 30% of this addition contained hexagonal platelets of SrAl$_{12}$O$_{19}$. Normally, Ce-TZP has as mechanical strength of about 400 MPa, while the materials made by Cutler et al. showed 725 MPa. Hardness increased as well from 950 to 1350, while fracture toughness decreased slightly from 12.5 MPa$\sqrt{m}$ to 11 MPa$\sqrt{m}$. Figure 36 shows the influence of the SrO amount, which is relevant for the platelet formation with respect to fracture toughness for two different amounts of alumina addition to the Ce-TZP matrix.

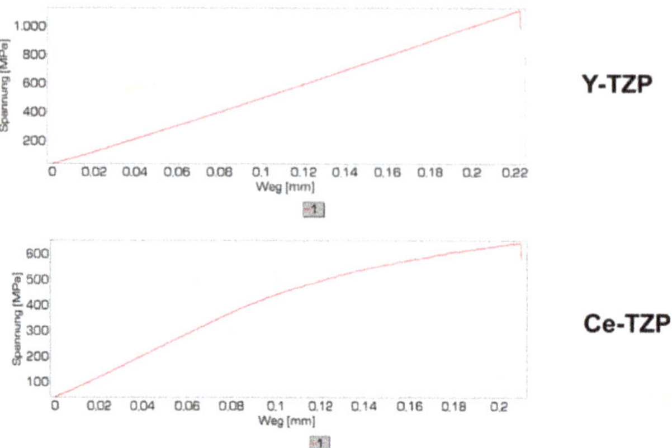

Figure 35. Stress-strain behaviour of Y-TZP and Ce-TZP.

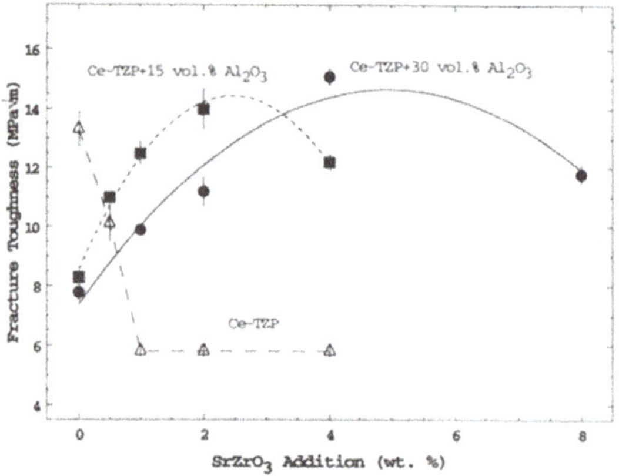

Fracture toughness of Ce-TZP with and without Al₂O₃ additions as a function of SrZrO₃ additions[22]. SrO additions to Ce-TZP lower the toughness due to a refinement in grain size while SrO additions to Ce-TZP/Al₂O₃ increase the toughness most likely due to modification of the transformation zone around the crack tip caused by *in-situ* SrO·6Al₂O₃ formation

Figure 36. Fracture toughness related to the addition of SrO and final platelet formation.

As already mentioned, mechanical strength properties of Ce-TZP are quite limited, but fracture toughness is quite high. Therefore, the approach has been made to again combine 90 wt-% of Ce-TZP with Alumina/Lanthanum oxide (ZCA10P) in order to realize an amount of about 90% of platelet formation during sintering. The materials properties achieved in this first approach are summarized in Table 7 and the corresponding microstructure is shown in Figure 37. From Table 7 it becomes obvious that also by this approach the mechanical strength increases significantly compared to pure Ce-TZP. Further optimisation of the amount of stabilizing ceria finally lead to a fracture toughness of 15 MPa√m at a reasonable mechanical strength of 760 MPa in the 4-point bending test [48].

Table 7. Mechanical properties of Ce-TZP/LaAl$_{11}$O$_{18}$ composite.

Property		
ρ_E [g/cm^3]		5.95
HV$_{0.5}$	Sintered	1200
σ_{3B} [MPa] [1]	Sintered	950
K$_{Ic}$ [MPa$\sqrt{m}$] [2]	Sintered	10.8

[1] 3-Pkt. (ISO 6872); [2] HV$_{10}$-indent.

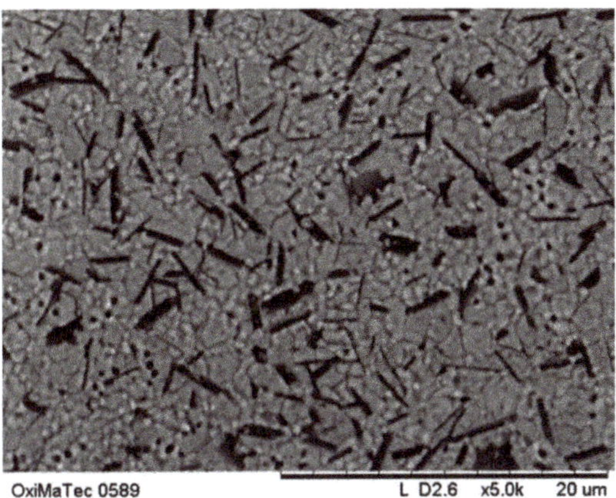

Figure 37. Microstructure of Ce-TZP/LaAl$_{11}$O$_{18}$ composite (ZCA10P).

Finally, we have investigated mixed stabilisation of zirconia by yttria and ceria in a system, which contains about 5 wt-% of alumina/lanthanum oxide. By this approach, the high mechanical strength properties can be retained and fracture toughness can be increased compared to Y-TZP. Initial results obtained are summarized in Table 8. Figure 38 shows the corresponding microstructure.

Meanwhile, the material composition has been optimized further. In addition, the raw material has been substituted by a nanoscaled zirconia powder, derived from direct synthesis of zircon-tetra-chloride with water steam. This optimisation work finally leads to a material with excellent mechanical strength of σ = 1100–1200 MPa measured with 4-point bending test, a fracture toughness k$_{Ic}$ = 12–14 MPa$\sqrt{m}$ and a hardness of HV$_{10}$ = 1250. The microstructure of this optimized material is shown in Figure 39. Within the zirconia matrix there are globular alumina particles and platelets homogeneously distributed.

Table 8. Mechanical properties of Y-Ce-TZP/Al$_2$O$_3$/LaAl$_{11}$O$_{18}$.

Property		ZA05P
ρ_E [g/cm^3]		5.96
HV$_{0.5}$	Sintered	1350
σ_{3B} [MPa] [1]	Sintered	1250
K$_{Ic}$ [MPa$\sqrt{m}$] [2]	Sintered	8.5

[1] 3-Pkt. (ISO 6872); [2] HV$_{10}$-indent.

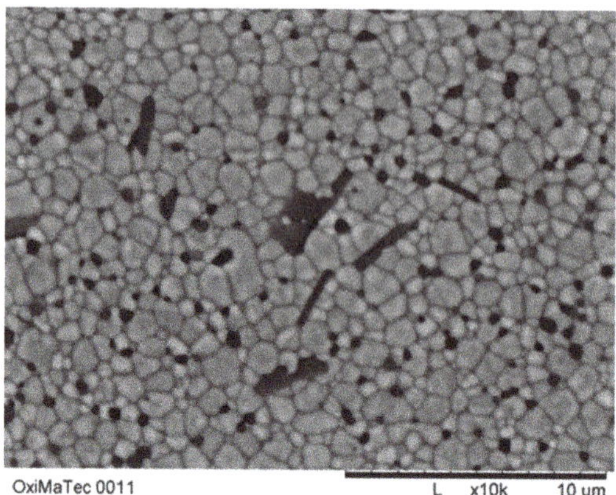

Figure 38. Microstructure of preliminary Y-Ce-TZP/Al$_2$O$_3$/LaAl$_{11}$O$_{18}$.

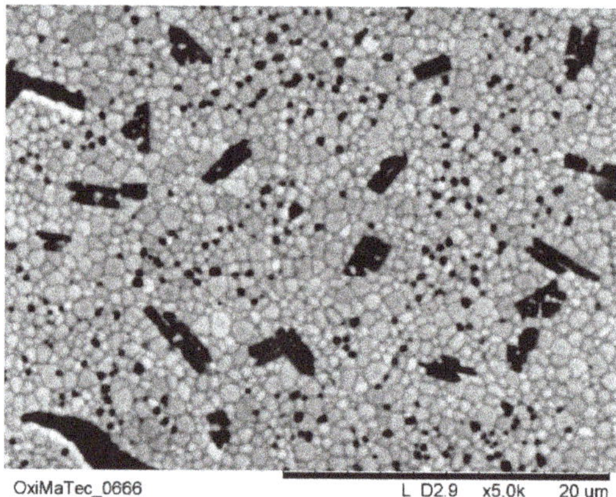

Figure 39. Microstructure of final Y-Ce-TZP/Al$_2$O$_3$/LaAl$_{11}$O$_{18}$ (ZA05P).

Both materials, ZCA10P and ZA05P, have finally been treated under hydrothermal conditions in the same way, just as ZY has been treated. Figure 40 shows the result obtained for ZCA10P (Ce-TZP/LaAl$_{11}$O$_{18}$) and Figure 41 the results obtained for ZA05P (Y-Ce-TZP/Al$_2$O$_3$/LaAl$_{11}$O$_{18}$). Opposite to ZY, which already has a significantly higher hydrothermal stability compared to conventional Y-TZP based on co-precipitated powders, there is no phase transformation.

So, it can be concluded that both materials don't show any hydrothermal decomposition reaction and therefore are preferred materials for bioceramic applications. Both materials have to be regarded as interesting alternative ceramic materials in total hip replacement systems, especially when mated against UHMWPE. Compared to the standard ZPTA (zirconia and platelet reinforced alumina) ceramics of today, these materials are less hard and have a significantly higher fracture toughness, which makes them even safer than the existing solution.

We expect that the excellent material properties of these platelet reinforced zirconia ceramics may play a very important role in dental industry in future. Because of their relatively low hardness (ZCA10P), they might become of interest for prosthetic applications, as well as for implants.

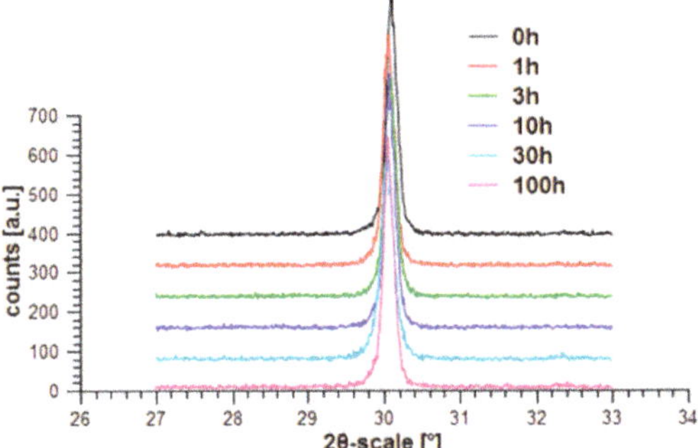

Figure 40. Hydrothermal treatment of ZCA10P (Ce-TZP/LaAl$_{11}$O$_{18}$).

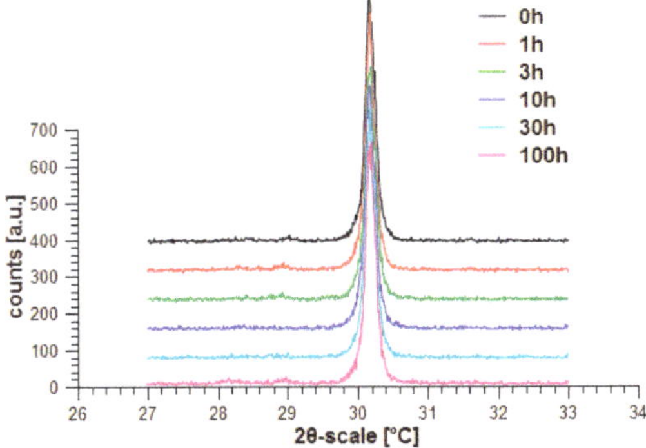

Figure 41. Hydrothermal treatment of ZA05P (Y-Ce-TZP/Al$_2$O$_3$/LaAl$_{11}$O$_{18}$).

6. Conclusions

Two different classes of composite ceramic materials have been described in more detail. One of them is based on an alumina matrix and the other on a zirconia matrix. In case a very high hardness in combination with high strength is required, the alumina matrix composites are preferred. In case a combination of high mechanical strength and fracture toughness is required, a zirconia matrix material is preferred.

The description of the two different composite systems are only a few representative materials which were developed over the past 15 years. However, one can assume that through intelligent additional dopants these two material classes can be improved further and finally tailored to the required properties.

For sure, it is much more difficult to handle these kinds of systems during body preparation compared to the use of commercially available yttria stabilized zirconia. However, in order to really have progress in development of new products, it is mandatory to develop useful technologies for body preparation. The deep understanding of the colloidal processes, which occur in a suspension with the different oxides is mandatory in order to finally come to a good and homogeneous product. Since there are no spectroscopic methods available, the approach for understanding the processes more in detail requires experience in chemistry and a deep understanding of the raw material powders, as well as their behaviour in a suspension. Finally, it has to be pointed out that commercially available dispersing agents are not useful in order to understand the processes. Only the use of defined chemicals help to understand the colloidal chemistry.

Optimisation of the inorganic composition in the alumina/zirconia system requires well educated solid state chemists, who understand crystallography and the chemical solid state reactions, which take place during sintering. Taking into account the amount of different ceramic compositions compared to metallic materials, in ceramics there are only a few different materials compared to the many different materials based on iron. Even in stainless steel there are more materials with different properties available than all ceramic compositions available on the market. To transfer the philosophy of steel into ceramics, a high flexibility within the different companies, who still manufacture their own ready-to-press powders, is required.

Finally, it is state-of-the-art that a good ceramic material requires excellent and completely reproducible raw material powders. Such kinds of powders nowadays are available on the market; i.e., alumina based on the alkoxide process or zirconia based on the chloro-carbonation process. But these raw materials cannot be handled like cheap raw materials such as alumina coming from the Bayer-process or zirconia coming from the thermal decomposition of zircon sand. Chemically derived powders are much more difficult in handling and also in the forming steps, like the pressing of green parts. Finally, sintering is not only a densification process; it is applied to solid state chemistry, by which the densification behaviour can be improved further.

Author Contributions: The work related to platelet reinforced alumina- and zirconia matrix composites has been made together by the authors. Recent results reported for Y-TZP are also based on a close collaboration among the authors. All authors have read and agreed to the published version of the manuscript.

Funding: Research work related to Y-TZP has been funded within the BriteEuram project BE5172 (BIOHPZ) and the most recent results by OxiMaTec GmbH. Research work for reinforced alumina matrix composites has been funded by European Project GRD1-199-10585 (HYPERCER) and also by OxiMaTec GmbH. Finally, most of the work related to zirconia matrix materials has been financed by OxiMaTec GmbH and a part of it by AiF-project ZU4433101KI7.

Conflicts of Interest: The authors declare no conflict of interest.

References and Notes

1. Dörre, E.; Hübner, H. *Alumina: Processing, Properties, and Applications*; Springer: Berlin/Heidelberg, Germany; New York, NY, USA; Tokyo, Japan, 1984; p. 5.
2. Dworak, U.; Olapinski, H.; Thamerus, G. Festigkeitssteigerung von mehrphasigen keramischen Werkstoffen am Beispiel der Systeme ZrO_2-ZrO_2/Al_2O_3-ZrO_2/Al_2O_3-TiC. *Ber. DKG* **1978**, *55*, 98.
3. Garvie, R.C.; Hannink, R.H.; Pascoe, R.T. Ceramic steel? *Nature* **1975**, *258*, 703. [CrossRef]
4. Dworak, U.; Olapinski, H. Sinterwerkstoff. Deutsches Patent DE27 44 700, 27 May 1987.
5. Christel, P. Zirconia: The second generation of ceramics for total hip replacement. *Bull. Hosp. Jt. Dis. Orthop. Inst.* **1989**, *49*, 170.
6. Piconi, C.; Rimondini, L.; Cerroni, L. *La Zirconia in Odontoiatria*; Elsevier: Amsterdam, The Netherlands, 2008.
7. Cockayne, B.; Jones, D.W. *Modern Oxide Materials*; Academic Press: London, UK, 1972.
8. Glizen, W.H. *Alumina as a Ceramic Material*; American Ceramic Society: Columbus, OH, USA, 1970.
9. Schüller, K.-H. Aluminiumoxid. In *Handbuch der Keramik*; Schmid: Freiburg, Germany, 1970.
10. Johnson, W.C.; Coble, R.I. A Test of the Second-Phase and Impurity-Segregation Models for MgO-Enhanced Densification of Sintered Alumina. *J. Am. Ceram. Soc.* **1978**, *61*, 110. [CrossRef]
11. Dörre, E.; (Feldmühle AG Plochingen, Düsseldorf, Deutschland). Personal communication, 1986.
12. Burger, W. *Maßgeschneiderte Hochleistungskeramiken für hochfeste und komplex Bauteile*; Innovation und Technik: Altena, Germany, 2007; Volume 8.
13. Willmann, G. Development in medical-grade alumina during the past two decades. *J. Mater. Process. Technol.* **1996**, *56*, 168. [CrossRef]
14. Curtis, C.E. Development of zirconia resistant to thermal shock. *J. Am. Ceram. Soc.* **1974**, *30*, 180. [CrossRef]
15. Grain, C.F. Phase Relations in the ZrO_2-MgO System. *J. Am. Ceram. Soc.* **1967**, *50*, 288. [CrossRef]
16. Geller, R.F.; Lang, S.M. *Decomposition of $ZrSiO_4$*; NIST: Gaithersburg, MD, USA, 1959.
17. Burger, W. Zirkonoxid in der medizintechnik. In *Technische Keramische Werkstoffe*; Kriegesmann, J., Ed.; Verlagsgrussp Deutscher Wirtschaftsdienst: Köln, Germany, 1996.
18. Dransfield, G.P.; Fothergill, K.A.; Egerton, T.A. *Euro Ceramics*; De, G., Terpstra, R., Metselaar, R., Eds.; Elsevier: London, UK, 1989; Volume 1, p. 275.
19. Blackburn, S.R.; Egerton, T.A.; Jones, A.G. Vapour phase synthesis of nitride ceramic powders using a DC plasma. *Br. Ceram. Prod.* **1991**, *47*, 87.
20. Garvie, R.C.; Hannink, R.H. Sub-eutectoid aged Mg-PSZ alloy with enhanced thermal up-shock resistance. *J. Mater. Sci.* **1982**, *17*, 2637.

21. Garvie, R.C.; Hannink, R.H.J.; Mekinnon, N.A. Partially Stabilized Zirconia Ceramics; Method of Making Said Ceramics, Dies Constructed of Said Ceramics Cutting Tools with Cutting Surface and Tappet Facings Formed of Said Ceramics. European Patent EP 0 013 599, 9 May 1984.
22. Olapinski, H.; Dworak, U.; Burger, W. *Ceramic Materials and Components for Engines*; Bunk, W., Hausner, H., Eds.; Wiley: Hoboken, NJ, USA, 1987; p. 537.
23. Dworak, U.; Olapinski, H.; Burger, W. Science and technology of zirconia III. In *Advances in Ceramics*; Heuer, A.H., Hobbs, L.W., Eds.; The American Ceramic Society: Columbus, OH, USA, 1988; Volume 24, p. 545.
24. Drennan, J.; Hannink, R.H.J. Effect of SrO Additions on the Grain-Boundary Microstructure and Mechanical Properties of Magnesia-Partially-Stabilized Zirconia. *J. Am. Ceram. Soc.* **1986**, *69*, 541. [CrossRef]
25. Burger, W.; Richter, H.G.; Piconi, C.; Vatteroni, R.; Cittadini, A.; Boccalari, M. Materials in medicine. *J. Mater. Sci.* **1997**, *8*, 113.
26. Burger, W.; Richter, H.G.; Slawik, G.; Willmann, G. Sintering Behaviour of Medical Grade Zirconia (Y-TZP) Powders. *Ceram. Trans.* **1995**, *48*, 43.
27. Richter, H.G.; Burger, W.; Willmann, G. Ceramic hip joint heads made from alumina and zirconia—A comparison. *Ceram. Trans.* **1995**, *48*, 73.
28. Chevalier, J.; Cales, B.; Drouin, M. Low-temperature aging of Y-TZP ceramics. *J. Am. Ceram. Soc.* **1999**, *82*, 2150. [CrossRef]
29. Zhang, F.; Venmeeusel, K.; Inokoshi, M.; Batuk, M.; Hadermann, J.; van Meerbeck, B.; Naert, I.; Vleugels, J. 3Y-TZP ceramics with improved hydrothermal degradation resistance and fracture toughness. *J. Eur. Ceram. Soc.* **2014**, *34*, 2453. [CrossRef]
30. Nakamura, K.; Kauno, T.; Milleding, P.; Ortengren, U. Zirconia as a dental abutment material: A systematic review. *Int. J. Prosthodont.* **2010**, 299.
31. Kern, F. *Non-Published Investigations on Several Zirconia Matrix Materials*; Stuttgart University: Stuttgart, Germany, 2020.
32. Chevalier, J.; Gremillard, L.; Deville, S. Low temperature degradation of zirconia and implication for biomedical implants. *Annu. Rev. Mater. Res.* **2007**, *37*, 1–31. [CrossRef]
33. Dörre, E.; (Feldmühle AG Plochingen, Düsseldorf, Deutschland). Personal communication, 1989.
34. Cevales, G. Das Zustandsdiagramm Al_2O_3-ZrO_2 und die Bestimmung einer neuen Hochtemperaturphase. *Ber. Dtsch. Keram. Ges.* **1968**, *45*, 217.
35. Claussen, N. Fracture toughness of Al_2O_3 with an unstabilized ZrO_2 dispersed phase. *J. Am. Ceram. Soc.* **1976**, *59*, 49. [CrossRef]
36. Lange, F.F. Transformation toughening; part 3: Experimental observations in the ZrO_2-Y_2O_3-system. *J. Mater. Sci.* **1982**, *17*, 247. [CrossRef]
37. Lange, F.F.; Hirlinger, M.M. Hinderance of grain growth in Al_2O_3 by ZrO_2 inclusions. *J. Am. Ceram. Soc.* **1984**, *67*, 164. [CrossRef]
38. Burger, W. Umwandlungs-und plateletverstärkte Aluminiumoxidmatrixwerkstoffe. *Keram. Z.* **1997**, *49*, 1067.
39. Burger, W. Umwandlungs-und plateletverstärkte Aluminiumoxidmatrixwerkstoffe. (Teil 2). *Keram. Z.* **1998**, *50*, 18.
40. Cutler, R.A.; Mayhow, R.J.; Rettyxmean, K.M.; Virkar, A.V. High-Toughness Ce-TZP/Al_2O_3 Ceramics with Improved Hardness and Strength. *J. Am. Ceram. Soc.* **1991**, *74*, 179. [CrossRef]
41. Yasuoka, M.; Hirao, K.; Brito, M.G.; Kanzaki, S. High-Strength and High-Fracture-Toughness Ceramics in the Al_2O_3/$LaAl_{11}O_{18}$ Systems. *J. Am. Ceram. Soc.* **1995**, *78*, 1853. [CrossRef]
42. Burger, W.; Kiefer, G.; Bellido, E.; Andersch, H. Sintered Shaped Body Reinforced with Platelets. U.S. Patent 6,452,957, 17 September 2002.
43. Burger, W. (coordinator) Results summarized in the final report of Brite Euram Project GRD1-199-10585, 2005.
44. Ceramic Dental Implants: Where Do We Stand? Available online: https://www.straumann.com/en/discover/youtooth/article/esthetics/2020/ceramic-dental-implants-where-do-we-stand.html (accessed on 22 June 2020).
45. EI Attaoui, H.; Saädaoui, M.; Chevalier, J.; Fantozzi, G. Static and cyclic crack propagation in Ce-TZP ceramics with different amounts of transformation toughening. *J. Eur. Ceram. Soc.* **2007**, *27*, 483. [CrossRef]
46. *Brochure "Nanozyr"*; Panasonic Electric Works Yokkaichi Co., Ltd.: Tokyo, Japan, 2011.
47. Burger, W. Oxidkeramik—Neue werkstoffe für die medizintechnik und industrielle anwendungen. In *Keramische Werkstoffe*; Hennicke, D., Ed.; Springer: Köln, Germany, 2012.
48. Non-published results, based on European Patent EP 2 086 908 (2006) and EP 2 086 909 (2006).

Journal of
Composites Science

Article

New Perspectives on Zirconia Composites as Biomaterials

Giuseppe Magnani, Paride Fabbri, Enrico Leoni, Elena Salernitano and Francesca Mazzanti *

ENEA, Laboratory of Materials Technologies Faenza, Via Ravegnana 186, 48018 Faenza, Italy;
giuseppe.magnani@enea.it (G.M.); paride.fabbri@enea.it (P.F.); enrico.leoni@enea.it (E.L.);
elena.salernitano@enea.it (E.S.)
* Correspondence: francesca.mazzanti@enea.it

Abstract: Zirconia–alumina composites couple the high toughness of zirconia with the peculiar properties of alumina, i.e., hardness, wear, and chemical resistance, so they are considered promising materials for orthopedic and dental implants. The design of high performance zirconia composites needs to consider different aspects, such as the type and amount of stabilizer and the sintering process, that affect the mechanics of toughening and, hence, the mechanical properties. In this study, several stabilizers (Y_2O_3, CuO, Ta_2O_5, and CeO_2) were tested together with different sintering processes to analyze the in situ toughening mechanism induced by the tetragonal–monoclinic (t–m) transformation of zirconia. One of the most important outcomes is the comprehension of the opposite effect played by the grain size and the tetragonality of the zirconia lattice on mechanical properties, such as fracture toughness and bending strength. These results allow for the design of materials with customized properties and open new perspectives for the development of high-performance zirconia composites for orthopedic implants with high hydrothermal resistance. Moreover, a near-net shape forming process based on the additive manufacturing technology of digital light processing (DLP) was also studied to produce ceramic dental implants with a new type of resin–ceramic powder mixture. This represents a new frontier in the development of zirconia composites thanks to the possibility to obtain a customized component with limited consumption of material and reduced machining costs.

Keywords: zirconia–alumina composite; stabilizing oxides; critical grain size; tetragonality; mechanical properties; fracture toughness; flexural strength; ceramic additive manufacturing; DLP

Citation: Magnani, G.; Fabbri, P.; Leoni, E.; Salernitano, E.; Mazzanti, F. New Perspectives on Zirconia Composites as Biomaterials. *J. Compos. Sci.* **2021**, *5*, 244. https://doi.org/10.3390/jcs5090244

Academic Editor: Francesco Tornabene

Received: 30 July 2021
Accepted: 7 September 2021
Published: 11 September 2021

Publisher's Note: MDPI stays neutral with regard to jurisdictional claims in published maps and institutional affiliations.

1. Introduction

Zirconia-toughened alumina (ZTA) and alumina-toughened zirconia (ATZ) composites have been studied for many decades to overcome some drawbacks of the tetragonal zirconia polycrystal (TZP) [1–3]. Zirconia–alumina composites have been used for several years as load-bearing biomaterials [4–6]. They combine the high toughness and strength of zirconia with the high hardness and stiffness of alumina, and they show also an increased hydrothermal stability of the tetragonal zirconia phase. It is well-known that the stress-induced tetragonal-to-monoclinic (t–m) transformation of zirconia results in fracture toughness improvement [7–11] due to energy-dissipative mechanisms and the inhibition of crack tip propagation [12]. Characteristics such as grain size, the type and amount of stabilizer, and the sintering process strongly affect the tetragonal zirconia transformability and the transformation toughening mechanism. In fact, the grain size of tetragonal zirconia has to be maintained below a critical size to reach a high value of fracture toughness [13].

Many oxides have already been tested as stabilizers to increase the metastability of the tetragonal phase by means of varying the c/a ratio of the elementary cell. The c/a ratio of the tetragonal phase is generally known as "tetragonality" and is an indicator of the distortion of the t-ZrO_2 unit-cell, hence the instability. On the other hand, alumina addition increases matrix stiffness and exerts a constraint on zirconia particles, maintaining them in the metastable tetragonal state [14,15] and acting as a "mechanical stabilizer".

J. Compos. Sci. **2021**, *5*, 244. https://doi.org/10.3390/jcs5090244

One of the main problems of zirconia-based compounds as biomaterials is the sensitivity of 3Y-TZP (3 mol% yttria tetragonal zirconia polycrystal) ceramics to low temperature degradation (LTD) when they are in contact with water that is already at human body temperature or water vapor [16]. The inherent presence of oxygen vacancies, generated when Y^{3+} replaces Zr^{4+} in the cationic sub-lattice, can be at the origin of aging, since they can be refilled by hydroxyl groups in the presence of water [17]. As a result of the LTD process, the t–m transformation of zirconia grains spontaneously occurs without any external applied stress. The correlated volume expansion results in the formation of microcracks that can catastrophically damage orthopedic or dental prostheses. In the literature, many data regarding the lifetime estimation of Y-TZP and ATZ or ZTA composites have been collected [18–22]. Accelerated aging tests in steam and hot water at low temperatures (e.g., 90–134 °C) are the accepted methods to simulate an in vivo aging behavior with the determination of activation energy value for environmentally driven t–m transformation. Fabbri et al. [21] studied a ZTA composite that showed a very low reactivity to the LTD compared to 3Y-TZP. This behavior of ZTA composites confirms that the presence of alumina grains can act as a barrier for the propagation of phase transformation to the neighboring zirconia grains, promoting the higher hydrothermal stability of the tetragonal phase [23–25]. Other studies have evaluated the possibility to significantly retard the hydrothermal degradation of Y-TZPs with small amounts of alumina addition. This result is attributed to the segregated Al^{3+} at the grain boundary of zirconia [24,26–28] without compromising the mechanical properties [17,29,30]. LTD is also influenced by the microstructure. Halmann et al. [31] showed that a finer microstructure had a beneficial effect on the LTD of Y-TZP. At the same time, a finer microstructure does not always affect the mechanical properties, such as flexural strength and fracture toughness, of zirconia-based materials in a positive manner [12,32–36]. In any case, all the previous studies confirmed that alumina–zirconia composites represent an improvement in terms of LTD resistance.

In last two decades, additive manufacturing (AM) technology has been brought from research or niche and expensive industrial applications to everyone thanks to the cost reduction of 3D printers. AM has been demonstrated to be effective in almost every material field and in multiple applications. The digital light processing (DLP) technique consists of the light-induced, layer-by-layer polymerization of a photocurable resin filled with ceramic powders. This technique allows for the manufacturing of relative dense ceramic components, with high degree of detail and surface finishing, that can be advantageously applied in, for instance, the biomedical field (bone scaffolds), the sector of metal-free dental restoration (endosseous implants and dental crowns), and microelectronics (sensors). 3D printing can be considered to be the most promising near net-shape forming technique for technical ceramics. In fact, it has opened the space for application in sectors where high manufacturing costs, connected to the machining costs (30–50% to the total manufacturing costs), usually prevent ceramic use [37]. In addition to the economical evaluation, we should also consider the important aspects related to the realization of parts with completely new designs and positive impacts on environmental sustainability due to the limited production of wastes and the sustainable use of raw materials. Finally, zirconia-based composites represent a new class of materials for applications with 3D printing technologies [38–42]: the need for the complex or customized shapes required in the field of biomaterials could be more easily satisfied by AM techniques. Additional studies are, however, required in order to demonstrate that AM can be conveniently applied to zirconia–alumina composites to produce reliable components.

In this paper, a comprehensive study of the effects of different parameters, i.e., type and amount of stabilizers, sintering thermal cycles, on the mechanical properties of zirconia-based materials, is described along with a demonstration of the applicability of the DLP AM technique for the manufacturing of zirconia–alumina-based dental elements.

2. Materials and Methods

2.1. Ceramic Mixtures and Sample Preparation

Yttria-stabilized zirconia (TZ-3YB, Tosoh, Tokyo, Japan), monoclinic zirconia (TZ0, Tosoh, Tokyo, Japan), alumina (Baikalox SM8, Baikowski Chimie, Poisy, France), chromium (III) oxide, tantalum (V) oxide, copper (II) oxide, and cerium (IV) oxide (99.9%, Carlo Erba, Milano, Italy) powders were mixed in the weight ratios reported in Table 1. The average particle was is 40 nm for both zirconia powders, 120 nm for alumina, and 0.7–1 μm for the other powders (Ta_2O_5, CeO_2, CuO, and Cr_2O_3).

Table 1. List of zirconia and zirconia composites considered in the present study.

Samples	ZrO_2-3Y (wt%)	ZrO_2-TZ0 (wt%)	Ta_2O_5 (wt%)	CeO_2 (wt%)	CuO (wt%)	Al_2O_3 (wt%)	Cr_2O_3 (wt%)
Zr3Y	100	-	-	-	-	-	-
Zr2Y	67.2	32.8	-	-	-	-	-
Zr3YTa	99.85	-	0.15	-	-	-	-
ZrCe	-	84	-	16	-	-	-
Zr3YCu	99.9	-	-	-	0.1	-	-
20803Y	80	-	-	-	-	20	-
50502Y	33.42	16.32	-	-	-	50	-
50502.5Y	41.8	8.2	-	-	-	50	-
50503Y	50	-	-	-	-	50	-
60402Y	26.8	13.08	-	-	-	59.52	0.60
60403Y	39.88	-	-	-	-	59.52	0.60

In case of ZTA or zirconia stabilized with oxides, a slurry (38.5 wt% of solid) was prepared using water as a solvent and 1 wt% of dolapix PC33 (Zschimmer & Schwarz, Lahnstein, Germany) as a dispersant; this slurry was homogenized with a Turbula mixer for 8 h in the presence of 3 mm zirconia spheres.

The slurry was dried with an IR lamp or by freeze-drying. The freeze-drying process was performed with an apparatus composed by a vacuum chamber paired with a vacuum pump through a cold trap filled with liquid nitrogen. The slurry (25 wt% of solid) was granulated in a liquid nitrogen bath with an ultrasonic nebulizer probe. The frozen granules were placed in the chamber under an active vacuum. The temperature of the frozen slurry was naturally maintained at about −20 °C by the heat removed during the water sublimation. The freeze-drying process ended when the powder naturally reached room temperature and the pressure decreased to 0.1 Pa.

The green samples were prepared by die pressing at 60–80 MPa, followed by cold isostatic pressing (CIP) at 100–150 MPa.

2.2. Ceramic Resin and 3D Printing

The ATZ resin was prepared by mixing liquid acrylate monomers (Sartomer, IGM Resins and Allnex), a 405 nm photo-initiator (IGM Resins, Waalwijk, The Netherlands), a commercial zirconia (TZ-3YS, Tosoh, Tokyo, Japan) with an average grain size of 90 nm, and the abovementioned commercial alumina powders with a weight ratio of 80/20. A dispersant (2 wt%) was added to the preparation before the high energy ball milling process to reduce the viscosity of the photocurable ceramic slurries [43]. A solid content in the range of 36–38 vol% of ceramic powder was reached in the slurries in order to obtain a viscosity lower than 1 Pa.s at 10 s^{-1} [44]. A DLP 3D printer (3DLPrinter-HD 2.0+, Robotfactory, Italy; construction volume of L 100 × W 56 × H 150 mm^3) was used and equipped with a projector using a UV–visible high pressure Hg lamp (250 W of power and 3000 lm of luminous flux). The layer height and exposure time for each layer were set in the ranges of 30–50 μm and 6–20 s, respectively. After printing and washing, a post-curing step with a UV lamp was also applied.

2.3. Sintering Process: Thermal Cycles

The green samples were debinded and pressureless solid state sintered (SSS) in flowing air (LINN Elektronik HT—1800 VAC, LINN HIGH THERM GMBH, Hirschbach, Germany). Samples were dewaxed with a cycle up to 800 °C (10 °C/h ramp) in flowing air and then pressureless sintered in flowing air in the range of 1450–1570 °C for different holding times (2–80 h) depending on the composition. The dewaxing and sintering steps for the 3D-printed green bodies were performed at 1550 °C for ATZ for 1 h after a debinding step performed at 800 °C.

In addition, the two-step sintering (TSS) process was also tested. In this case, T_1 and T_2 were in the ranges of 1400–1500 and 1350–1450 °C, respectively, with zirconia and ATZ.

2.4. Physical, Microstructural and Mechanical Characterization

Sintered density was determined by Archimedes' method.

Diffraction patterns were collected by using a Philips X-ray powder diffractometer with Bragg–Brentano geometry and Cu Kα radiation (40 kV and 35 mA) to identify the crystalline phases in the sintered samples and to evaluate the tetragonality of the tetragonal phase.

Viscosity measurements were performed using a Malvern Kinexus Pro+ rheometer (Kinexus pro+, Malvern Instruments, Ltd., Worcestershire, UK) at 25 °C with cone-plate geometry (4°, 40 mm) in shear rate control from 0.1 to 300 s^{-1}.

The microstructural analysis of both the surface and cross sections of sintered bodies was performed with a scanning electron microscope (SEM, LEO 438 VP).

The flexural strength was determined at room temperature with four-point bending tests (five tests for each composition). Samples, in the form of 2 × 2.5 × 25 mm bars, were prepared and tested in accordance with the standard ENV843-1:2004 (cross head speed of 0.5 mm/min and support span of 20 mm). Hardness (H_v) was determined by means of Vickers indentation with a load of 9.8 N, while fracture toughness (K_{IC}) was determined by means of Vickers indentation with a load ranging from 9.8 to 98 N. To calculate fracture toughness, the formula proposed by Niihara [45] for Palmqvist cracks was used (Equation (1)):

$$K_{IC} = \frac{0.035 \left(H_v a^{0.5}\right) \left(3\frac{E}{H_v}\right)^{0.4} \left(\frac{l}{a}\right)^{-0.5}}{3} \quad \text{for } 0.25 < \frac{l}{a} < 2.5 \tag{1}$$

where a is the indent half-diagonal, E is the Young's modulus, and l is the Palmqvist crack length.

3. Results and Discussion

3.1. Stabilization of Zirconia: Variables That Influence Transformability

3.1.1. Type of Stabilizer

Different zirconia-based materials were produced and characterized to study the effect of the type of the stabilizer on the t–m phase transformation. The dopants could be classified according to their oxidation state (Cu^{2+}, Y^{+3}, Ce^{4+}, and Ta^{5+}). More precisely, they are stabilizers of the tetragonal phase (Y_2O_3 and CeO_2) and toughening oxides (Ta_2O_5 and CuO) [46].

Cations' valence and size affect the stabilization mechanism of the tetragonal phase [47–50], even if the correlation is not univocal, as suggested by Yoshimura et al. [51].

The phase composition and crystallographic parameters were evaluated for each mixture (lattice constants c and a and their c/a ratio, namely "tetragonality") of doped tetragonal zirconia. The values of fracture toughness and hardness were also determined (Table 2).

Table 2. Properties of zirconia-based materials doped with different stabilizers and toughening oxides. DR is the relative density, c/a is the tetragonality, K_{IC} the fracture toughness, and H_V the microhardness.

Samples	DR (%)	Tetragonal ph. (vol%)	Monoclinic ph. (vol%)	c/a	K_{IC} (MPa m$^{1/2}$)	H_V (GPa)
ZrCe	98	100	-	1.0191	18.4 ± 0.9	8.7 ± 0.2
Zr2Y	98	83	17	1.0166	9.4 ± 0.4	12.5 ± 0.5
Zr3Y	99	100	-	1.0159	4.0 ± 0.1	11.0 ± 0.3
Zr3YTa	99	100	-	1.0173	9.4 ± 0.5	12.7 ± 0.4
Zr3YCu	98	100	-	1.0161	4.6 ± 0.2	12.5 ± 0.4

Yttrium oxide (Y_2O_3) is the most common stabilizer of the tetragonal phase, and Y-TZP is widely used due to its strong mechanical properties [26,52]. The addition of different amounts of yttria influenced the c/a ratio, which indicated the transformability of the available tetragonal phase. In comparing the values in Table 2 for the mixtures of Zr2Y and Zr3Y, it is clear that a higher yttria content (3 vs. 2 mol%) led to a greater stabilization of the tetragonal phase, which corresponded to a decrease in the c/a ratio (1.0159 vs. 1.0166) and a lower toughness (4.0 vs. 9.4 MPa m$^{1/2}$).

Cerium oxide (CeO_2) is another well-known stabilizer of zirconia, and ceria-doped zirconia exhibits very high values of fracture toughness [53]. The ZrCe sample in our study showed toughness value four times higher than that of Zr3Y (18.4 vs. 4.0 MPa m$^{1/2}$, respectively) as indicated in Table 2. CeO_2 is a stabilizer as Y_2O_3, but its c/a ratio is higher; this means that the tetragonal phase is less stabilized, so its transformation is easier, thus leading to an increase in fracture toughness. On the other hand, as described in the literature [54], CeO_2 does not allow one to obtain high values of mechanical resistance due to its limited capability to contain grain growth during sintering. Indeed, ZrCe grains are wider (ca. 2.0 μm) than Y-TZP ones (ca. 0.5–0.8 μm) [27]. As the oxidation state is the same of Zr^{4+}, Ce^{4+} does not generate oxygen vacancies inside the ZrO_2 cell, so, in a humid environment, the t–m spontaneous transformation is not promoted and CeO_2-stabilized zirconia shows significantly high resistance to LTD [17,54,55].

Tantalum oxide (Ta_2O_5) is known in the literature for its toughening effect when added to 3Y-TZP [56,57]. In our study, the addition of Ta_2O_5 led to a higher value of fracture toughness than 3Y-TZP (9.4 vs. 4.0 MPa m$^{1/2}$, respectively), as shown in Table 2. The addition of Ta_2O_5 to 3Y-TZP increased the c/a ratio (1.0173 vs. 1.0159, respectively) such that the chemical driving force for the t–m transformation was enhanced, and this led to a higher value of fracture toughness. On the other hand, the stabilizing effect of Y_2O_3 was contrasted by the addition of Ta_2O_5, which is a toughening oxide that increases the t–m martensitic transformation temperature [50], resulting in a toughening effect.

Copper oxide (CuO) was also tested as toughening agent for Y-TZP. The results reported in Table 2 show that the addition of CuO only led to a slight increase in the fracture toughness of the 3Y-TZP (4.6 vs. 4.0 MPa m$^{1/2}$, respectively). This result is in contrast with the results reported by Ramesh et al. [58], where a different Y-TZP powder was used.

After comparing the fracture toughness values (Table 2) as function of the tetragonality, a linear correlation was obtained, as shown in Figure 1. If the c/a ratio of the tetragonal phase was near 1 (i.e., the c/a value of the cubic phase), the tetragonal phase was more stable and hence the t–m transformation became more difficult and the fracture toughness decreased. On the contrary, if the c/a ratio of the tetragonal phase increased up to 1.022 (which is the b/a value of the monoclinic phase), the t–m transformation was favored and the fracture toughness increased.

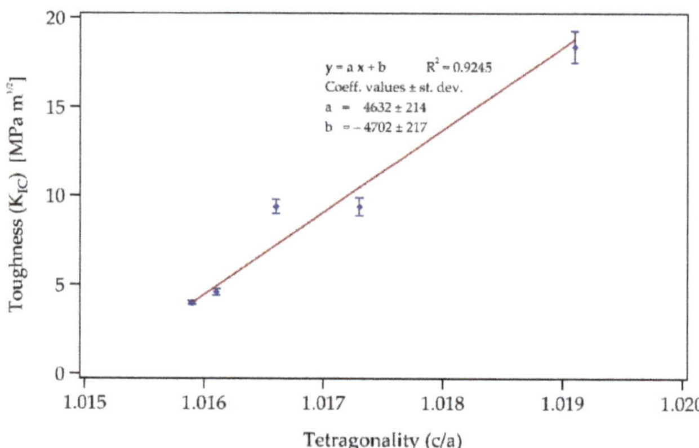

Figure 1. Fracture toughness vs. tetragonality of the different stabilizers in zirconia-based materials.

3.1.2. Stabilizer Content

The effect of different contents of stabilizer (Y_2O_3) was studied in ZTA composites with a 60/40 alumina/zirconia weight ratio. The results reported in Table 3 show that the fracture toughness reached a maximum value of 6.2 MPam$^{1/2}$ with the lowest amount of stabilizer (60402Y). The same results were previously observed in ZTA composites with a 50/50 alumina/zirconia weight ratio, as reported in Table 3 [59].

Table 3. Properties of ZTA materials doped with different amounts of stabilizer. DR is the relative density, c/a is the tetragonality, K_{IC} the fracture toughness, H_V the microhardness, and MOR is the four-point flexural strength.

Samples	DR (%)	c/a	Tetragonal ph. (vol%)	Cubic ph. (vol%)	Monoclinic ph. (vol%)	K_{IC} (MPa m$^{1/2}$)	H_V (GPa)	MOR (MPa)
60403Y	99.9	1.0168	76.3	23.7	0	5.1 ± 0.3	17.7 ± 0.3	660 ± 23
60402Y	98.8	1.0169	98.0	0	2	6.2 ± 0.2	18.0 ± 0.3	794 ± 98
50503Y [1]	99.3	1.0165	98.7	1.3	0	6.0 ± 0.1	14.9 ± 0.5	-
50502.5Y [1]	99.9	1.0170	94.6	5.4	0	5.6 ± 0.2	15.7 ± 0.3	-
50502Y [1]	99.2	1.0175	100	0	0	8.1 ± 0.1	15.4 ± 0.2	-

[1] Reprinted with permission from ref. [59]. Copyright © 2021 Elsevier Ltd.

This behavior can be explained by the analysis of the variation of the tetragonal phase amount and the tetragonality with stabilizer content. In fact, 60403Y had a lower amount of tetragonal phase (less than 80%) and a lower tetragonality than those of 60402Y. This means that a lower quantity of tetragonal phase was available to the toughening t–m transformation in the 60403Y composite. Furthermore, in the same sample, the lower tetragonality enhanced the stability of the tetragonal phase, which caused a decrease in the fracture toughness. These observations are also in line with the study of Yoshimura et al. [60], which reported the dependence of the c/a ratio on stabilizer content.

3.1.3. Critical Grain Size

3Y-TZP was sintered in six different conditions in order to highlight the effect of the grain size variation on tetragonality and, consequently, fracture toughness. The experimental results are reported in the Table 4.

After increasing the sintering time to 60 h at 1550 °C, the fracture toughness and grain size increased up to maximum values of 7.7 MPa m$^{1/2}$ and 1.19 μm, respectively (Figure 2). Furthermore, a strong dependence between the fracture toughness and tetragonality was observed at the microstructural level. Indeed, with the increase in sintering time, tetragonality increased, i.e., the tetragonal cell instability grew. This instability, caused by the distortion of the cell, promoted the t–m transformation and a consequent increase in fracture toughness.

Table 4. Properties of zirconia-based (3Y-TZP) samples sintered with different thermal cycles. DR is the relative density, Dm the average grain size, c/a is the tetragonality, K_{IC} the fracture toughness, and H_V the microhardness.

Thermal Cycle	DR (%)	Dm (μm)	c/a	K_{IC} (MPa m$^{1/2}$)	H_V (GPa)
1500 °C—2 h	98.9	0.53	1.0159	5.0 ± 0.1	11.0 ± 0.2
1550 °C—20 h	99.9	0.78	1.0164	5.3 ± 0.1	12.1 ± 0.2
1550 °C—30 h	99.6	0.87	1.0165	5.4 ± 0.1	11.5 ± 0.2
1550 °C—40 h	99.9	0.95	1.0165	6.6 ± 0.2	11.9 ± 0.2
1550 °C—60 h	99.6	1.19	1.0167	7.7 ± 0.1	11.0 ± 0.2
1550 °C—80 h	95.8	-	1.0162	-	-

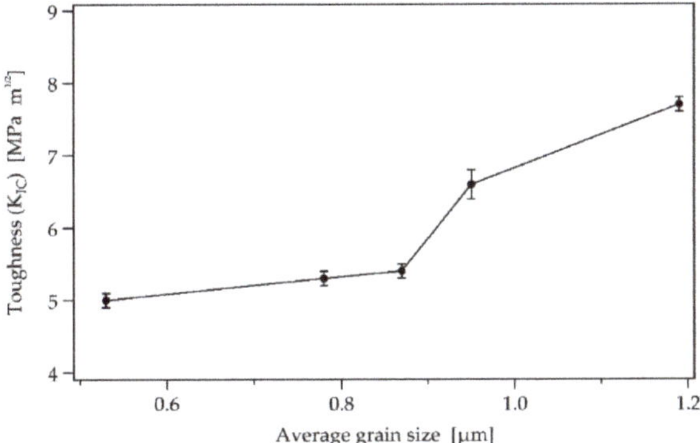

Figure 2. Fracture toughness vs. average grain size of the different sintered zirconia-based materials.

The sample sintered at 1550 °C for 80 h was characterized by the lowest sintered density due to the formation of the monoclinic phase, and it showed many cracks. In fact, XRD analysis confirmed that all the samples were mainly constituted by the tetragonal phase with traces of the cubic phase, while the sample sintered at 80 h showed an increase in monoclinic phase content.

According to these data, the critical grain size can be estimated to be equal or greater than 1.19 μm for this 3Y-TZP material. This value is in agreement with the critical grain observed by Lange [32].

3.2. Parameters That Influence Mechanical Properties

The relationships between the microstructure and mechanical properties of Y-TZP ceramics have been extensively studied over the past four decades, and different effects have been identified.

It was demonstrated that the toughening effect, related to the t–m transformation mechanism in Y-TZP ceramics, is promoted by larger grain sizes [12,32–36]. On the other hand, some mechanical properties, including flexural strength, are known to be enhanced by fine microstructures [36,61,62].

Indeed, as grain size coarsens, the critical defect enlarges, thus leading to a strength decrease [63]. According to the Griffith (Equation (2)), strength (σ_R), fracture toughness (K_{IC}), and failure origin size (c) are strictly connected and their control is necessary to obtain reliable structural ceramic materials.

$$\sigma_R \sim \frac{K_{IC}}{\sqrt{\pi c}} \tag{2}$$

Unfortunately, the best conditions (composition, grain size, transformability, etc.) to reach ceramic strength in zirconia-based materials are not the same for maximizing fracture toughness, so the reliability of these ceramics comes from the compromise of these two properties [64].

Hardness is also influenced by microstructure [65]. Generally, hardness is strictly related to density, but no univocal correlation between hardness and grain size has been proven. The hardness values of 3Y-TZP do not show the influence of the grain dimension in the submicrometric range [66].

A typical method to obtain ceramics with fine microstructures and improved mechanical properties (flexural strength) is based on the application of innovative sintering processes that limit grain growth. Among the best known sintering methods to refine ceramic microstructures, the spark plasma sintering (SPS) [67,68] and microwave sintering (MWS) [69] methods are the most efficient.

A simple and cost-effective method for industrial applications to obtain near full dense ceramics with controlled grain growth is TSS (two-step sintering) [70], in which the sample is first heated to a higher temperature to achieve an intermediate density and then cooled down and held at a lower temperature until it is fully dense. This sintering method has been successfully applied for ZTA composites [71,72].

The effect of TSS on the 3Y-TZP and ZTA samples (Table 5) was studied and compared to that of classic SSS. In the case of 3Y-TZP, TSS showed an advantageous effect on grain size (almost halved), as shown in Figure 3a,b. However, TSS seemed to have no effect on the fracture toughness. This was probably due to two opposite and concomitant effects of TSS that compensate for each other. The grain size refinement contrasted with the toughening effect achieved when the grain size approached the critical value. On the other hand, the tetragonal phase obtained with the TSS was more transformable, as evidenced by the slight increase in the tetragonality. It is probable that the longer holding time at the higher temperature promoted the migration of the stabilizer (Y^{3+}) [52]; hence, the yttria concentration within the tetragonal phase decreased and enhanced transformability.

Table 5. Properties of 3Y-TZP and ZTA materials sintered with the single step (SSS) or two-step cycles (TSS). DR is the relative density, K_{IC} the fracture toughness, H_V the microhardness, MOR is the four-point flexural strength, Dm is the average grain size (A refers to alumina and Z to zirconia grains), and c/a is the tetragonality.

Samples	Thermal Cycle	DR (%)	Dm A/Z (μm)	c/a	K_{IC} (MPa m$^{1/2}$)	H_V (GPa)	MOR (MPa)
3Y-TZP SSS	1500 °C—1 h	99.7	0.33	1.0154	5.0 ± 0.1	13.3 ± 0.3	1095 ± 75
3Y-TZP TSS	1400/1350 °C—30 h	99.8	0.18	1.0157	4.9 ± 0.1	13.4 ± 0.2	1102 ± 85
60402Y SSS	1550 °C—1 h	98.8	0.71/0.44	1.0169	6.2 ± 0.2	18.0 ± 0.3	794 ± 98
60402Y TSS	1500/1450 °C—30 h	98.6	0.58/0.35	-	5.5 ± 0.3	16.1 ± 0.4	660 ± 89
60402Y TSS FD	1500/1450 °C—30 h	98.7	0.59/0.34	-	-	-	872 ± 47
60403Y SSS	1550 °C—1 h	99.9	0.70/0.43	1.0168	5.1 ± 0.3	17.7 ± 0.3	660 ± 23
60403Y TSS	1500/1450 °C—30 h	99.8	0.58/0.36	-	4.8 ± 0.2	16.0 ± 0.2	700 ± 57

Again, the flexural strength values were very similar despite the halved grain size. It is probable that the grain refinement obtained with TSS did not contribute to a decrease in critical defect size. In fact, as observed by Xiong et al. [73], the TSS method could yield the formation of thermodynamically stable large pores, thus showing its limit in eliminating last residual porosity (1–2%). The effects of grain size refinement and critical defect dimension compensate for each other, thus leaving the strength value unaltered (as also described by Trunec [62]).

In the case of the ZTA composites, the TSS method effectively limited grain growth (Figure 3c,d). Comparing two samples with the same stabilizer content, the grain size refinement resulted in a lower toughness, probably due to the average grain dimension being too far from the critical grain size. The strength values of the 60403Y samples were found to be similar, likely because the increase in the critical defect size was not sufficiently compensated for by the refinement of the microstructure, as suggested by Trunec [62]. For the 60402Y samples, dynamic pore coalescence occurred in the second step of TSS, which did not aid the elimination of residual porosity and had detrimental effects on bending strength [73].

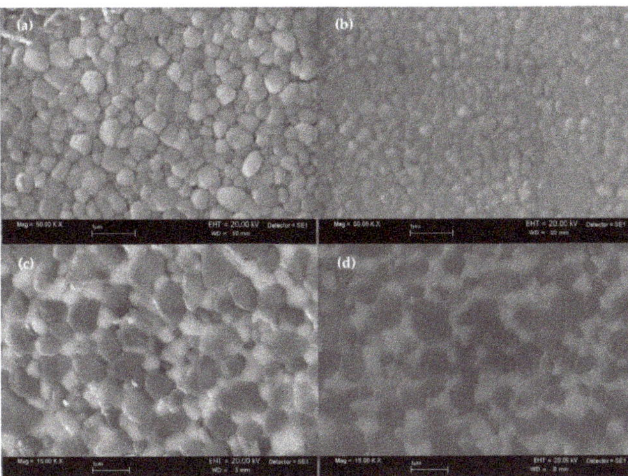

Figure 3. SEM micrographs of zirconia samples 3Y-TZP SSS (**a**) and 3Y-TZP TSS (**b**), as well as of ZTA samples 60403Y SSS (**c**) and 60403Y TSS (**d**).

Finally, the bending strength was also influenced by the powder preparation technique. Using the freeze-drying technique to dry the slurry, the production of a homogeneous granulate without aggregates was achieved (Figure 4). This granulation process strongly influences the quality of a green and sintered body [74]; in our study, higher values of bending strength were obtained (872 ± 47 MPa for 60402Y TSS-FD; see Table 5).

Figure 4. SEM micrograph of 60402Y powder prepared with freeze-dry granulation.

3.3. New Manufacturing Techniques: 3D Printing

The production of ceramic components via the DLP technique is strictly connected to the availability of a suitable ceramic slurry. Nowadays, the most important producers of vat polymerization printers commercialize feedstocks for their 3D printer models with limited possibility to access to other resins available on the market. Another problem for the AM of ceramics with the DLP technique is the low disposability of printable slurries filled with desired ceramic powders.

For the preparation of new resin–ceramic powder mixtures, one of the main problems related to the addition of a high content of ceramic powder to the photopolymeric resin is the increase in the viscosity of the mixture. This drawback was solved here by wisely selecting monomers with different functionalities and molecular weights. The shear thickening behavior that is commonly observed in high solid loaded suspensions was reduced by the use of an appropriate surfactant and a zirconia powder with a lower surface area $(7 \pm 2 \text{ m}^2/\text{g})$. In this way, high content ceramic photocurable resins (see Materials and Methods section) with low viscosity, suitable for the DLP printing process, were prepared.

The shear viscosity for two ATZ resins is reported in Figure 5.

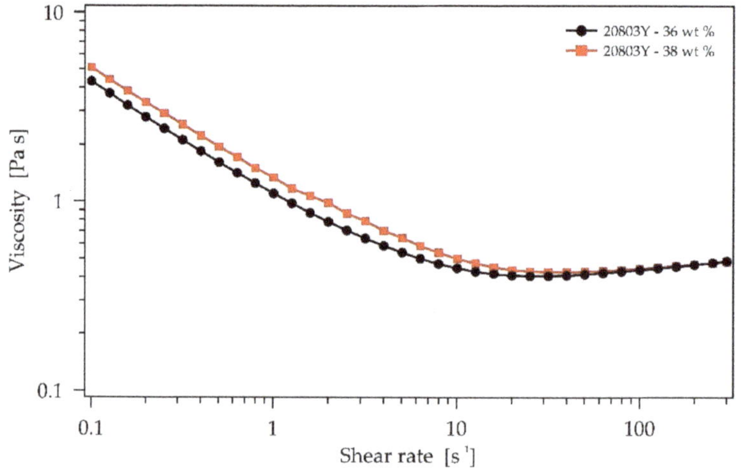

Figure 5. Photocurable slurry viscosity.

Prototypal dental endosseous implants were obtained via the DLP technique with the developed ZTA resin (Figure 6), which was sintered up to 1550 °C for 1 h and reached a final density of 96.8%.

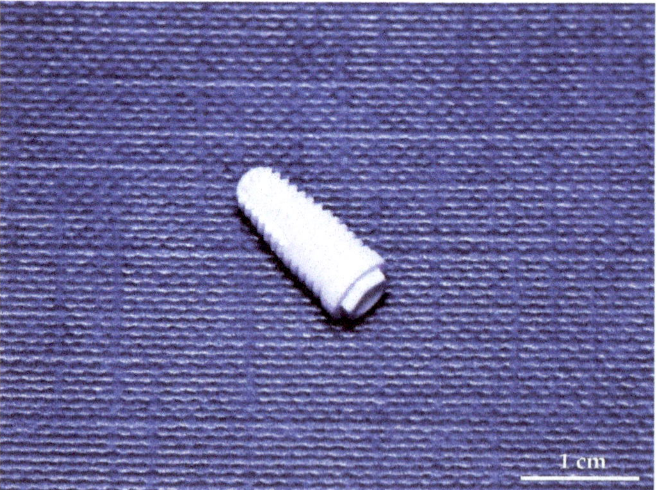

Figure 6. Endosseous dental implant in ATZ printed with the DLP technique.

More complex shapes, as the lattice structure shown in Figure 7, were successfully printed with a final relative density of 98%. Layer-by-layer deposition is highlighted in Figure 8. SEM observations revealed a regular lattice structure profile, where the overlapping layers and their homogeneity in thickness were clearly visible. The slicing value was set to 50 μm and fell to 35 μm after sintering shrinkage. Nevertheless, the layer adhesion could be further enhanced to completely avoid the delamination defects partially present in these items.

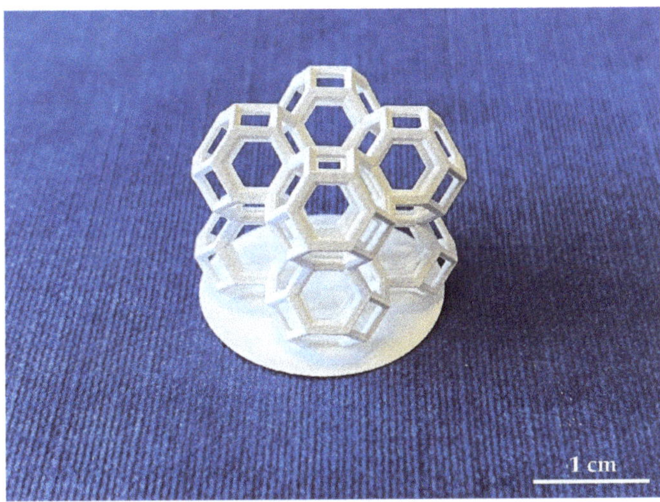

Figure 7. Lattice structure in ATZ printed by the DLP technique.

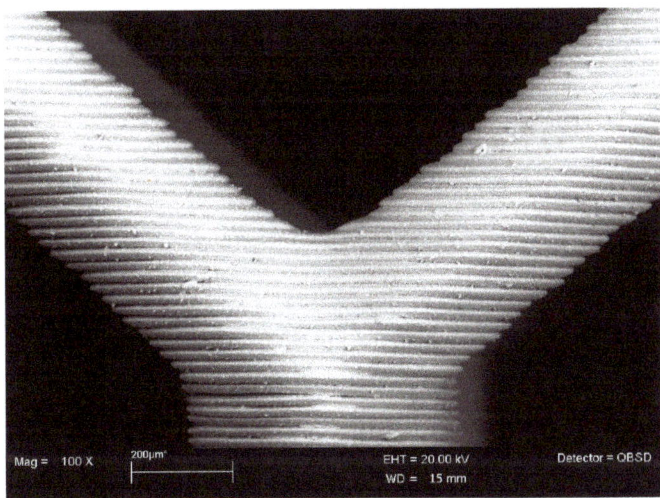

Figure 8. Micrograph of the lattice structure profile.

These preliminary outcomes highlighted the possibility to develop resins with the required ceramic material and the feasibility to print ceramic materials with low cost and widely available DLP printers.

4. Conclusions

Alumina–zirconia composites emphasize the unique properties of zirconia and show many positive aspects that encourage their applications as biomaterials.

The type of stabilizer of the tetragonal phase and the use of toughening oxides with different oxidation states were found to strongly influence the value of the tetragonality, which is the c/a ratio of the lattice parameters of the tetragonal cell. The existence of a linear correlation between tetragonality and fracture toughness was verified. The c/a ratio revealed the instability of the tetragonal cell and, therefore, its tendency to transform, with a consequent toughness increase.

It was also observed that the amount of stabilizer influenced the c/a ratio; in particular, the lower the stabilizer content, the higher the tetragonality and, therefore, the fracture toughness.

The relationships between microstructure and mechanical properties were investigated. Despite this effect not being completely clear in some cases, it was generally proven that as the average grain size grew, the fracture toughness increased until it approached the critical grain size. On the other hand, flexural strength was not significantly affected by the grain size refinement, probably because of the presence of larger critical defects when TSS was applied instead of SSS.

These experimental results could constitute a scientific base to design new high-performance ZTA composites that applicable in orthopedic and dental implants with high hydrothermal resistance.

Furthermore, an innovative forming technique based on additive manufacturing technology—DLP technique—was successfully tested to produce dental components with zirconia–alumina composites. This technique represents a very interesting perspective for the development of highly customized devices with lower waste and reduced cost that is suitable for small batch production in the biomedical field.

Author Contributions: Conceptualization, G.M.; methodology, G.M.; investigation, G.M., P.F., E.L., F.M., and E.S.; all authors analyzed the data and wrote the paper. All authors have read and agreed to the published version of the manuscript.

Funding: This research received no external funding.

Data Availability Statement: All relevant data are represented in this publication. All raw data and evaluated data are available from the authors upon request.

Conflicts of Interest: The authors declare no conflict of interest.

References

1. Claussen, N. Fracture Toughness of Al_2O_3 with an unstabilized ZrO_2 dispersed phase. *J. Am. Ceram. Soc.* **1976**, *59*, 49–51. [CrossRef]
2. Wang, J.; Stevens, R. Zirconia-toughened alumina (ZTA) ceramics. *J. Mater. Sci.* **1989**, *24*, 3421–3440. [CrossRef]
3. Maji, A.; Choubey, G. Microstructure and mechanical properties of alumina toughened zirconia (ATZ). *Mater. Today Proc.* **2018**, *5*, 7457–7465. [CrossRef]
4. Wimmer, M.A.; Pacione, C.; Yuh, C.; Chan, Y.-M.; Kunze, J.; Laurent, M.P.; Chubinskaya, S. Articulation of an alumina-zirconia composite ceramic against living cartilage—An in vitro wear test. *J. Mech. Behav. Biomed. Mater.* **2020**, *103*, 103531. [CrossRef]
5. Solarino, G.; Piconi, C.; De Santis, V.; Piazzolla, A.; Moretti, B. Ceramic Total knee arthroplasty: Ready to go? *Joints* **2017**, *05*, 224–228. [CrossRef]
6. Piconi, C.; Sprio, S. Zirconia implants: Is there a future? *Curr. Oral Health Rep.* **2018**, *5*, 186–193. [CrossRef]
7. Tsukuma, K.; Ueda, K.; Shimada, M. Strength and fracture toughness of isostatically hot-pressed composites of Al_2O_3 and Y_2O_3-partially-stabilized ZrO_2. *J. Am. Ceram. Soc.* **1985**, *68*, C4–C5. [CrossRef]
8. Tuan, W.H.; Chen, R.Z.; Wang, T.C.; Cheng, C.H.; Kuo, P.S. Mechanical properties of Al_2O_3/ZrO_2 composites. *J. Eur. Ceram. Soc.* **2002**, *22*, 2827–2833. [CrossRef]
9. Shin, Y.-S.; Rhee, Y.-W.; Kang, S.-J.L. Experimental evaluation of toughening mechanisms in alumina-zirconia composites. *J. Am. Ceram. Soc.* **1999**, *82*, 1229–1232. [CrossRef]
10. Hannink, R.H.J.; Kelly, P.M.; Muddle, B.C. Transformation toughening in zirconia-containing ceramics. *J. Am. Ceram. Soc.* **2000**, *83*, 461–487. [CrossRef]
11. Tsukuma, K.; Takahata, T.; Shiomi, M. Strength and fracture toughness of Y-TZP, Ce-TZP, Y-TZP/Al_2O_3, and Ce-TZP/Al_2O_3. In *Advances in Ceramics: Science and Technology of Zirconia III*; Sōmiya, S., Yamamoto, N., Yanagida, H., Eds.; The American Ceramic Society: Columbus, OH, USA, 1988; Volume 24B, pp. 721–728, ISBN 9780916094874.
12. Lange, F.F. Transformation toughening: Part 1 Size effects associated with the thermodynamics of constrained transformations. *J. Mater. Sci.* **1982**, *17*, 225–234. [CrossRef]

13. Garvie, R.C. The occurrence of metastable tetragonal zirconia as a crystallite size effect. *J. Phys. Chem.* **1965**, *69*, 1238–1243. [CrossRef]
14. Gutknecht, D.; Chevalier, J.; Garnier, V.; Fantozzi, G. Key role of processing to avoid low temperature ageing in alumina zirconia composites for orthopaedic application. *J. Eur. Ceram. Soc.* **2007**, *27*, 1547–1552. [CrossRef]
15. Zadorozhnaya, O.Y.; Khabas, T.A.; Tiunova, O.V.; Malykhin, S.E. Effect of grain size and amount of zirconia on the physical and mechanical properties and the wear resistance of zirconia-toughened alumina. *Ceram. Int.* **2020**, *46*, 9263–9270. [CrossRef]
16. Cotič, J.; Jevnikar, P.; Kocjan, A.; Kosmač, T. Complexity of the relationships between the sintering-temperature-dependent grain size, airborne-particle abrasion, ageing and strength of 3Y-TZP ceramics. *Dent. Mater.* **2016**, *32*, 510–518. [CrossRef]
17. Chevalier, J.; Gremillard, L.; Virkar, A.V.; Clarke, D.R. The tetragonal-monoclinic transformation in zirconia: Lessons learned and future trends. *J. Am. Ceram. Soc.* **2009**, *92*, 1901–1920. [CrossRef]
18. Schneider, J.; Begand, S.; Kriegel, R.; Kaps, C.; Glien, W.; Oberbach, T. Low-temperature aging behavior of alumina-toughened zirconia. *J. Am. Ceram. Soc.* **2008**, *91*, 3613–3618. [CrossRef]
19. Chevalier, J.; Grandjean, S.; Kuntz, M.; Pezzotti, G. On the kinetics and impact of tetragonal to monoclinic transformation in an alumina/zirconia composite for arthroplasty applications. *Biomaterials* **2009**, *30*, 5279–5282. [CrossRef]
20. Deville, S.; Chevalier, J.; Fantozzi, G.; Bartolomé, J.F.; Requena, J.; Moya, J.S.; Torrecillas, R.; Díaz, L.A. Low-temperature ageing of zirconia-toughened alumina ceramics and its implication in biomedical implants. *J. Eur. Ceram. Soc.* **2003**, *23*, 2975–2982. [CrossRef]
21. Fabbri, P.; Piconi, C.; Burresi, E.; Magnani, G.; Mazzanti, F.; Mingazzini, C. Lifetime estimation of a zirconia-alumina composite for biomedical applications. *Dent. Mater.* **2014**, *30*, 138–142. [CrossRef]
22. Lughi, V.; Sergo, V. Low temperature degradation -aging- of zirconia: A critical review of the relevant aspects in dentistry. *Dent. Mater.* **2010**, *26*, 807–820. [CrossRef] [PubMed]
23. Deville, S.; Chevalier, J.; Dauvergne, C.; Fantozzi, G.; Bartolomé, J.F.; Moya, J.S.; Torrecillas, R. Microstructural investigation of the aging behavior of (3Y-TZP)–Al$_2$O$_3$ composites. *J. Am. Ceram. Soc.* **2005**, *88*, 1273–1280. [CrossRef]
24. Tsubakino, H.; Nozato, R.; Hamamoto, M. Effect of alumina addition on the tetragonal-to-monoclinic phase transformation in zirconia–3 mol% yttria. *J. Am. Ceram. Soc.* **1991**, *74*, 440–443. [CrossRef]
25. Tolkachev, O.S.; Dvilis, E.S.; Alishin, T.R.; Khasanov, O.L.; Miheev, D.A.; Chzhan, T. Assessment of the hydrothermal resistance of Y-TZP ceramics by the degree of tetragonality of major phases. *Lett. Mater.* **2020**, *10*, 416–421. [CrossRef]
26. Zhang, F.; Vanmeensel, K.; Inokoshi, M.; Batuk, M.; Hadermann, J.; Van Meerbeek, B.; Naert, I.; Vleugels, J. Critical influence of alumina content on the low temperature degradation of 2–3 mol% yttria-stabilized TZP for dental restorations. *J. Eur. Ceram. Soc.* **2015**, *35*, 741–750. [CrossRef]
27. Kohorst, P.; Borchers, L.; Strempel, J.; Stiesch, M.; Hassel, T.; Bach, F.W.; Hübsch, C. Low-temperature degradation of different zirconia ceramics for dental applications. *Acta Biomater.* **2012**, *8*, 1213–1220. [CrossRef]
28. Nogiwa-Valdez, A.A.; Rainforth, W.M.; Zeng, P.; Ross, I.M. Deceleration of hydrothermal degradation of 3Y-TZP by alumina and lanthana co-doping. *Acta Biomater.* **2013**, *9*, 6226–6235. [CrossRef]
29. Zhang, F.; Vanmeensel, K.; Inokoshi, M.; Batuk, M.; Hadermann, J.; Van Meerbeek, B.; Naert, I.; Vleugels, J. 3Y-TZP ceramics with improved hydrothermal degradation resistance and fracture toughness. *J. Eur. Ceram. Soc.* **2014**, *34*, 2453–2463. [CrossRef]
30. Gremillard, L.; Chevalier, J.; Epicier, T.; Deville, S.; Fantozzi, G. Modeling the aging kinetics of zirconia ceramics. *J. Eur. Ceram. Soc.* **2004**, *24*, 3483–3489. [CrossRef]
31. Hallmann, L.; Mehl, A.; Ulmer, P.; Reusser, E.; Stadler, J.; Ha, C.H.F.; Zenobi, R.; Stawarczyk, B.; Mutlu, O. The influence of grain size on low-temperature degradation of dental zirconia. *J. Biomed. Mater. Res. B Appl. Biomater.* **2012**, *100*, 447–456. [CrossRef]
32. Lange, F.F. Transformation toughening: Part 3 Experimental observations in the ZrO$_2$-Y$_2$O$_3$ system. *J. Mater. Sci.* **1982**, *17*, 240–246. [CrossRef]
33. Becher, P.F.; Swain, M. V Grain-size-dependent transformation behavior in polycrystalline tetragonal zirconia. *J. Am. Ceram. Soc.* **1992**, *75*, 493–502. [CrossRef]
34. Suresh, A.; Mayo, M.J.; Porter, W.D.; Rawn, C.J. Crystallite and grain-size-dependent phase transformations in yttria-doped zirconia. *J. Am. Ceram. Soc.* **2003**, *86*, 360–362. [CrossRef]
35. Bravo-Leon, A.; Morikawa, Y.; Kawahara, M. Fracture toughness of nanocrystalline tetragonal zirconia with low yttria content. *Acta Mater.* **2002**, *50*, 4555–4562. [CrossRef]
36. Eichler, J.; Rödel, J.; Eisele, U.; Hoffman, M. Effect of Grain size on mechanical properties of submicrometer 3Y-TZP: Fracture strength and hydrothermal degradation. *J. Am. Ceram. Soc.* **2007**, *90*, 2830–2836. [CrossRef]
37. SmarTech Markets Publishing. *Ceramics Additive Manufacturing Markets 2017–2028. An Opportunity Analysis and Ten-Year Market Forecast*; SmarTech Analysis: Crozet, VA, USA, 2018. Available online: https://www.smartechanalysis.com (accessed on 10 September 2020).
38. SmarTech Analysis. *Ceramics Additive Manufacturing Production Markets: 2019–2030*; SmarTech Analysis: Crozet, VA, USA, 2020. Available online: http://www.smartechanalysis.com (accessed on 10 September 2020).
39. Sun, Y.; Wang, C.; Zhao, Z. ZTA Ceramic materials for DLP 3D printing. *IOP Conf. Ser. Mater. Sci. Eng.* **2019**, *678*, 012020. [CrossRef]

40. Wu, H.; Liu, W.; Lin, L.; Li, L.; Li, Y.; Tian, Z.; Zhao, Z.; Ji, X.; Xie, Z.; Wu, S. Preparation of alumina-toughened zirconia via 3D printing and liquid precursor infiltration: Manipulation of the microstructure, the mechanical properties and the low temperature aging behavior. *J. Mater. Sci.* **2019**, *54*, 7447–7459. [CrossRef]

41. Coppola, B.; Lacondemine, T.; Tardivat, C.; Montanaro, L.; Palmero, P. Designing alumina-zirconia composites by DLP-based stereolithography: Microstructural tailoring and mechanical performances. *Ceram. Int.* **2021**, *47*, 13457–13468. [CrossRef]

42. Zhang, X.; Wu, X.; Shi, J. Additive manufacturing of zirconia ceramics: A state-of-the-art review. *J. Mater. Res. Technol.* **2020**, *9*, 9029–9048. [CrossRef]

43. Li, X.; Zhong, H.; Zhang, J.; Duan, Y.; Bai, H.; Li, J.; Jiang, D. Dispersion and properties of zirconia suspensions for stereolithography. *Int. J. Appl. Ceram. Technol.* **2020**, *17*, 239–247. [CrossRef]

44. de Camargo, I.L.; Morais, M.M.; Fortulan, C.A.; Branciforti, M.C. A review on the rheological behavior and formulations of ceramic suspensions for vat photopolymerization. *Ceram. Int.* **2021**, *47*, 11906–11921. [CrossRef]

45. Niihara, K. A fracture mechanics analysis of indentation-induced Palmqvist crack in ceramics. *J. Mater. Sci. Lett.* **1983**, *2*, 221–223. [CrossRef]

46. Li, P.; Chen, I.-W.; Penner-Hahn, J.E. Effect of dopants on zirconia stabilization–An X-ray absorption study: II, tetravalent dopants. *J. Am. Ceram. Soc.* **1994**, *77*, 1281–1288. [CrossRef]

47. Nisticò, R. Zirconium oxide and the crystallinity hallows. *J. Aust. Ceram. Soc.* **2021**, *57*, 225–236. [CrossRef]

48. Jang, J.W.; Kim, D.J.; Lee, D.Y. Size effect of trivalent oxides on low temperature phase stability of 2Y-TZP. *J. Mater. Sci.* **2001**, *36*, 5391–5395. [CrossRef]

49. Zhang, F.; Batuk, M.; Hadermann, J.; Manfredi, G.; Mariën, A.; Vanmeensel, K.; Inokoshi, M.; Van Meerbeek, B.; Naert, I.; Vleugels, J. Effect of cation dopant radius on the hydrothermal stability of tetragonal zirconia: Grain boundary segregation and oxygen vacancy annihilation. *Acta Mater.* **2016**, *106*, 48–58. [CrossRef]

50. Tien, T.Y. Toughened Ceramics. U.S. Patent U.S. 4,886,768A, 12 December 1987.

51. Yoshimura, M. Phase stability of zirconia. *Am. Ceram. Soc. Bull.* **1988**, *67*, 1950–1955.

52. Matsui, K.; Horikoshi, H.; Ohmichi, N.; Ohgai, M.; Yoshida, H.; Ikuhara, Y. Cubic-formation and grain-growth mechanisms in tetragonal zirconia polycrystal. *J. Am. Ceram. Soc.* **2003**, *86*, 1401–1408. [CrossRef]

53. Shimozono, T.; Ikeda, J.; Pezzotti, G. Evaluation of transformation zone around propagating cracks in zirconia biomaterials using raman microprobe spectroscopy. In Proceedings of the Bioceramics 18, Kyoto, Japan, 8–12 December 2005; Nakamura, T., Yamashita, K., Neo, M., Eds.; Trans Tech Publications Ltd.: Bäch, Switzerland, 2006; Volume 309–311, pp. 1207–1210. [CrossRef]

54. Matsumoto, R.L.K. Aging Behavior of Ceria-Stabilized Tetragonal Zirconia Polycrystals. *J. Am. Ceram. Soc.* **1988**, *71*, C128–C129. [CrossRef]

55. Chevalier, J.; Gremillard, L. Ceramics for medical applications: A picture for the next 20 years. *J. Eur. Ceram. Soc.* **2009**, *29*, 1245–1255. [CrossRef]

56. Kim, D.-J. Effect of Ta_2O_5, Nb_2O_5, and HfO_2 alloying on the transformability of Y_2O_3-stabilized tetragonal ZrO_2. *J. Am. Ceram. Soc.* **1990**, *73*, 115–120. [CrossRef]

57. Kim, D.-J.; Tien, T.-Y. Phase Stability and Physical Properties of Cubic and Tetragonal ZrO_2 in the System ZrO_2-Y_2O_3-Ta_2O_5. *J. Am. Ceram. Soc.* **1991**, *74*, 3061–3065. [CrossRef]

58. Ramesh, S.; Gill, C.; Lawson, S. Effect of copper oxide on sintering, microstructure, mechanical properties and hydrothermal ageing of coated 2.5Y-TZP ceramics. *J. Mater. Sci.* **1999**, *34*, 5457–5467. [CrossRef]

59. Magnani, G.; Brillante, A. Effect of the composition and sintering process on mechanical properties and residual stresses in zirconia-alumina composites. *J. Eur. Ceram. Soc.* **2005**, *25*, 3383–3392. [CrossRef]

60. Yoshimura, M.; Yashima, M.; Noma, T.; Sōmiya, S. Formation of diffusionlessly transformed tetragonal phases by rapid quenching of melts in ZrO_2-$RO_{1.5}$ systems (R = rare earths). *J. Mater. Sci.* **1990**, *25*, 2011–2016. [CrossRef]

61. Rice, R.W. Ceramic tensile strength-grain size relations: Grain sizes, slopes, and branch intersections. *J. Mater. Sci.* **1997**, *32*, 1673–1692. [CrossRef]

62. Trunec, M. Effect of grain size on mechanical properties of 3Y-TZP ceramics. *Ceram. Silikaty* **2008**, *52*, 165–171.

63. Casellas, D.; Alcalá, J.; Llanes, L.; Anglada, M. Fracture variability and R-curve behavior in yttria-stabilized zirconia ceramics. *J. Mater. Sci.* **2001**, *36*, 3011–3025. [CrossRef]

64. Chevalier, J.; Liens, A.; Reveron, H.; Zhang, F.; Reynaud, P.; Douillard, T.; Preiss, L.; Sergo, V.; Lughi, V.; Swain, M.; et al. Forty years after the promise of «ceramic steel?»: Zirconia-based composites with a metal-like mechanical behavior. *J. Am. Ceram. Soc.* **2020**, *103*, 1482–1513. [CrossRef]

65. Sadowski, T.; Łosiewicz, K.; Boniecki, M.; Szutkowska, M. Assessment of mechanical properties by nano- and microindentation of alumina/zirconia composites. *Mater. Today Proc.* **2021**, *45*, 4196–4201. [CrossRef]

66. Krell, A. Load dependence of hardness in sintered submicrometer Al_2O_3 and ZrO_2. *J. Am. Ceram. Soc.* **1995**, *78*, 1417–1419. [CrossRef]

67. Chintapalli, R.; Mestra, A.; García Marro, F.; Yan, H.; Reece, M.; Anglada, M. Stability of nanocrystalline spark plasma sintered 3Y-TZP. *Materials* **2010**, *3*, 800–814. [CrossRef]

68. Yao, W.; Liu, J.; Holland, T.B.; Huang, L.; Xiong, Y.; Schoenung, J.M.; Mukherjee, A.K. Grain size dependence of fracture toughness for fine grained alumina. *Scr. Mater.* **2011**, *65*, 143–146. [CrossRef]

69. Borrell, A.; Salvador, M.D.; Peñaranda-Foix, F.L.; Cátala-Civera, J.M. Microwave sintering of zirconia materials: Mechanical and microstructural properties. *Int. J. Appl. Ceram. Technol.* **2013**, *10*, 313–320. [CrossRef]
70. Chen, I.-W.; Wang, X.-H. Sintering dense nanocrystalline ceramics without final-stage grain growth. *Nature* **2000**, *404*, 168–171. [CrossRef]
71. Loong, T.H.; Soosai, A.; Muniandy, S. Effect of temperature and holding time on zirconia toughened alumina (ZTA) prepared by two-stage sintering. *Mater. Sci. Forum* **2021**, *1030*, 11–18. [CrossRef]
72. Sivanesan, S.; Loong, T.H.; Namasivayam, S.; Fouladi, M.H. Two-stage sintering of alumina-Y-TZP (Al$_2$O$_3$/Y-TZP) composites. *Key Eng. Mater.* **2019**, *814*, 12–18. [CrossRef]
73. Xiong, Y.; Hu, J.; Shen, Z. Dynamic pore coalescence in nanoceramic consolidated by two-step sintering procedure. *J. Eur. Ceram. Soc.* **2013**, *33*, 2087–2092. [CrossRef]
74. Bergstrom, L. Colloidal Processing of Ceramics. In *Handbook of Applied Surface and Colloid Chemistry*; Holmberg, K., Ed.; John Wiley & Sons, Ltd.: Chichester, UK, 2001; Volume 1, pp. 201–218.

Journal of
Composites Science

MDPI

Article

Is Surface Metastability of Today's Ceramic Bearings a Clinical Issue?

Alessandro Alan Porporati [1,2], Laurent Gremillard [3], Jérôme Chevalier [3], Rocco Pitto [4] and Marco Deluca [5,*]

[1] CeramTec GmbH, Medical Products Division, Ceramtec-Platz 1-9, 73207 Plochingen, Germany; a.porporati@ceramtec.de

[2] Department of Engineering and Architecture, University of Trieste, Via Alfonso Valerio 6/1, 34127 Trieste, Italy

[3] Univ Lyon, INSA Lyon, UCBL, CNRS, MATEIS, UMR 5510, F-69621 Villeurbanne, France; laurent.gremillard@insa-lyon.fr (L.G.); jerome.chevalier@insa-lyon.fr (J.C.)

[4] Department of Orthopaedic Surgery, Middlemore Hospital, University of Auckland, South Auckland Clinical Campus, Auckland 93311, New Zealand; r.pitto@auckland.ac.nz

[5] Materials Center Leoben Forschung GmbH, Roseggerstr. 12, 8700 Leoben, Austria

* Correspondence: marco.deluca@mcl.at; Tel.: +43-3842-45922-530

Abstract: Recent studies on zirconia-toughened alumina (ZTA) evidenced that in vivo aged implants display a much higher monoclinic zirconia content than expected from in vitro simulations by autoclaving. At the moment, there is no agreement on the source of this discrepancy: Some research groups ascribe it to the effect of mechanical impact shocks, which are generally not implemented in standard in vitro aging or hip walking simulators. Others invoke the effect of metal transfer, which should trigger an autocatalytic reaction in the body fluid environment, accelerating the kinetics of tetragonal-to-monoclinic transformation in vivo. Extrapolations of the aging kinetics from high (autoclave) to in vivo temperature are also often disputed. Last, Raman spectroscopy is by far the preferred method to quantify the amount of monoclinically transformed zirconia. There are, however, many sources of errors that may negatively affect Raman results, meaning that the final interpretation might be flawed. In this work, we applied Raman spectroscopy to determine the monoclinic content in as-received and in vitro aged ZTA hip joint implants, and in one long-term retrieval study. We calculated the monoclinic content with the most used equations in the literature and compared it with the results of X-ray diffraction obtained on a similar probe depth. Our results show, contrary to many previous studies, that the long-term surface stability of ZTA ceramics is preserved. This suggests that the Raman technique does not offer consistent and unique results for the analysis of surface degradation. Moreover, we discuss here that tetragonal-to-monoclinic transformation is also necessary to limit contact damage and wear stripe extension. Thus, the surface metastability of zirconia-containing ceramics may be a non-issue.

Keywords: zirconia-toughened alumina; phase transformation; Raman spectroscopy

Citation: Porporati, A.A.; Gremillard, L.; Chevalier, J.; Pitto, R.; Deluca, M. Is Surface Metastability of Today's Ceramic Bearings a Clinical Issue? *J. Compos. Sci.* **2021**, *5*, 273. https://doi.org/10.3390/jcs5100273

Academic Editor: Francesco Tornabene

Received: 1 August 2021
Accepted: 8 October 2021
Published: 14 October 2021

Publisher's Note: MDPI stays neutral with regard to jurisdictional claims in published maps and institutional affiliations.

1. Introduction

The current trend in total hip arthroplasty (THA) is to gradually prefer ceramic-based implants over metallic implants due to their excellent biocompatibility [1], both in bulk and particulate form, and high long-term survival [2]. Nowadays, the ceramic of choice for THA is zirconia-toughened alumina (ZTA), often in the form of the BIOLOX®*delta* material, the most commercially successful material. BIOLOX®*delta* (Delta) was developed in the early 2000s by CeramTec GmbH and is composed of 17 vol.% yttria-stabilized tetragonal zirconia particles (Y-TZP) embedded into an alumina matrix. The function of Y-TZP is to improve the toughness of the prosthetic material, through a mechanism called phase transformation toughening: under stress, a fraction of the tetragonal zirconia grains may undergo a phase transition to a monoclinic phase (t–m transformation). This phase has

67

a larger volume and thus sets the surrounding material under compressive stress. This increases the crack propagation resistance of the material. The alumina phase, on the other hand, provides hardness and wear resistance [3]. Furthermore, Delta also contains 0.3 w.% chromia, which imparts the pink color to the material, and about 3 vol.% of strontium hexaaluminate platelets, which further increases the fracture toughness of the ceramic composite. The stability of alumina matrix material bearings in the body of patients is still being discussed, despite the best long-term clinical outcomes shown by the arthroplasty registries when compared to alternative bearings [4–8]. Before being introduced into the medical market, the Delta material was thoroughly tested according to the available ASTM F2345-03 standard for in vitro testing, which states that a one-hour exposure at 134 °C under 2 bars water steam (in autoclave) corresponds to two years in vivo [9]. Extrapolations based on those tests predicted a very slow transformation (less than 5% increase over the first 10 years) [10]. However, investigations on revised Delta implants after few years (in some cases, months) in vivo [3,11–14], where the causes of revision were unrelated to the ceramic material, revealed worn areas [12,14,15], the presence of metal transfer [11–14], and a much higher monoclinic content than expected according to in vitro simulations [11,13]. This monoclinic content was always necessarily reported as the difference between worn and non-worn areas [11,15] because the initial monoclinic phase amount cannot be measured prior to implantation, and this value may vary slightly due to the manufacturer's batch processing [11].

Many material research groups have attempted to clarify this discrepancy between in vitro and in vivo results, but up to now, no consensus on the real causes has been found. Pezzotti et al. proposed one hypothesis [11,16–18] that ascribes this discrepancy to the effects of metal transfer. In fact, in correspondence with metal transfer, high wear values and high monoclinic contents in zirconia were measured [11–13]. According to Pezzotti et al. [11], in the aqueous body fluid environment, the autocatalytic dissociation of water molecules is promoted at the transferred metal's surface; this apparently causes an annihilation of oxygen vacancies first in the alumina phase and then in the tetragonal zirconia phase of Delta ceramics. This latter oxygen vacancy reduction should then trigger the tetragonal-to-monoclinic transformation in zirconia. This conclusion was supported by cathodoluminescence and X-ray photoelectron spectroscopies [11,16], and similar mechanisms were also observed in yttria-stabilized zirconia [19]. Although this explanation may sound plausible, especially in the case of metal transfer, it is not clear whether the initial data from the retrievals are consistent: the measured high wear could be due to the surface asperities produced by metal smearing, and not the ceramic. In addition, more likely, the presence of metal affects the quantification of the monoclinic phase by Raman spectroscopy because of the influence of metal on the intensity of Raman peaks due to plasmonic effects [20]. A poor signal-to-noise ratio could also affect spectra collected in the presence of chemisorbed proteins [21], which highlights the need for a thorough surface cleanup before Raman analysis.

Another explanation has been proposed by Perrichon et al. [15,22,23], who ascribe the discrepancy to the effect of shocks due to microseparation, which are not implemented in the ASTM in vitro standard, or in hip walking simulators. Perrichon et al. were able to demonstrate that using a specific in vitro test route that includes hip walking simulations, shock tests, and environmental aging tests, the discrepancy with in vivo studies can be reduced [23]. In this case, the monoclinic content on worn areas was much higher than on non-worn zones, in agreement with observations on retrieved implants. This shows that the phase transformation toughening mechanism was activated under stresses to limit mechanical damage. In other words, t–m transformation due to shock is not related to degradation due to aging.

The last explanation is related to the model used to relate the simulated aging kinetics obtained in the autoclave to in vivo aging. It is postulated that aging follows an Arrhenius law, and thus knowledge of its activation energy enables establishing a time–temperature equivalence. However, two main limits exist for this model: First, data at a low temperature take a long time to obtain and thus are hardly available; thus, it is not completely certain that the Arrhenian behavior is still valid around body temperature. Second, small uncertainties in the activation energy may lead to large variability in the time–temperature equivalence, while the ASTM standard tacitly considers a single activation energy for all zirconia-containing materials.

However, more importantly, it has been recently shown that the in vivo zirconia transformation in Delta does not affect the mechanical performance of the total hip arthroplasty components [24]. It follows that differences in the t–m transformation of zirconia in the Delta material between in vitro tests and ex vivo components may not be clinically relevant; therefore, solving the problem is conducted only for the sake of knowledge. In the following, we will try to explain the differences due to the measurement method.

Although it is clear that in vitro testing standards should be reviewed in order to better approach the conditions encountered by the implant inside the patient's body, one aspect may be overlooked: comparisons between in vitro and in vivo studies are based mainly on the value of V_m, the volume fraction of monoclinic zirconia, which is measured by Raman spectroscopy. However, currently, there is neither a standard that exists regarding how V_m should be measured from Raman spectra, nor sufficient explanations reported in the previous literature about how data collection and treatment have been performed. Very likely, each research group uses a different procedure for V_m quantification, which (at least) makes the comparison of V_m results obtained by different groups questionable. In addition, some data analysis procedures might lead to severe artefacts, with the consequence that wrong values of V_m are calculated, leading to flawed interpretations.

In this work, we show that the use of the Clarke/Adar equation—deemed unreliable by some research groups—is indeed the best choice for our measurement setup for Delta ceramics. This is confirmed by a control procedure using X-ray diffraction (XRD). Furthermore, we demonstrate—by analyzing a retrieval affected by metal transfer—that the effect of metal on the monoclinic transformation is negligible. Lastly, we show that specific choices in data analysis, such as the use of the absolute/integrated intensity, the choice of a baseline, or the overall signal-to-noise ratio of the spectrum, have a large impact on the obtainable results in terms of V_m. These results clearly suggest that a standard for V_m quantification using Raman spectroscopy (including sample preparation, spectroscopic procedures, and data treatment) should be promptly put in place.

2. V_m Quantification by Raman Spectroscopy

Raman spectroscopy probes the inelastic light scattering from vibrational motions of atoms in a solid, and as such, it is sensitive to any change in the way atoms vibrate, as caused, for instance, by the presence of a different phase. In other words, the Raman spectrum is a fingerprint of the state a solid assumes. In zirconia ceramics, the tetragonal and monoclinic polymorphs present very different spectra [25], and in mixed phases, a superposition between those two spectra appears, the extent of which depends on the volume fraction of the monoclinic phase, V_m, in the investigated area. Various researchers have derived an expression to quantify V_m from the intensity of Raman peaks belonging to tetragonal and monoclinic phases, building upon equations already available for XRD analyses [26]. The equation has the following form [26,27]:

$$V_m = \frac{I_m^{181} + I_m^{190}}{k\left(I_t^{147} + \delta I_t^{265}\right) + I_m^{181} + I_m^{190}} \tag{1}$$

and it differs among the available approaches only for the values of the δ and k coefficients. $I_{m,t}^i$ is the intensity of Raman peaks (at position i—in cm^{-1}) belonging to monoclinic and tetragonal phases. The two most used equations for the determination of V_m in THA implants are the one derived by Clarke and Adar ($\delta = 1$, $k = 0.97$) [27] and the one derived by Katagiri et al. ($\delta = 0$, $k = 2.2 \pm 0.2$) [26], whereby the latter has been used in the majority of recent studies concerning the Delta material. Tabares and Anglada [26] recently carried out a systematic study with both Raman and XRD using bulk mixtures of tetragonal and monoclinic zirconia powders with from 0% to 100% of monoclinic phase content. They calculated V_m using both the Clarke/Adar and Katagiri equations and demonstrated that while Katagiri's equation correctly reproduced the monoclinic content, the Clarke/Adar equation largely underestimated it. XRD results were also in better accordance with the Katagiri equation. They suggested that this discrepancy is related to the localization of the monoclinic polymorph (i.e., the V_m profile) in the material used for calibration by Clarke and Adar: fracture surfaces in ZrO_2/Y_2O_3 specimens, where the monoclinic phase is expected to be present only near the surface. In this case, the penetration depth of X-rays depends on the angle of incidence, and thus it can be suggested that the discrepancy in the V_m calculated with Clarke and Adar's formula is due to a different angle of incidence that Tabares and Anglada used for their XRD measurements, compared to the one used by Clarke and Adar [26]. Hence, according to Tabares and Anglada, the Katagiri equation seems to have universal validity because it has been obtained on bulk mixtures of tetragonal and monoclinic zirconia powders, where the monoclinic content is homogeneous across the whole probed volume by both Raman and XRD. Its validity, however, has neither been systematically demonstrated in materials where a sharp gradient in the monoclinic phase is present, nor in sintered materials. Both aspects clearly apply to in vitro and in vivo aged Delta [3,9,11,28].

Apart from the choice of the equation to calculate V_m, there are many other aspects that could lead to errors and discrepancies between the V_m values reported in the literature:

- Raman spectra need to be fitted (using mathematical expressions) to obtain intensity values of the respective tetragonal and monoclinic peaks. Spectra with different qualities (i.e., different signal-to-noise ratios, SNRs) might lead to different V_m values because of fitting errors. Factors influencing the quality of spectra are the optical system, the laser, the time used for collection/accumulation of spectra, and the quality of the investigated surface.
- Spectra are often affected by a background due to elastic scattering or to the presence of fluorescence (particularly true for Delta). In these cases, a baseline is generally subtracted to avoid the influence of the background on the final result [26]. However, the choice of the baseline might affect the final result as well.
- It is not clear whether the integrated or the absolute intensity of Raman peaks should be used in Equation (1). With Equation (1) being an intensity ratio, this question may seem unimportant; however, the absolute intensity might not fully represent the monoclinic content, especially by low V_m values [29].
- Each spectrometer used for Raman analysis has different characteristics (e.g., the focal length, the number of gratings, the confocal pinhole width) affecting the SNR ratio and the spectral resolution, which could lead to different results if the same material is probed by different equipment.

A comparison between the Clarke/Adar and Katagiri equations using both Raman and XRD on Delta has not yet been reported in the literature, and also a thorough analysis of the aforementioned error sources (even partly) has never been attempted.

3. Materials and Methods

3.1. In Vitro Aging Study Samples

Ten Delta heads and ten Delta cup inserts (CeramTec GmbH, Plochingen, Germany) were analyzed by both XRD and Raman spectroscopy in order to independently quantify the monoclinic content. The areas investigated corresponded to the head apex in the heads (polished), and to the center of the bottom (opposite of the cup) in the inserts (ground). The two different surface finishes were selected with the intent to attempt to cover as much of the V_m range, from 0% to 100%, as possible, this way mimicking non-wear and wear zones in real implants, respectively. The aforementioned total hip arthroplasty implant components were tested both before and after extreme hydrothermal aging in an autoclave at 134 °C and 2.2 bars for 150 h, which would correspond to more than 300 years in vivo according to the ASTM standard.

3.2. Retrieval Sample

The studied retrieval was constituted by a fully Delta ceramic-on-ceramic (CoC) bearing couple. The total hip replacement (THR) components were a DePuy Pinnacle cup 60 mm and a Summit stem size 3 high offset. The 28 mm Delta ceramic head taper had a +8.5 mm neck length offset.

The patient was informed that the data concerning the case would be submitted for publication, and she provided consent.

The patient underwent complex total hip arthroplasty in 2004 (diagnosis: secondary osteoarthritis following developmental dysplasia of the hip). After twelve years of pain-free normal function, the patient presented with periprosthetic joint infection (caused by Klebsiella pneumoniae). After failure of debridement/irrigation and two dislocation events (managed with closed reduction), the patient underwent two-stage revision surgery in late 2016. The retrieved Delta 28 mm CoC bearing appeared intact, with titanium metal stripes in the femoral head caused by recurrent dislocation events. Areas with metal transfer were investigated before and after a cleaning procedure to remove the metal, which consisted in a 10 h bath at 60 °C with 30% aqueous H_2SO_4 solution.

The patient was 50 years old at the time of index surgery; The body weight of the patient was 78 kg, and the patient's body mass index (BMI) was 31. The patient was a housewife with a part-time administration job, and she performed no sports activities.

3.3. Characterization Methods

X-ray diffraction (XRD) analyses were carried out on a Bruker AXS D8 advance diffractometer (Bruker, Karlsruhe, Germany) using the Bragg–Brentano configuration. The excitation of the tube was fixed at 30 KV and 20 mA, the slit was fixed at 0.6 mm, and the probe size was around 6 × 12 mm. A position-sensitive detector (LynxEye, Bruker) was used to collect the data between 10° and 70° (2θ), with a 0.015° step size and a 0.4 s/step acquisition speed. With this configuration, and considering the peaks of interest (the (−111) and (111) monoclinic peaks and the (101) tetragonal peak, respectively, located around 28.3°, 31.5°, and 30.1°), 90% of the XRD signal comes from the first 17 μm below the surface. The monoclinic fraction was determined from the integrated intensities of the XRD peaks after subtracting a linear baseline, using Garvie and Nicholson's equation [30].

Raman spectra were collected with a single spectrograph (Horiba Jobin Yvon LabRAM HR800) with a grating of 1800 gr/mm and Ar+ laser excitation at 514.5 nm wavelength. The laser power on the sample was maintained at ~2 mW with a 100× long-working distance objective to avoid excessive laser-induced heating. With the chosen optical configuration, the laser had a lateral resolution of ~1 μm and a penetration depth of 4.2 μm or 15 μm when the confocal pinhole was fixed at 100 μm or 1000 μm, respectively (intended as the depth from which 90% of the signal comes from). These values were determined following the procedure outlined in Pezzotti et al. [31]. On each specimen, three adjacent points (>10 μm apart) were measured. The collected spectra were fitted with Gaussian–Lorentzian functions after subtracting a linear baseline; the integrated intensity values of monoclinic and tetragonal peaks were used to calculate V_m with both the Clarke/Adar and Katagiri equations and compared with the results of the XRD analyses. An example of the fitted spectrum after baseline subtraction is shown in Figure 1.

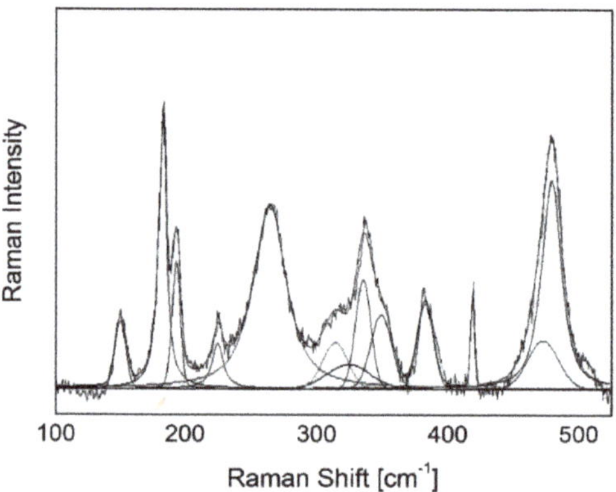

Figure 1. Example of a fitted Raman spectrum of Delta using the procedure followed in this paper. The spectrum was taken on a head specimen in a region with a high monoclinic fraction.

4. Results

4.1. In Vitro Aging Study

Figure 2 and Table 1 present the results of V_m measurements by Raman spectroscopy carried out on Delta femoral heads and inserts, both as received and after the aging procedure. The values of V_m were calculated from the integrated intensity of peaks belonging to the monoclinic and tetragonal phase after the fitting procedure described in Section 3. As it can clearly be seen, the values obtained with the through-focus configuration are smaller than those obtained with the confocal one; in the latter case, due to the smaller probe depth, the volume closer to the surface of the sample was analyzed. Hence, this result shows that the monoclinic fraction is higher in the vicinity of sample surfaces. Aged samples reveal a higher monoclinic content (up to a factor of 2 and higher), as expected, and the difference from the pristine state is larger near the sample surface. Moreover, inserts have a higher monoclinic content due to the raw (grinded backside) surface finish.

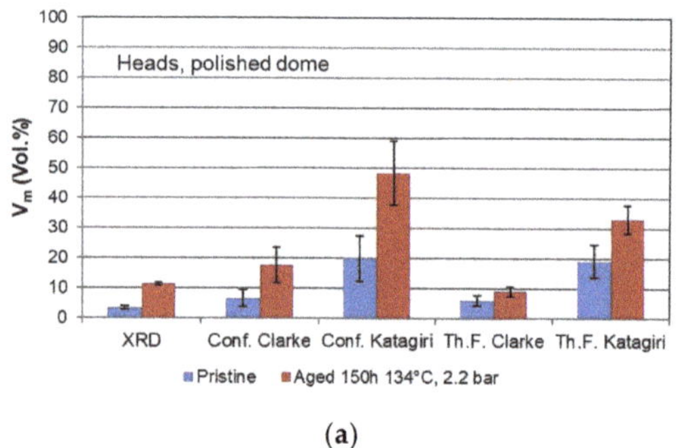

(**a**)

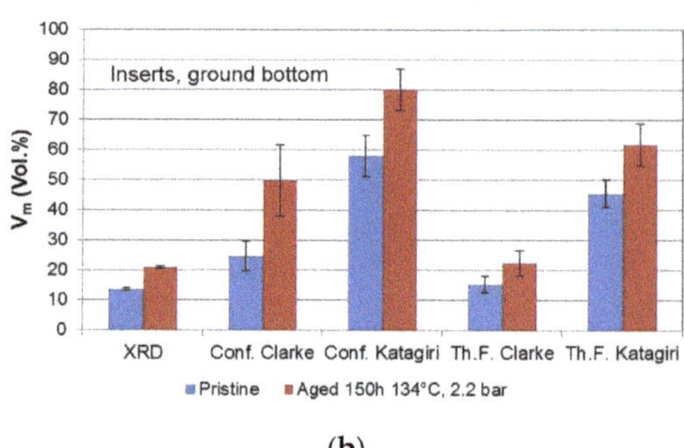

(**b**)

Figure 2. Values of V_m measured by Raman spectroscopy and XRD on both non-aged and in vitro aged Delta heads (**a**) and inserts (**b**). Conf. = confocal pinhole closed down to 100 μm (penetration depth of Raman signal: 4.2 μm); Th. F. = confocal pinhole fully open (1000 μm)—penetration depth of Raman signal: 15 μm.

Comparing values obtained with the Clarke/Adar and Katagiri equations, it is evident that a higher monoclinic content results from the Katagiri equation. This is in line with the findings of Tabares and Anglada [24], who concluded that the Clarke/Adar formula underestimated the monoclinic content for powders. However, a direct comparison between the through-focus Raman results and the XRD results (which have a very similar penetration depth of ~15 μm and 17 μm, respectively) shows that, indeed, it is the

Clarke/Adar equation that provides the best correspondence with the XRD measurements. This is valid on both sample types and for both pristine and aged specimens. This is more evident from Figure 3, where a direct comparison between V_m by XRD and Raman is provided for all samples both in the (a) confocal and (b) through-focus configurations. In Figure 3a, both equations overestimate the V_m by Raman, and this is due to the difference in the volume probed by the two techniques (with confocal Raman, the probe depth is much smaller). For the through-focus case (Figure 3b), where a direct comparison between Raman and XRD is more pertinent due to the very similar penetration depth, the Katagiri equation clearly overestimates (by a factor of 2.5) the monoclinic content, whereas the Clarke/Adar equation provides only a slightly lower V_m than XRD. This latter equation seems thus more suitable for the determination of V_m in the case of aged femoral heads, where the monoclinic content is not constant over the probed depth, keeping in mind that the obtained value is then a weighted average over 15 μm under the surface, as with V_m obtained by XRD.

Table 1. Values of V_m measured by Raman spectroscopy and XRD on both non-aged and *in vitro* aged Delta heads and inserts. Confocal Raman results belong to a depth up to ~4 μm below the surface (pinhole diameter: 100 μm). Through-focus Raman data correspond to a fully opened confocal pinhole (1000 μm) and thus encompass a depth of 15 μm. The penetration depth of XRD is 17 μm.

	Raman CONFOCAL				Raman THROUGH-FOCUS				XRD	
	Clarke/Adar		Katagiri		Clarke/Adar		Katagiri			
	V_m	STD	V_m	STD	V_m	STD	V_m	STD	V_m	STD
Heads										
non-aged	7	5	20	13	6	2	19	7	4	1
aged	18	4	49	9	9	2	33	6	11	1
Inserts										
non-aged	25	8	58	8	15	3	46	6	14	1
aged	50	16	80	10	23	8	62	11	21	0

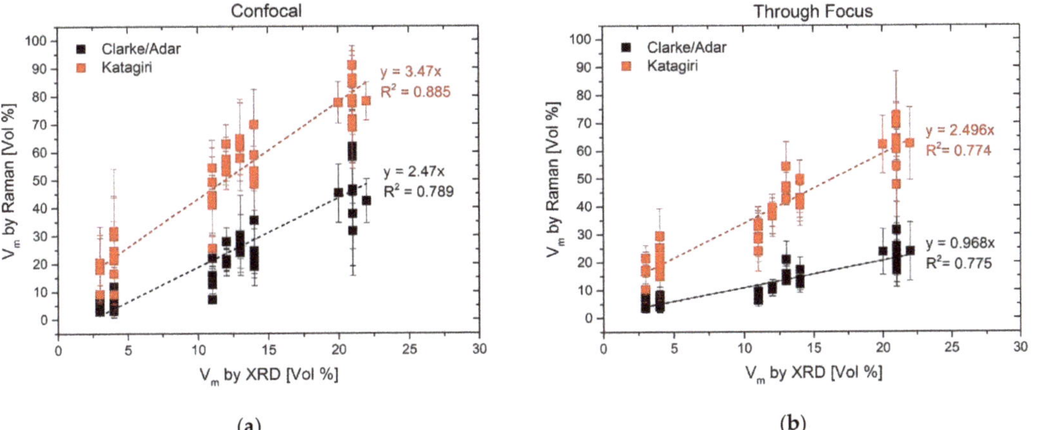

Figure 3. Comparison between all V_m values measured by XRD and Raman spectroscopy and on both non-aged and *in vitro* aged Delta heads and inserts. (**a**) Confocal = confocal pinhole closed down to 100 μm (penetration depth of Raman signal: 4.2 μm); (**b**) Through Focus = confocal pinhole fully open (1000 μm)—penetration depth of Raman signal: 15 μm. The Through Focus measurement mode (with the Clarke/Adar equation) best reproduces the XRD results due to the similar probed volume (penetration depth of XRD: 17 μm.

4.2. In Vivo Aging Study

The explanted femoral head presented significant metal transfer across the whole implant except on apex areas, caused by the two dislocation events with consequent closed reduction procedures. Roughness values measured before and after chemical attack revealed that in the metal transfer area, approximately half of the measured roughness (0.154 μm vs. 0.079 μm in the cleaned sample) was due to the metal smearing and not eventual ceramic surface wear nor scratches on the ceramic surface. Hence, the roughness results reported by other groups in metal transfer areas (without removing the metal) may be questioned [11]. A picture of the retrieved head before and after the cleaning and identification of zones is provided in Figure 4. Zones A, B, C, D, and E are defined as stripe wear, transition area, main wear, metal transfer, and no wear (control area).

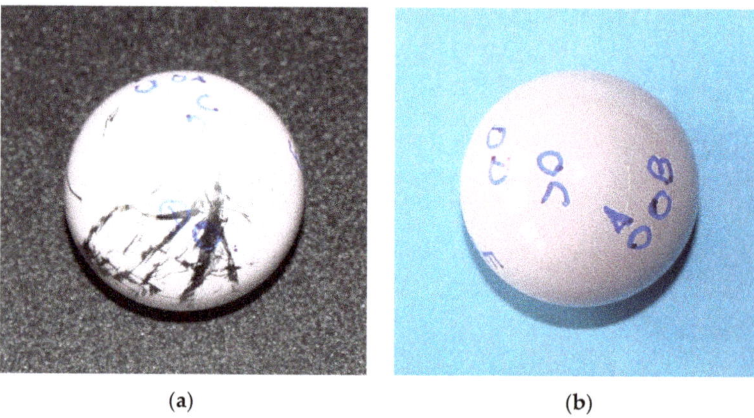

(a) (b)

Figure 4. Picture of retrieved head before (**a**) and after (**b**) the cleaning and identification of zones.

No evidence was found of an increased monoclinic content led by metal transfer, whereas wear seemed to be more critically related to the monoclinic content: in wear areas, we found a higher monoclinic content (especially at the surface)—cf. Table 2. This result supports the interpretation that the discrepancy between in vitro and in vivo is related to shocks contributing to wear and stress-induced phase transformation rather than to metal transfer [15,22,23].

Table 2. Roughness values and values of V_m by XRD; Raman V_m values are determined as difference from E (non-wear case).

		Roughness (μm)		V_m, Raman (%)				V_m, XRD (%)
Area	Description	Metal	Metal Removed	Confocal (Clarke/Adar)	Confocal (Katagiri)	Through-Focus (Clarke/Adar)	Through-Focus (Katagiri)	
A	Stripe wear	0.035	0.036	5	18	3	5	
B	Transition area	0.012	0.017	16	36	8	19	
C	Main wear	0.009	0.008	40	49	20	28	17
D	Metal transfer	0.154	0.079	20	35	7	13	17
E	No wear	0.008	0.007	-	-	-	-	16

5. Discussion

5.1. Use of Clarke/Adar and Katagiri Equations

Our study clearly demonstrates that in the investigated materials (both for in vitro and in vivo aged specimens), the Clarke/Adar equation, and not the Katagiri equation, produced results that are in better accordance with the XRD measurements. This contradicts the current trend in the literature and suggests that the validity of Raman data in the literature is questionable. There is, in particular, a discrepancy with Tabares and Anglada's work [26], where on the basis of Raman and XRD measurements on several monoclinic/tetragonal powder mixtures, Katagiri's equation was deemed more suitable, whereas the Clarke/Adar equation underestimated the results. Tabares and Anglada explained this result with an intrinsic difference residing in the experimental procedure followed by Clarke and Adar: They used fracture surfaces of sintered samples in which the monoclinic phase was confined to a thin surface layer. Consequently, Clarke and Adar's specimens were affected by a concentration gradient in the depth direction, which caused the value of V_m measured by XRD to depend on the wavelength and the angle of incidence of the radiation. In other words, since Tabares and Anglada used different XRD settings for their calibration, the value of 0.97 for the k coefficient in Equation (1) is not valid in their case, and the Clarke/Adar equation underestimates V_m.

This, however, should also apply to our case. Interestingly, it is the Clarke/Adar equation that performs better in our case. One possible explanation is the fact that in our work, we carried out all measurements on sintered samples. It may be envisaged that the coefficient $k = 2.2$ obtained by Katagiri and confirmed by Tabares/Anglada for the Katagiri equation is valid only on powder mixtures, whereas the functional form including the tetragonal peak at ~265 cm^{-1} (and $k = 0.97$) has to be taken for a sintered material, which is the case of the calibration performed by Clarke and Adar. Another possible explanation is the fact that both Clarke/Adar and Tabares/Anglada worked on monolithic zirconia (thus with a much lower penetration depth for XRD: around 5 μm).

A further proof that the Clarke/Adar equation, and not the Katagiri equation, has to be used for our setting is provided in Figure 5. The upper (blue) spectrum in Figure 6 belongs to an area (named area A) with a low V_m located at the apex of non-aged polished Delta head domes, whereas the lower spectrum (red) corresponds to regions with a high Vm (named area B) at the center of the ground bottom of aged heads and inserts. The spectrum in area A is associated with a V_m of 10.2% or 30.5% if calculated with the Clarke/Adar or Katagiri equation, respectively. The spectrum in area B corresponds to a V_m of 66.7% (Clarke/Adar) or 90.1% (Katagiri). Such a high monoclinic content as obtained from the Katagiri equation seems unlikely given the still very strong intensity of the tetragonal peak at ~265 cm^{-1}. In a fully monoclinic material, the 265 cm^{-1} peak is, in fact, absent [18].

The main intrinsic limitation of the Katagiri equation is evident from its functional form displayed in Equation (1): For the calculation of V_m, it considers two monoclinic peaks and only one tetragonal peak. Consequently, if the coefficient k is not correct for the investigated material, the contribution of the monoclinic peaks is disproportionately high. Very likely, for the investigated sintered material, the coefficient k should be higher. Based on a comparison with the V_m obtained here using the Clarke/Adar equation, a coefficient of $k = 4.7$ for the Katagiri equation is probably more realistic in the present case. The coefficient k is probably not only dependent on the materials used for the calibration but also on the type of Raman spectrometer used and on the depth profile of the monoclinic fraction. A careful analysis of the available literature, in fact, suggests that the Katagiri equation performs better on triple spectrometers [11,24,32], whereas the Clarke/Adar equation performs better on single spectrographs [15,21]. This might be explained by differences in the measured relative intensities by the different equipment.

5.2. Spectral Quality and Fitting

Further aspects that could lead to differences in the values of Vm published by various research groups are (i) the overall quality (in terms of the SNR) of the collected spectra and (ii) the procedure used for data regression. Let us first investigate the former aspect. Figure 6 reports two spectra collected on the same polished spot of a Delta head. One spectrum was taken with shorter acquisition times and less repetitions in order to obtain two spectra with very different SNRs. The low-SNR spectrum mimics the case in which a spectrum was taken focusing through the metal in an area affected by metal transfer on a retrieved implant (cf. Figure 3c in [13]). The high-SNR spectrum (black line) corresponds to a V_m of 14% or 34% (with the Clarke/Adar or Katagiri equation, respectively), whereas the low-SNR spectrum corresponds to a V_m of 19% or 41% (Clarke/Adar or Katagiri equation, respectively). Therefore, despite those spectra belonging to the same area, a difference of

~20% was obtained. In other words, using spectra with a low SNR (such as the ones taken in a metal transfer area without removing the metal) may produce an overestimation of the monoclinic content of about 20%.

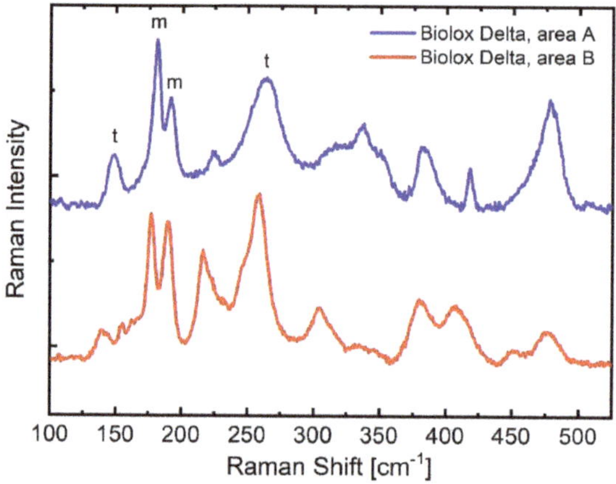

Figure 5. Comparison of Raman spectra of Delta that underwent a low monoclinic transformation (blue line—area A) and a high monoclinic transformation (red line—area B). Peaks belonging to the monoclinic (m) and tetragonal (t) phases of the area used in the analysis are labeled on the upper (area A) spectrum. The area A spectrum is associated with V_m = 10% and 31% with the Clarke/Adar and Katagiri equations, respectively. In area B, the monoclinic content amounts to 67% and 90% (according to the Clarke/Adar and Katagiri equations, respectively). Such a high monoclinic content as obtained from the Katagiri equation seems unlikely given the still very strong intensity of the tetragonal peak at ~265 cm^{-1}.

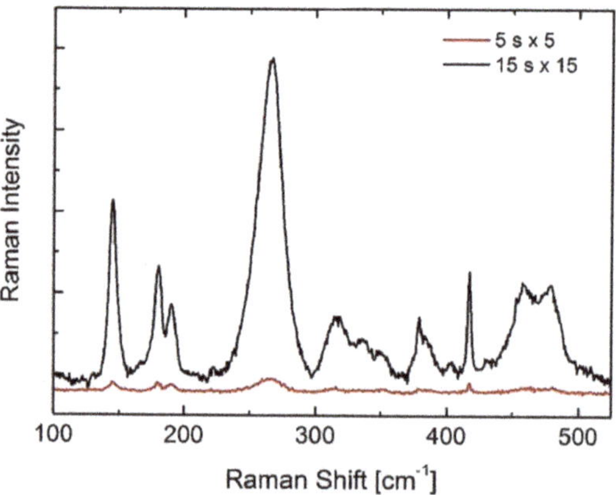

Figure 6. Raman spectra collected on Delta with different acquisition times in order to obtain different SNRs. The spectra were collected on the same point of a polished specimen surface, but the low-SNR spectrum mimics the case of a spectrum collected through a metal layer in correspondence with metal transfer.

Another issue that is often overlooked in the literature is the use of absolute or integrated intensities. In general, integrated intensities should be more suitable in low-V_m cases [27]. Nevertheless, the use of absolute intensities may seem attractive in cases in which a large fluorescence background is present. According to our analysis, using absolute instead of integrated intensities causes an overestimation of the monoclinic content amounting to 26% for the Katagiri equation and up to 60% for the Clarke equation. Hence, use of integrated intensities is mandatory.

The latter result highlights the intrinsic weakness of the Clarke/Adar equation with respect to variations in the overall background of the spectrum, such as in the case of fluorescent emission. To highlight this aspect, we carried out a study in which we fitted the spectra using two different baselines, both in inserts and heads belonging to the investigated Delta implant components. Figure 7 shows the different baselines used in spectra collected on the rear of an insert (a, b) and on the apex of a head (c, d). These cases will be called Cups A and B and Heads A and B, henceforth. From Cup A, values of V_m of 53% and 84% were measured with the Clarke/Adar and Katagiri equations, respectively. The V_m for Cup B amounted to 64% with the Clarke/Adar equation and 84% with the Katagiri equation. Concerning the head, spectrum A had a V_m of 10% with the Clarke/Adar equation and 35% with the Katagiri equation. Head B produced a V_m of 15% with the Clarke/Adar equation and 34% with the Katagiri equation. Hence, choosing a different baseline brings about an error as high as 33% for the Clarke/Adar equation, whereas the Katagiri equation is more stable (maximum error ~3%). For the sake of clarity, we mention that in all spectra used for comparison in Figure 3 above, we used a 15 s acquisition time, 15 repetitions, and the same baseline used for background subtraction as that reported in Figure 7a.

Despite performing better than the Katagiri equation in the investigated samples, the Clarke/Adar equation thus seems to be more prone to errors. The reason resides in the use of the large tetragonal peak at ~265 cm^{-1}, which is strongly influenced by the background and is largely affected by changes in the choice of the baseline for data regression. Hence, Katagiri's choice of excluding this peak from the analysis is not at all wrong; however, we demonstrate here that in this case, the coefficient k has to be recalibrated every time a new material (e.g., sintered instead of powders) or a new instrument is used. Moreover, any modification in both the data collection and treatment procedures risks introducing sources of errors that are non-negligible even in the case of the Katagiri equation. This suggests that the V_m values obtained by different research groups using different equipment and different data treatments can hardly be compared. The only way out of this issue is to define a standard procedure for the analysis of the monoclinic content in zirconia via Raman spectroscopy.

5.3. Proposed Standard Procedure

In the following, we suggest a standard procedure that would allow obtaining V_m values on implants which can be compared between different laboratories:

- First, a series of standard, sintered zirconia samples with a large span of monoclinic content should be prepared in a single batch by the same laboratory or company. These samples should serve as a reference for the calibration of all Raman equipment worldwide.
- Each laboratory should carry out a defined calibration procedure on the standard samples in order to determine the value of the coefficient k for the Katagiri equation that is valid in that specific laboratory.
- The procedure for data treatment, including a minimum SNR, a defined baseline subtraction, and a fitting procedure, should be defined.
- A standard procedure for cleaning the surface of retrievals on all areas, in order to obtain spectra with comparable SNRs over the whole implant, should be defined.

A standardization procedure of this type should be attempted and defined within a round robin study with the participation of a large number of scientific institutions worldwide. However, it should be kept in mind that such a standard procedure will have the limitation—intrinsic within both the Raman and XRD techniques—that the measurement result is the average monoclinic content at several micrometers depth and thus reveals neither the bulk composition nor the spatial distribution of the monoclinic phase. The penetration depth can be varied in Raman by changing the width of the confocal pinhole, and in XRD by modifying the angle of incidence of X-rays. Since spectra and diffractograms will be different for each modification of those parameters (for example, with a different SNR), the standard procedure should be repeated for each penetration depth selected.

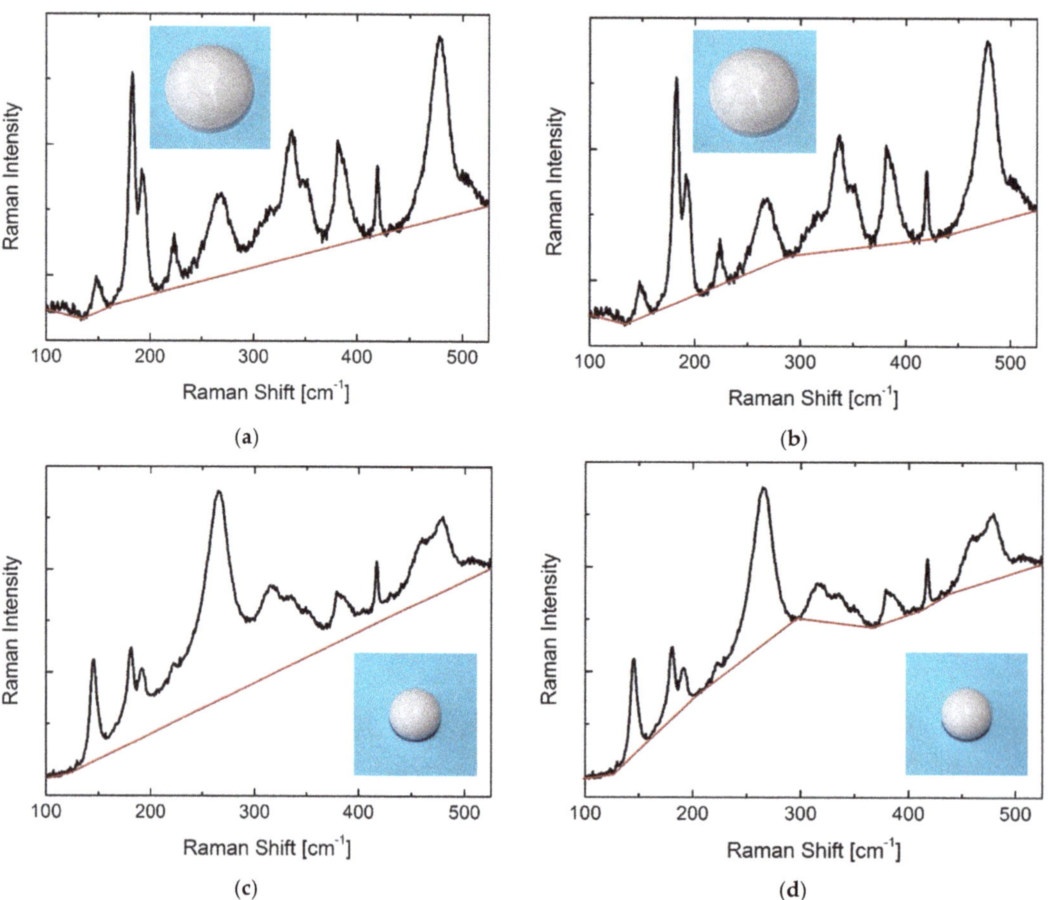

Figure 7. Different baseline choices on Raman spectra of Delta with (**a**,**b**) high and (**c**,**d**) low monoclinic contents. The spectrum shown in (**a**,**b**) was collected on the rear of an insert (rough surface), whereas the spectrum shown in (**c**,**d**) was collected on the apex of a head (polished surface).

A noteworthy alternative to the use of equations is the use of hyperspectral imaging and related statistical analyses (e.g., principal component analyses), as recently applied in surface-enhanced Raman spectroscopy [33].

5.4. Significance of V_m

Section 5.3 proposes a standard procedure for the evaluation of V_m from Raman spectrometry data. With this tool, one can now properly assess the amount of monoclinic phase on the surface of zirconia-toughened alumina hip prosthesis components. However, to determine whether the measured V_m has an influence on the performances of the components, one must also consider the origin of the transformation. Indeed, the origin can be twofold. First, the monoclinic phase can be formed by hydrothermal aging, after a spontaneous t–m transformation due to the presence of water. In this case, the t–m transformation is, in itself, a degradation mechanism. Second, the t–m transformation can occur as a response to high mechanical stresses (phase transformation toughening), as can occur in or around a wear stripe [15], or during shocks. In the latter case, the t–m transformation is necessary to limit the damage. This is visible, for example, from the smaller width of wear stripes measured on ZTA than on alumina (that presents a comparable hardness but no phase transformation toughening) in vitro [34–36].

Stress-induced phase transformation is therefore required to obtain good crack and wear resistance, and the monoclinic content per se should not be considered as an indicator of degradation.

6. Conclusions

In our work, by measurements on both in vitro and in vivo aged BIOLOX®*delta* specimens, we determined that the Clarke/Adar equation is the most suitable equation to quantify the monoclinic content in the investigated material with the used experimental setup. Furthermore, we confirmed that metal transfer is not necessarily related to a high monoclinic content; previous studies showing the contrary might be affected by measurement artefacts leading to an exaggerated monoclinic content. This suggests that the discrepancy between in vitro and in vivo aged implants is rather ascribable to the effect of shocks than to the influence of metal transfer. Moreover, it must be considered that metastability of the tetragonal phase, to a certain extent, is necessary to guarantee good mechanical properties.

In addition, we demonstrated that Raman spectroscopy is a delicate procedure that is very much prone to errors. Critical aspects are associated with the used equation for the calculation of V_m, and with the definition of the related calibration coefficients. Other important issues are related to the spectral quality and data regression procedures. Our study demonstrates that there is a lack of standards concerning the quantification of the monoclinic content in zirconia by Raman spectroscopy. Such standards should be promptly put in place in order to avoid misinterpretations that could ultimately affect the well-being of THA patients.

Author Contributions: Conceptualization, A.A.P. and M.D.; methodology, A.A.P. and M.D.; software, M.D.; validation, A.A.P., M.D. and L.G.; formal analysis, M.D. and L.G.; investigation, A.A.P., M.D. and R.P.; resources, A.A.P., M.D. and R.P.; data curation, M.D. and L.G.; writing—original draft preparation, M.D. and A.A.P.; writing—review and editing, A.A.P., J.C., L.G. and R.P.; visualization, M.D. All authors have read and agreed to the published version of the manuscript.

Funding: This research received no external funding.

Institutional Review Board Statement: Not applicable.

Informed Consent Statement: Not applicable.

Conflicts of Interest: A.A.P. is an employee of CeramTec GmbH (Plochingen, Germany). M.D., L.G., J.C. and R.P. declare no conflict of interest.

References

1. Rahaman, M.N.; Yao, A.; Bal, B.S.; Garino, J.P.; Ries, M.D. Ceramics for prosthetic hip and knee joint replacement. *J. Am. Ceram. Soc.* **2007**, *90*, 1965–1988. [CrossRef]
2. Bierbaum, B.E.; Nairus, J.; Kuesis, D.; Morrison, J.C.; Ward, D. Ceramic-on-ceramic bearings in total hip arthroplasty. *Clin. Orthop. Relat. Res.* **2002**, *405*, 158–163. [CrossRef]
3. Taddei, P.; Pavoni, E.; Affatato, S. Raman and Photoemission Spectroscopic Analyses of Explanted Biolox® Delta Femoral Heads Showing Metal Transfer. *Materials* **2017**, *10*, 744. [CrossRef]
4. National Joint Registry for England, Wales. Northern Ireland and the Isle of Man, 17th Annual Report 2020. Surgical Data to 31 December 2019, 2020. Available online: https://reports.njrcentre.org.uk/Portals/0/PDFdownloads/NJR%2017th%20Annual%20Report%202020.pdf (accessed on 22 July 2021).
5. Australian Orthopaedic Association National Joint Replacement Registry (AOANJRR). *Hip, Knee & Shoulder Arthroplasty: 2018 Annual Report*; Data Period 1 September 1999–31 December 2017; AOA: Adelaide, Australia, 2018; Available online: https://aoanjrr.sahmri.com/annual-reports-2020 (accessed on 22 July 2021).
6. Peters, R.M.; Van Steenbergen, L.N.; Stevens, M.; Rijk, P.C.; Bulstra, S.K.; Zijlstra, W.P. The effect of bearing type on the outcome of total hip arthroplasty. *Acta Orthop.* **2018**, *89*, 163–169. [CrossRef]

7.   Report of R.I.P.O. Regional Register of Orthopaedic Prosthetic Implantology Overall Data Hip, Knee and Shoulder Arthroplasty in Emilia-Romagna Region (Italy) 2000–2018. 2020. Available online: https://ripo.cineca.it/authzssl/Reports.html (accessed on 22 July 2021).

8.   Sharplin, P.; Wyatt, M.C.; Rothwell, A.; Frampton, C.; Hooper, G. Which is the best bearing surface for primary total hip replacement? A New Zealand Joint Registry study. *Hip. Int.* **2018**, *28*, 352–362. [CrossRef]

9.   Arita, M.; Takahashi, Y.; Pezzotti, G.; Shishido, T.; Masaoka, T.; Sano, K.; Yamamoto, K. Environmental stability and residual stresses in zirconia femoral head for total hip arthroplasty: In vitro aging versus retrieval studies. *Biomed. Res. Int.* **2015**, *2015*, 638502. [CrossRef] [PubMed]

10. Kuntz, M. Live-Time Prediction of BIOLOX®delta. In *Bioceramics and Alternative Bearings in Joint Arthroplasty*, 1st ed.; Chang, J.-D., Billau, K., Eds.; Steinkopff: Darmstadt, Germany, 2007; pp. 281–288.

11. Pezzotti, G.; Affatato, S.; Rondinella, A.; Yorifuji, M.; Marin, E.; Zhu, W.; McEntire, B.; Bal, S.B.; Yamamoto, K. In vitro versus in vivo phase instability of zirconia-toughened alumina femoral heads: A critical comparative assessment. *Materials* **2017**, *10*, 466. [CrossRef]

12. Affatato, S.; Ruggiero, A.; De Mattia, J.S.; Taddei, P. Does metal transfer affect the tribological behaviour of femoral heads? Roughness and phase transformation analyses on retrieved zirconia and Biolox® Delta composites. *Compos. Part B Eng.* **2016**, *92*, 290–298. [CrossRef]

13. Zhu, W.; Pezzotti, G.; Boffelli, M.; Chotanaphuti, T.; Khuangsirikul, S.; Sugano, N. Chemistry-driven structural alterations in short-term retrieved ceramic-on-metal hip implants: Evidence for in vivo incompatibility between ceramic and metal counterparts. *J. Biomed. Mater. Res. B Appl. Biomater.* **2017**, *105*, 1469–1480. [CrossRef] [PubMed]

14. Elpers, M.; Nam, D.; Boydston-White, S.; Ast, M.P.; Wright, T.M.; Padgett, D.E. Zirconia phase transformation, metal transfer, and surface roughness in retrieved ceramic composite femoral heads in total hip arthroplasty. *J. Arthroplast.* **2014**, *29*, 2219–2223. [CrossRef] [PubMed]

15. Perrichon, A.; Reynard, B.; Gremillard, L.; Chevalier, J.; Farizon, F.; Geringer, J. Effects of in vitro shocks and hydrothermal degradation on wear of ceramic hip joints: Towards better experimental simulation of in vivo ageing. *J. Tribol. Int.* **2016**, *100*, 410–419. [CrossRef]

16. Pezzotti, G.; Bal, B.S.; Zanocco, M.; Marin, E.; Sugano, N.; McEntire, B.J.; Zhu, W. Reconciling in vivo and in vitro kinetics of the polymorphic transformation in zirconia-toughened alumina for hip joints: III. Molecular scale mechanisms. *Mater. Sci. Eng. C* **2017**, *71*, 552–557. [CrossRef] [PubMed]

17. Pezzotti, G.; Bal, B.S.; Zanocco, M.; Marin, E.; Sugano, N.; McEntire, B.J.; Zhu, W. Reconciling in vivo and in vitro kinetics of the polymorphic transformation in zirconia-toughened alumina for hip joints: I. Phenomenology. *Mater. Sci. Eng. C* **2017**, *72*, 252–258. [CrossRef] [PubMed]

18. Pezzotti, G.; Bal, B.S.; Zanocco, M.; Marin, E.; Sugano, N.; McEntire, B.J.; Zhu, W. Reconciling in vivo and in vitro kinetics of the polymorphic transformation in zirconia-toughened alumina for hip joints: II. Theor. *Mater. Sci. Eng. C* **2017**, *71*, 446–451. [CrossRef] [PubMed]

19. Zhu, W.; Nakashima, S.; Marin, E.; Gu, H.; Pezzotti, G. Microscopic mapping of dopant content and its link to the structural and thermal stability of yttria-stabilized zirconia polycrystals. *J. Mater. Sci.* **2019**, *55*, 524–534. [CrossRef]

20. Otto, A.; Mrozek, I.; Grabhorn, H.; Akemann, W. Surface-enhanced Raman scattering. *J. Phys. Condens. Matter.* **1992**, *4*, 1143. [CrossRef]

21. Rufaqua, R.; Vrbka, M.; Hemzal, D.; Choudhury, D.; Rebenda, D.; Krupka, I.; Hartl, M. Analysis of Chemisorbed Tribo-Film for Ceramic-on-Ceramic Hip Joint Prostheses by Raman Spectroscopy. *J. Funct. Biomater.* **2021**, *12*, 29. [CrossRef]

22. Perrichon, A.; Liu, B.; Chevalier, J.; Gremillard, L.; Reynard, B.; Farizon, F.; Liao, J.-D.; Geringer, J. Ageing, shocks and wear mechanisms in ZTA and the long-term performance of hip joint materials. *Materials* **2017**, *10*, 569. [CrossRef] [PubMed]

23. Perrichon, A.; Reynard, B.; Gremillard, L.; Chevalier, J.; Farizon, F.; Geringer, J. A testing protocol combining shocks, hydrothermal ageing and friction, applied to Zirconia Toughened Alumina (ZTA) hip implants. *Mech. Behav. Biomed. Mater.* **2017**, *65*, 600–608. [CrossRef]

24. Nguyen, T.M.; Weitzler, L.; Esposito, C.I.; Porporati, A.A.; Padgett, D.E.; Wright, T.M. Zirconia phase transformation in zirconia-toughened alumina ceramic femoral heads: An implant retrieval analysis. *J. Arthroplast.* **2019**, *34*, 3094–3098. [CrossRef]

25. Kim, B.K.; Hahn, J.W.; Han, K.R. Quantitative phase analysis in tetragonal-rich tetragonal/monoclinic two phase zirconia by raman spectroscopy. *J. Mater. Sci. Lett.* **2017**, *16*, 669–671. [CrossRef]

26. Munoz Tabares, J.A.; Anglada, M.J. Quantitative analysis of monoclinic phase in 3Y-TZP by raman spectroscopy. *J. Am. Ceram. Soc.* **2010**, *93*, 1790–1795. [CrossRef]

27. Clarke, D.R.; Adar, F. Measurement of the crystallographically transformed zone produced by fracture in ceramics containing tetragonal zirconia. *J. Am. Ceram. Soc.* **1982**, *65*, 284–288. [CrossRef]

28. Pezzotti, G.; Yamada, K.; Sakakura, S.; Pitto, R.P. Raman spectroscopic analysis of advanced ceramic composite for hip prosthesis. *J. Am. Ceram. Soc.* **2008**, *91*, 1199–1206. [CrossRef]

29. Vega, M.M.; Bonifacio, A.; Lughi, V.; Sergo, V. Low-Level Monoclinic Content Detection in Zirconia Implants Using Raman Spectroscopy. In *Nano-Structures for Optics and Photonics. NATO Science for Peace and Security Series B: Physics and Biophysics*, 1st ed.; Di Bartolo, B., Collins, J., Silvestri, L., Eds.; Springer: Dordrecht, The Netherlands, 2015; pp. 539–540.

30. Garvie, R.C.; Nicholson, P.S. Phase analysis in zirconia systems. *J. Am. Ceram. Soc.* **1972**, *55*, 303–305. [CrossRef]

31. Pezzotti, G.; Munisso, M.C.; Lessnau, K.; Zhu, W. Quantitative assessments of residual stress fields at the surface of alumina hip joints. *J. Biomed. Mater. Res. Part B Appl. Biomater.* **2010**, *95*, 250–262. [CrossRef]
32. Tateiwa, T.; Marin, E.; Rondinella, A.; Ciniglio, M.; Zhu, W.; Affatato, S.; Pezzotti, G.; Bock, R.M.; McEntire, B.J.; Bal, B.S.; et al. Burst Strength of BIOLOX®delta Femoral Heads and Its Dependence on Low-Temperature Environmental Degradation. *Materials* **2020**, *13*, 350. [CrossRef]
33. Vega, M.M.; Alois, B.; Vanni, L.; Sergo, V. Fine determination of monoclinic phase in zirconia-based implants: A surface-enhanced Raman spectroscopy (SERS) study. *J. Nanosci. Nanotechnol.* **2020**, *20*, 2430–2435. [CrossRef]
34. Uribe, J.; Geringer, J.; Gremillard, L.; Reynard, B. Degradation of alumina and zirconia toughened alumina (ZTA) hip prostheses tested under microseparation conditions in a shock device. *Tribol. Int.* **2013**, *63*, 151–157. [CrossRef]
35. Piconi, C.; Porporati, A.A.; Streicher, R.M. Ceramics in THR bearings: Behavior under off-normal Conditions. *Key Eng. Mater.* **2014**, *631*, 3–7. [CrossRef]
36. De Fine, M.; Terrando, S.; Hintner, M.; Porporati, A.A.; Pignatti, G. Pushing Ceramic-on-Ceramic in the most extreme wear conditions: A hip simulator study. *Orthop. Traumatol. Surg. Res.* **2021**, *107*, 102643. [CrossRef] [PubMed]

Journal of
Composites Science

Citation: Solarino, G.; Spinarelli, A.; Virgilio, A.; Simone, F.; Baglioni, M.; Moretti, B. Outcomes of Ceramic Composite in Total Hip Replacement Bearings: A Single-Center Series. *J. Compos. Sci.* **2021**, *5*, 320. https://doi.org/10.3390/jcs5120320

Academic Editor: Francesco Tornabene

Received: 25 October 2021
Accepted: 6 December 2021
Published: 8 December 2021

Publisher's Note: MDPI stays neutral with regard to jurisdictional claims in published maps and institutional affiliations.

Article

Outcomes of Ceramic Composite in Total Hip Replacement Bearings: A Single-Center Series

Giuseppe Solarino *, Antonio Spinarelli, Antonio Virgilio *, Filippo Simone, Marco Baglioni and Biagio Moretti

Department of Basic Medical Sciences, Neuroscience and Organs of Sense, School of Medicine, AOU Policlinico Consorziale, Università di Bari "Aldo Moro", 70124 Bari, Italy; antonio.spinarelli@gmail.com (A.S.); filsimo1993991@gmail.com (F.S.); baglioni.m86@gmail.com (M.B.); biagio.moretti@uniba.it (B.M.)
* Correspondence: giuseppe.solarino@uniba.it (G.S.); antoniovirg@yahoo.it (A.V.);

Abstract: Despite the fact that total hip replacement is one of the most successful surgical procedures for treatment of a variety of end-stage hip diseases, the process of osteolysis and implant loosening remains a significant problem, especially in young and high-demand patients. More than 40 years ago, ceramic bearings were introduced due to their mechanical advantage in order to obtain a reduction in wear debris, and due to the conviction that it was possible to minimize friction and wear owing to their mechanical hardness, high chemical stability, surface lubrication by fluids and low friction coefficient. Together with excellent mechanical properties, ceramics have a biological inertness: eventual ceramic debris will lead to a reactive response with a high predominance of fibrocystic cells, rather than macrophagic cells, and absence of giant cells, which is ideal from a biological perspective. As a consequence, they will not trigger the granulomatous reaction necessary to induce periprosthetic osteolysis, and this clearly appears to be of great clinical relevance. In recent years, tribology in manufacturing ceramic components has progressed with significant improvements, owing to the development of the latest generation of ceramic composites that allow for an increased material density and reduced grain size. Currently, ceramic-on-ceramic bearings are considered the attractive counterparts of ceramic- or metal-on-polyethylene ones for patients with a long life expectancy. The aim of this paper is to report the results of total hip replacements performed with a ceramic-on-ceramic articulation made from a ceramic composite in a single center, focusing on its usefulness in specific preoperative diagnosis.

Keywords: hip; alumina matrix composite; AMC; hip prosthesis; prosthesis; case series; ceramic-on-ceramic

1. Introduction

Currently, one of the most successful operations in the orthopedic surgery field is prosthetic hip replacement (total hip arthroplasty, THA), owing to the excellent immediate results, and in the long term, it is possible to obtain a rapid recovery with great satisfaction from patients.

Over the years, there has been continuous research in order to obtain an improvement in the components of prostheses in order to reduce complications and maximize results.

From a tribological point of view, there has been a continuous development of materials and a continuous search for the best combinations of materials, especially in terms of duration and complications. The choice of materials is, in fact, able to influence the functional recovery, to reduce the risk of complications including the formation of debris and to reduce the number of revisions, thus influencing the long-term duration of prostheses.

In 1962, the introduction of ultra-high molecular weight (UHMW) polyethylene as a bearing material by Sir John Charnley was the first important push toward the modern THA procedure. Some years later, in 1971, Pierre Boutin introduced the ceramic alumina as a coupling in hip implants, realizing the alumina-on-alumina THA procedure [1].

Among the various characteristics of alumina, there is its hydrophilic property which therefore allows for excellent lubrication of prostheses [2].

Subsequently, over the years, there has been a continuous evolution of ceramics, owing to the progressive reduction in the dimensions of particles that compose them, and hot pressurization techniques, resulting in a progressive increase in flexural resistance. Several generations of alumina have followed one another, of which the third generation of ceramics is called BIOLOX® forte [3].

The most important improvement was obtained through the creation of the alumina matrix composite (AMC), obtained by joining alumina (82%) to zirconia oxide, strontium aluminate and chromium oxide particles; in this way, the density of the material and its three-dimensional composition can be modified, obtaining a much harder and more resistant compound. This material is the fourth generation of alumina and is called BIOLOX® delta [4]. Important features of the AMC are its resistance in the event of impacts and malpositioning and, above all, the possibility to use large-diameter heads while maintaining a very thin insert, in order to ensure maximum mobility and minimize the risk of dislocation [5]. Among the advantages of using the alumina matrix composite (AMC), there is the low frequency of complications [6–8]. Osteolysis related to wear debris is one of the major reasons for failure of hip prostheses and is therefore capable of negatively affecting the survival of THA implants, especially in young patients [9–12]. Although there is a risk of osteolysis, this occurrence is still very rare. Kim et al. in 2016, in the case of implants with a delta ceramic head and polyethylene liner (HXLPE), did not detect any case of osteolysis in patients under 50 years and with an average follow-up of 11.9 years [13]. Depending on the used materials, the possible complications related to debris are variable. In the case of use of metal-on-metal prostheses, there is an excretion of ions in the urine, toxicity, neuropathies and the possibility of developing a pseudotumor. The ceramic-on-ceramic (CoC) combination is able to guarantee optimal tribological properties including exceptional hardness, excellent wettability and lubrication and, above all, a very low release of particles, which, in any case, have a high level of biocompatibility [14,15]. It should also be considered that over the course of various generations of ceramics, the risk of ceramic fracture associated with this material has enormously decreased, owing to improvements from the point of view of design and manufacturing processes, and it should stand at values between 0.001% and 0.021% [16]. According to a 2017 evaluation, new composite ceramics have lowered the risk of femoral head fracture to 0.002%; however, the risk of liner fracture has increased somewhat (0.02%). Ceramic bearings provide substantial advantages over traditional bearings, according to the authors, given that the components are introduced without impingement [17]. Howard et al. used data from the National Joint Registry (NJR) for England, Wales, Northern Ireland and the Isle of Man to show that there is good evidence that the latest generation of ceramic has greatly reduced the risk of head fracture, and that a small head size and a high patient BMI are associated with an increased risk of ceramic bearing fracture [18]. Yoon et al. performed a meta-analysis on 10,571 THAs from 45 trials and found that the AMC performed better than third-generation ceramic components [19].

Buttaro et al. retrospectively reviewed 939 cases (880 patients) of primary total hip arthroplasty with fourth-generation delta ceramic-on-ceramic bearings and found that the mean survival rate was 99.3% at 2–10 years, when revision or impending revision in relation to the bearing surface was considered the end point. They emphasized that the few problems they discovered were due to technical mistakes that surgeons should avoid when utilizing this type of surface [20]. According to Lee et al., who analyzed 2.78 million AMC ceramic ball heads implanted globally, the incidence of clinical fractures of contemporary AMC femoral heads is 1 in 100,000 (0.0010%). The majority of implant failures happened within 48 months after surgery, and they were typically caused by particular events such as trauma, mismatched components and dislocations [21]. Ceramics have a biological inertness: fibrocystic cells—rather than giant and macrophagic cells—are involved with high predominance in response as a reaction to eventual ceramic particles being worn out, which will lead to an ideal response from a biological perspective; in fact, an undesired consequence, i.e., periprosthetic osteolysis due to a granulomatous reaction, will not be induced and this obviously has enormous clinical relevance [22,23].

2. Our Experience

We looked at 62 cementless AMC-on-AMC THAs conducted on 54 patients in a row. Patients were operated on bilaterally in eight cases, with THAs performed on separate dates. We received clearance from our hospital's review board, and 32 men and 22 women, ranging in age from 14 to 68, were recruited in this retrospective study. In 34 cases, the right hip was operated on, while the left was operated on in 28 cases. Diagnoses were as follows: osteoarthritis, either primary or secondary, was diagnosed in 32 cases (51.6%), avascular necrosis of the femoral head in 12 cases (19.35%), intracapsular displaced fracture of the femoral neck in 12 cases (19.35%), secondary osteoarthritis as a sequela of Legg–Calve–Perthes disease/developmental congenital dysplasia of the hip/slipped capital femoral epiphysis in 4 cases (6.45%) and rheumatic arthritis in two cases (3.22%). Two fellowship-trained orthopedic surgeons performed preoperative templating to determine the right size of the components. All of the procedures were conducted by senior registrars with expertise in joint arthroplasty surgery; no orthopedic trainee, even under supervision, was ever engaged as the initial surgeon in any THA. All of the THAs were primary operations that were implanted in a traditional turbulent flow theater utilizing a direct lateral approach. After the acetabulum was reamed, the acetabular cup was set with a desired inclination angle of 40–45° and a planned anteversion of 15–20°. Only cementless hemispherical cups were implanted, and their outer diameter was as follows: 54 in 16 cases (26.6%), 50 in 12 (20%), 52 in 8 (13,3%), 56 in 8 (13.3%), 58 in 6 (10%), 60 in 4 (6.6%), 46 in 4 (6.6%), 48 in 2 (3.22%) and 62 in 2 (3.22%). Additional acetabular screws were utilized to complete fixation based on the patient's bone quality and the surgeon's choice, but there was no apparent link to the preoperative diagnosis. Before the ceramic inlay can be placed and positioned, the metal shell must be clean and dry on the inside. Because they cannot be crushed, every little piece of soft or bone tissue, any fluid and any remnant of cement must be removed from the metal shell. It is critical to confirm that the right seat of the ceramic insert has been attained by touching the cup rim with the fingers: the ceramic and metal rims must lay flat to one another. The biggest broach that would fill the metaphysis and leave minimal cancellous bone left was utilized to implant all cementless femoral components using a press fit method. Thorough cleaning and drying of the taper on the stem side are also required because any foreign material reduces the fracture strength of the ceramic femoral ball head and impairs its force transmission. Because relative motions between the components are minimized, a slight rotating motion when placing the ceramic head will ensure a safe locking. Finally, ensuring the femoral ball head is placed securely requires a moderate hammer blow on the impactor in the axial direction of the stem taper. The shuck test and examination of the main arc range of motion were used to examine the stability of the hip implant before it was closed in layers. The diameter of the femoral head utilized was at the discretion of the surgeon: a 36 mm femoral head was implanted in 44 hips (70.96%), a 32 mm femoral head in 14 hips (22.5%) and a 40 mm femoral head in 4 cases (6.45%). All patients received the same perioperative care: antibiotic and thromboembolic prophylaxis with heparin administration and compression stockings, passive motion exercises with a therapist immediately after the operation, removal of the single intra-articular suction drain on the second postoperative day and patients free to walk with two supports after three days for about six weeks. Preoperatively and postoperatively, patients were assessed at 6 weeks, 3 months, 1 year and every 2 years thereafter. Clinically, a history of squeaking and all postoperative problems were noted at each visit, and patient activity was measured using the Harris Hip Score (HHS); the preoperative HHS averaged 36.7 (range 15–58). The same observer who had not been engaged in the procedures for tilting or migration of the cup, radiolucent lines and osteolysis in acetabular component zones acquired and examined an anteroposterior pelvic and axial view of the affected hip radiographically. A cup migration of more than 3 mm, an angular rotation of more than 3° and/or a continuous radiolucent line wider than 2 mm were all considered loosening. Around the femoral component, there were indications of stem subsidence or displacement, as well as osteolysis. Our cutoff for vertical migration was set at 3 mm, and osteolysis was defined as a sharply demarcated lucent area adjacent to the acetabular and/or femoral component that was not visible on the immediate postoperative radiographs. All local problems, such as periprosthetic infection, dislocation, intraoperative and postoperative periprosthetic fracture and liner and/or head breakages, were documented.

Due to the death or loss of two patients with two THAs at the last follow-up at the study census date, 60 THAs were eligible for the current study. The mean HHS increased to 92.1 points (range 61–100) among them, with no significant differences between groups in terms of the preoperative diagnosis that led to the procedure. None of the implants' components failed due to mechanical failure, and none of them dislocated. The patients never mentioned squeaking.

3. Discussion

3.1. AMC-on-AMC Total Hip Replacement in Avascular Necrosis of the Femoral Head

Avascular necrosis (AVN) or osteonecrosis of the femoral head is caused by a lack of blood supply in the region; in the later stages of the disease, bone is reabsorbed and the epiphysis collapses: it is obvious that hip replacement with THA, which provides immediate pain relief and excellent functional outcomes, should be considered the preferred treatment. Nevertheless, previous reports have underlined that avascular necrosis represents a risk factor of biomechanical failure of implants [24,25] due to the pathogenesis of periprosthetic osteolysis, strictly connected to wear debris of polyethylene. Such a cause of failure of a THA remains one of the most prevalent in young and active patients, and thus alternative hard materials have been developed as bearing surfaces in clinical practice to reduce the detrimental effects of wear on prosthesis survival. As a result, there has been a revived interest in hard components as surfaces for total joint arthroplasty, including ceramic-on-ceramic and metal-on-metal bearings. However, despite their remarkable tribological characteristics (friction, lubrication and wear), specific difficulties must be accounted for with such materials because metallic ions or ceramic particles may be generated; furthermore, the biologic cascade resulting from debris created by these alternative bearings has yet to be completely understood. Current metal-on-metal bearings are self-polishing and made of cobalt-chromium-molybdenum alloys with varying carbon contents; undoubtedly, debris from such bearings is to blame for increased metal ion levels in urine and blood, eventual toxicity, capsular aggregation of lymphocytes and pseudotumor growth, all of which can cause neuropathy and delayed-type hypersensitivity. Park et al. [26] have already demonstrated a hypersensitivity reaction to cobalt in patients with early osteolysis after contemporary metal-on-metal THAs using skin patch tests, whereas Savarino et al. [15] have demonstrated high levels of cobalt and chromium with metal-on-metal articulations and negligible serum metal ion contents in ceramic-on-ceramic THAs using skin patch tests; they concluded that, because of the increased ion release, metal-on-metal coupling should be used with caution, particularly in young patients. We feel that when THAs are conducted in patients with AVN, certain critical precautions should be taken into account because this condition is considered a risk factor. According to Ortiguera et al. [27], the dislocation rate in osteonecrosis patients is expected to be much higher than in osteoarthritic patients because subjects with AVN have potentially much less stiffness than patients with osteoarthritis and thus are able to achieve a wider range of hip movement after surgery, making them more prone to joint instability. Berry et al. [28] examined the cumulative incidence of dislocation in a cohort of over 6000 prosthetic hips and found it to be more than doubled (14.1% compared to 6.4%) in patients with osteonecrosis than in those with osteoarthritis; their research also found that dislocation rates rise with extended durations of follow-up, possibly reflecting the impact of neuromuscular degeneration and, in particular, wear at the metal-on-polyethylene articulating surface.

Ceramic hip arthroplasty, we feel, is the most reliable treatment to suggest in active, youthful patients with a long life expectancy, such as those with AVN. Wear is often higher in this demographic. Intracellular ceramic wear debris, which may be observed in the soft periarticular tissues, will create an inflammatory reaction that the organism will tolerate much better than the one caused by metal-on-metal articulation, and as a result, it will be less aggressive. Ceramics have excellent biocompatibility due to their favorable tribologic properties, biologic inertness, extremely good surface finish and unique hardness, which allows surgeons to use femoral heads with diameters greater than 28 mm while also providing an increased range of motion and decreased dislocation, as well as a very low friction coefficient and high wear resistance.

Furthermore, other authors [29] comparing 654 THAs conducted for AVN and performed in 327 patients with an average age of 42 years operated bilaterally, with a CoC THA on one side and a CoPE THA on the contralateral side, demonstrated that the risk of a femoral fracture around the stem was strictly related to the bearing components used during prosthetic implantation. They discovered that late postoperative periprosthetic fractures increased in quantity with follow-up and occurred 30 times more commonly on the side with the PE cup than on the side with the CoC bearing. They came to the conclusion that because of the lack of wear and osteolysis, CoC couples reduce the incidence of late periprosthetic fractures.

In Figure 1, the case of the youngest patient in our group is shown: a 14-year-old girl suffering for multifocal AVN due to treatment for acute lymphoblastic leukemia. She was complaining of severe pain and considerable functional impairment in her left hip: a 48 mm cup with 32 mm AMC–AMC bearings on a standard stem was used.

3.2. Alumina-on-Alumina Total Hip Replacement in Developmental Dysplasia of the Hip

THA in sequelae of developmental dysplasia of the hip (DDH) is a challenging and difficult procedure in terms of restoring the normal biomechanics. In case of dysplastic patients, the pre- and intraoperative process of decision making is also complex due to the underdeveloped acetabulum, the contracture of the surrounding soft tissues and often the high riding head; on the pelvic side, the choice is frequently linked to the need of using small-diameter acetabular cups because DDH causes a lack of bone under the prosthesis, and sometimes, it may not be possible to place a large acetabular cup in the anatomical position with a medialization close to the radiological tear drop because of the hypoplasic acetabulum and thus the risk of miscoverage of the shell (Figure 2). We are convinced that obtaining the correct location of the acetabular cup, and thus providing the best function of the artificial hip, is dependent on the outer diameter of the shell because, typically, a cup with a small diameter is used, and as a result, the thickness of the acetabular insert decreases, the two components being inextricably linked. We are aware of manufacturer information stating that a 32 mm HXLPE inlay may only be utilized with a cup diameter of 52 mm or more [30]. The presence of small-diameter cups affects the size of the insert which will be proportionally smaller. The use of the AMC in these patients is essential because it allows the combination of very thin inserts even in acetabular cups with a reduced diameter, combining them with large diameter heads [31]. Inadequate polyethylene thickness is identified as the source of plastic particle-mediated osteolysis, not only in conventional polyethylene [32] but also in vitamin E-diffused HXLPE, and hence significant liner thinning is not advised [33]: Higher peak contact loads and smaller contact areas result from a decrease in polyethylene liner thickness or headliner conformance, resulting in reduced biomechanical wear factors. On the other hand, since larger femoral heads have been widely advocated to improve implant stability and range of motion, especially in patients with spinopelvic alignment, such as patients with hip osteoarthritis secondary to DDH, prosthetic heads must be enlarged, implying a reduction in liner thickness [34–40].

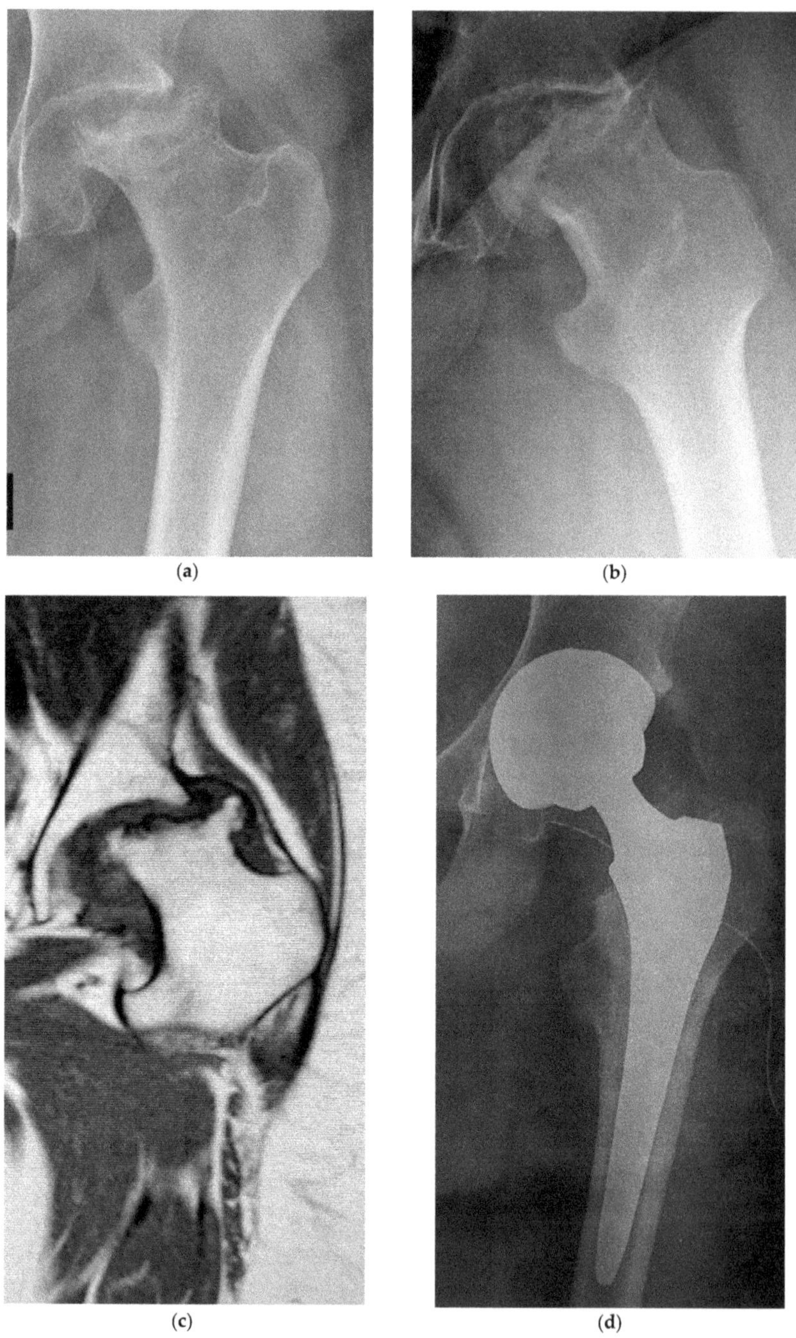

Figure 1. Female, 14 years old, osteonecrosis of the femoral head in her left hip: preoperative radiological features in AP (**a**), axial view (**b**) and magnetic resonance imaging (**c**). Postoperative AP X-ray (**d**) of THA with a 32 mm AMC-on-AMC coupling.

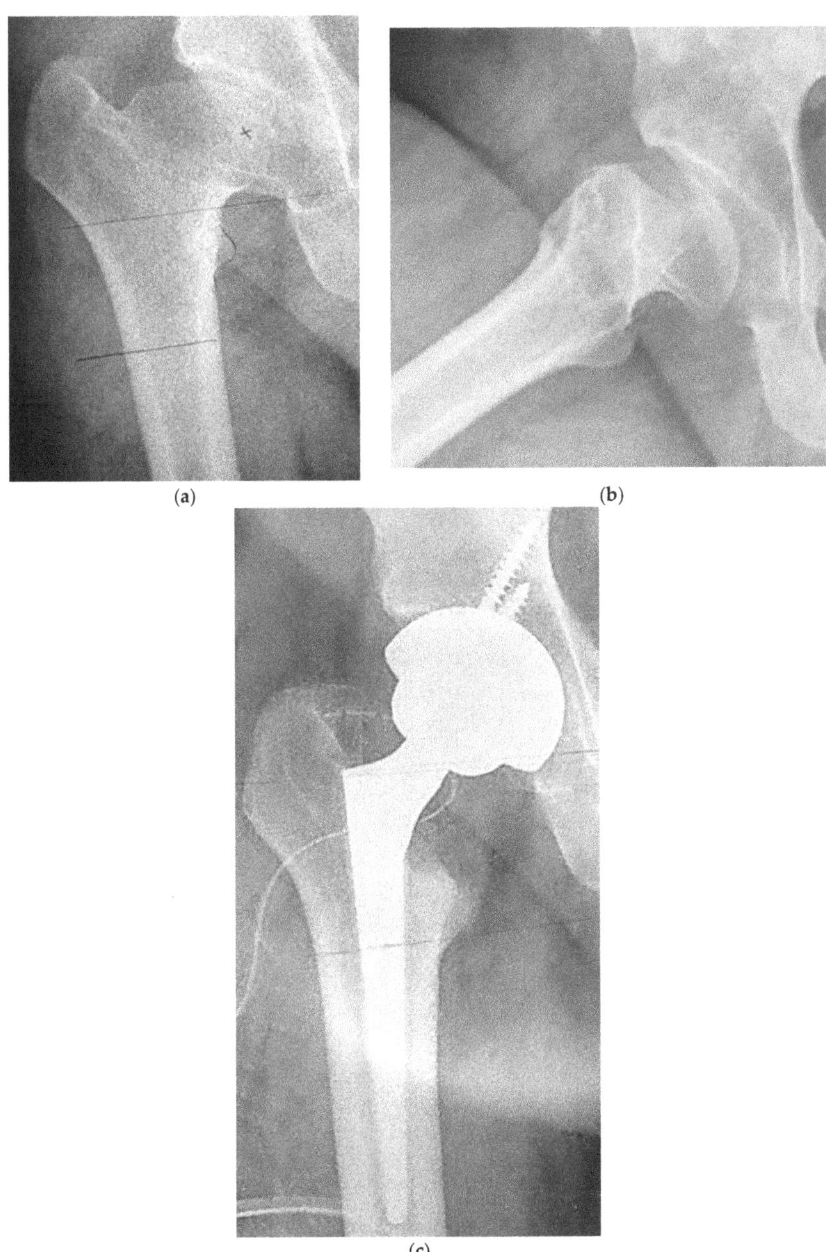

Figure 2. Male, 55 years old, right hip osteoarthritis secondary to development dysplasia: preoperative AP and axial radiographs (**a,b**), and postoperative X-ray (**c**) of the AMC-on-AMC THA using screws for implementation of cup fixation and a conical stem below it.

3.3. Alumina-on-Alumina Total Hip Replacement for Femoral Neck Fracture

Because internal fixation has a significantly greater failure probability, resulting in more pain for these patients, primary arthroplasty stands out as the best option for displaced Garden 3 and 4 femoral neck fractures (FFN). When a THA is performed, the surgeon must consider implant dislocation as a possible complication, which is said to be more common after a hip fracture, with the posterior surgical approach, in elderly patients with soft-tissue laxity due to sarcopenia and thus poor muscular strength, and the attempt to regain the full range of motion before the injury [41,42]. Even with heads bigger than 28 and 22.2 mm, the ceramic-on-ceramic connection delivers little friction and wear. We believe that after the use of ceramic-on-ceramic bearings in THA, the risk of dislocations can be influenced because, while the risk of fracture should be higher, it will actually be lower after use of heads larger than 28 mm, as in this series: the prevalence of fibrotic response in the tissues surrounding an implant with a CoC articulation may provide a thicker and more resistant capsule in the long term. The periprosthetic retroacetabular bone should not be regarded as a source of stress shielding: the titanium shell is thought to function as a shock absorber between the high stiffness of the alumina and the likely porotic bone, addressing the problem of socket fixation described when a cup of bulky alumina was cemented into the acetabulum [36]. In Bystrom S et al. [42], the femoral head size was shown to be a significant risk factor for prosthesis luxation in a retrospective study of 42,987 primary operations: 22 mm heads performed as well as or better than 28 mm heads, while 28 mm heads led to revision four times more often than 32 mm heads. According to the Norwegian Arthroplasty Register, femoral head size is a risk factor for total hip luxation, with 28 mm heads leading to revision substantially more frequently than 32 mm heads, and 26 mm heads leading to revision significantly more frequently than 30 mm heads. The preoperative diagnosis, i.e., an FFN, was also a significant determinant in the luxation revision rate.

In Figure 3, the case of a young man who suffered an FFN after a traffic accident (fall from an electric scooter) is documented: a THA with 36 mm AMC-on-AMC articulation was implanted.

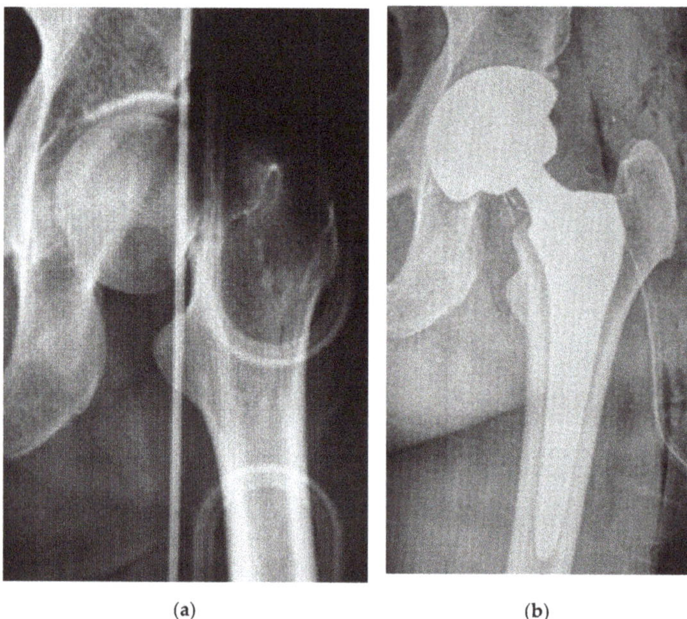

(a)   (b)

Figure 3. Male, 42 years old, displaced fracture of his left femoral neck: preoperative AP X-ray of the involved hip (**a**), and immediate postoperative X-ray (**b**) of the AMC-on-AMC THA with a 36 mm head and a stem with a modular neck.

4. Conclusions

Surgeries of joint arthroplasties are among the most successful operations in the entirety of medicine. Since the early 1960s, hip replacement surgical techniques and technology improvements have increased the effectiveness of this procedure. Because the range of motion grows and the jump distance to subluxation and/or dislocation becomes greater, there is significant interest in adopting larger-diameter heads. If we switch to hard-on-hard couplings, we may utilize heads larger than 32 mm: AMC-on-AMC bearing surfaces are appealing alternative bearing surfaces that remove or substantially minimize the issues associated with PE wear debris. A new low-frictional torque arthroplasty theory can be established owing to their sliding characteristics and better wettability than PE: with ceramic components (inlay and head), due to a lower coefficient of friction, it is possible to enlarge the femoral head diameter without increasing the frictional torque and to grant a wider range of movement that will last because ceramic–ceramic bearings avoid wear. When a PE liner develops wear, the center of rotation migrates centrally and/or cranially, and the deeper the head, the more restricted the range of motion becomes; in fact, late dislocation can be the first clinical sign of wear, with a cumulative risk of first-time dislocation that is 2-fold and 3-fold at 10 and 20 years, respectively, when compared to the 1-year follow-up [14,30–43]. Breakage of ceramic components is a ceramic-specific issue; we have not seen any fractures in our series.

This may be explained by the better manufacturing process, the exact contact surface geometry, including appropriate clearance, and the greater resistance to fracture when employing heads with a diameter of 32 mm or bigger. According to Santavirta S [44], the component fracture risk for currently available ceramic goods is essentially non-existent, as demonstrated in clinical trials at 4 and 5 years, even with a 28 mm head [45,46]. If the diameter is even bigger, more benefits in terms of both strength and function are obtained: in the first three months following surgery, the rate of dislocation was 0.88% for 36 mm heads and 4.64% for 28 mm heads [47].

Around the world, total hip arthroplasty is becoming more prevalent in younger patients [48]. Ceramic bearing implants for THA surfaces are now extensively utilized across the world to reduce wear debris and aseptic loosening. When utilizing ceramic components, the surgical technique must be precise [49].

Our findings show that AMC-on-AMC couplings provide outstanding outcomes in THA conducted in high-demand patients, with no material-related side effects, no ceramic fracture and no mechanical loosening of implant components. The danger of dislocation should be reduced if the head has a diameter of 32 mm or more. Overall, we believe AMC-on-AMC to be the bearing of choice for younger and more active patients due to its dependability and longevity.

Author Contributions: Writing—original draft, G.S., A.S., A.V., F.S., M.B. and B.M. All authors have read and agreed to the published version of the manuscript.

Funding: This research received no external funding.

Conflicts of Interest: The authors declare no conflict of interest.

References

1. Bizot, P.; Nizard, R.; Lerouge, S.; Prudhommeaux, F.; Sedel, L. Ceramic/ceramic total hip arthroplasty. *J. Orthop. Sci.* **2000**, *5*, 622–627. [CrossRef] [PubMed]
2. Hannouche, D.; Hamadouche, M.; Nizard, R.; Bizot, P.; Meunier, A.; Sedel, L. Ceramics in total hip replacement. *Clin. Orthop. Relat. Res.* **2005**, *430*, 62–71. [CrossRef] [PubMed]
3. Piconi, C.; Streicher, R.M. Forty years of ceramic-on-ceramic THR bearings. *Semin. Arthroplast.* **2013**, *24*, 188–192. [CrossRef]
4. Sedel, L. Evolution of alumina-on-alumina implants: A review. *Clin. Orthop. Relat. Res.* **2000**, *379*, 48–54. [CrossRef]
5. Lerouge, S.; Yahia Sedel, L. Alumina ceramic in total joint replacement. In *Hip Surgery*; Martin Dunitz Ltd.: London, UK, 1998; pp. 31–40.
6. Masson, B. Emergence of the alumina matrix composite in total hip arthroplasty. *Int. Orthop.* **2007**, *33*, 359–363. [CrossRef] [PubMed]

7.   Kuntz, M.; Usbeck, S.; Pandorf, T.; Heros, R. *Tribology in Total Hip Arthroplasty*; Springer: Heidelberg, Germany, 2011; pp. 25–40.
8.   Kim, Y.H.; Park, J.W.; Kim, J.S. Long-term results of third-generation ceramicon- ceramic bearing cementless total hip arthroplasty in young patients. *J. Arthroplast.* **2016**, *31*, 2520–2524. [CrossRef] [PubMed]
9.   Solarino, G.; Zagra, L.; Piazzolla, A.; Morizio, A.; Vicenti, G.; Moretti, B. Results of 200 Consecutive Ceramic-on-Ceramic Cementless Hip Arthroplasties in Patients Up To 50 Years of Age: A 5-24 Years of Follow-Up Study. *J. Arthroplast.* **2019**, *34*, S232–S237. [CrossRef] [PubMed]
10.   Kang, B.-J.; Ha, Y.-C.; Ham, D.-W.; Hwang, S.-C.; Lee, Y.-K.; Koo, K.-H. Third-Generation Alumina-on-Alumina Total Hip Arthroplasty: 14 to 16-Year Follow-Up Study. *J. Arthroplast.* **2015**, *30*, 411–415. [CrossRef]
11.   Solarino, G.; Piazzolla, A.; Notarnicola, A.; Moretti, L.; Tafuri, S.; De Giorgi, S.; Moretti, B. Long-term results of 32-mm alumina-on-alumina THA for avascular necrosis of the femoral head. *J. Orthop. Traumatol.* **2012**, *13*, 21–27. [CrossRef]
12.   Engh, C.A.; Stepniewski, A.S.; Ginn, S.D.; Beykirch, S.E.; Sychterz-Terefenko, C.J.; Hopper, R.H. A Randomized Prospective Evaluation of Outcomes After Total Hip Arthroplasty Using Cross-linked Marathon and Non–cross-linked Enduron Polyethylene Liners. *J. Arthroplast.* **2006**, *21*, 17–25. [CrossRef]
13.   Kim, Y.-H.; Kim, J.-S.; Park, J.-W.; Joo, J.-H. Contemporary total hip arthroplasty with and without cement in patients with oste-onecrosis of the femoral head. A concise follow-up, at a minimum of seventeen years of a previous report. *J. Bone Jt. Surg.* **2011**, *93*, 1806–1810. [CrossRef] [PubMed]
14.   Kim, Y.-H.; Kim, J.-S.; Park, J.-W.; Joo, J.-H. Periacetabular Osteolysis is the Problem in Contemporary Total Hip Arthroplasty in Young Patients. *J. Arthroplast.* **2012**, *27*, 74–81. [CrossRef] [PubMed]
15.   Kim, Y.-H.; Park, J.-W.; Kim, J.-S. Alumina Delta-on-Highly Crosslinked-Remelted Polyethylene Bearing in Cementless Total Hip Arthroplasty in Patients Younger than 50 Years. *J. Arthroplast.* **2016**, *31*, 2800–2804. [CrossRef] [PubMed]
16.   Solarino, G.; Abate, A.; Morizio, A.; Vicenti, G.; Moretti, B. Should we use ceramic-on-ceramic coupling with large head in total hip arthroplasty done for displaced femoral neck fracture? *Semin. Arthroplast.* **2013**, *24*, 255–260. [CrossRef]
17.   Savarino, L.; Padovani, G.; Ferretti, M.; Greco, M.; Cenni, E.; Perrone, G.; Greco, F.; Baldini, N.; Giunti, A. Serum ion levels after ceramic-on-ceramic and metal-on-metal total hip arthroplasty: 8-year minimum follow-up. *J. Orthop. Res.* **2008**, *26*, 1569–1576. [CrossRef]
18.   Macdonald, N.; Bankes, M. Ceramic on ceramic hip prostheses: A review of past and modern materials. *Arch. Orthop. Trauma Surg.* **2014**, *134*, 1325–1333. [CrossRef]
19.   Sentuerk, U.; von Roth, P.; Perka, C. Ceramic on ceramic arthroplasty of the hip: New materials confirm appropriate use in young patients. *Bone Jt. J.* **2016**, *98-B*, 14–17. [CrossRef]
20.   Howard, D.P.; Wall, P.; Fernandez, M.A.; Parsons, H.; Howard, P.W. Ceramic-on-ceramic bearing fractures in total hip arthroplasty. An analysis of data from the National Joint Registry. *Bone Jt. J.* **2017**, *99*, 1012–1019. [CrossRef]
21.   Yoon, B.-H.; Park, J.-W.; Cha, Y.-H.; Won, S.-H.; Lee, Y.-K.; Koo, K.-H. Incidence of Ceramic Fracture in Contemporary Ceramic-on-Ceramic Total Hip Arthroplasty: A Meta-analysis of Proportions. *J. Arthroplast.* **2019**, *35*, 1437–1443.e3. [CrossRef]
22.   Buttaro, M.A.; Zanotti, G.; Comba, F.M.; Piccaluga, F. Primary Total Hip Arthroplasty With Fourth-Generation Ceramic-on-Ceramic: Analysis of Complications in 939 Consecutive Cases Followed for 2-10 Years. *J. Arthroplast.* **2016**, *32*, 480–486. [CrossRef]
23.   Lee, G.-C.; Kim, R.H. Incidence of Modern Alumina Ceramic and Alumina Matrix Composite Femoral Head Failures in Nearly 6 Million Hip Implants. *J. Arthroplast.* **2016**, *32*, 546–551. [CrossRef] [PubMed]
24.   Tsikandylakis, G.; Overgaard, S.; Zagra, L.; Kärrholm, J. Global diversity in bearings in primary THA. *EFORT Open Rev.* **2020**, *5*, 763–775. [CrossRef]
25.   Zagra, L.; Gallazzi, E. Bearing surfaces in primary total hip arthroplasty. *EFORT Open Rev.* **2018**, *3*, 217–224. [CrossRef]
26.   Zustin, J.; Sauter, G.; Morlock, M.M.; Ruther, W.; Amling, M. Association of osteonecrosis and failure of hip resurfacing arthro-plasty. *Clin. Orthop. Relat. Res.* **2010**, *468*, 756–761. [CrossRef]
27.   Kannan, A.; Malhotra, R. Letter to the Editor: Association of Osteonecrosis and Failure of Hip Resurfacing Arthroplasty. *Clin. Orthop. Relat. Res.* **2010**, *468*, 902–903. [CrossRef]
28.   Park, Y.S.; Moon, Y.W.; Lim, S.J.; Yang, J.M.; Ahn, G.; Choi, Y.L. Early osteolysis following second-generation metal-on-metal hip re-placement. *J. Bone Jt. Surg.* **2005**, *87*, 1515–1521.
29.   Ortiguera, C.J.; Pulliam, I.T.; Cabanela, M.E. Total hip arthroplasty for osteonecrosis: Matched-pair analysis of 188 hips with long-term follow-up. *J. Arthroplast.* **1999**, *14*, 21–28. [CrossRef]
30.   Berry, D.J.; Harmsen, W.S.; Cabanela, M.E.; Morrey, B.F. Twenty-five-year survivorship of two thousand consecutive primary Charnley total hip replacements: Factors affecting survivorship of acetabular and femoral components. *J. Bone Jt. Surg.* **2002**, *84*, 171–177. [CrossRef] [PubMed]
31.   Hernigou, P.; Auregan, J.C.; Bastard, C.; Housset, V.; Lachaniette, C.-H.F.; Dubory, A. Higher prevalence of periprosthetic fractures with ceramic on polyethylene hip bearing compared with ceramic on ceramic on the contralateral side: A forty year experience with hip osteonecrosis. *Int. Orthop.* **2018**, *42*, 1457–1461. [CrossRef]
32.   Geesink, R. Prevention and treatment of recurrent dislocation after total hip arthroplasty. *Eur. Instr. Course Lect.* **2005**, *7*, 127–142.
33.   Solarino, G.; Moretti, B. Letter to the Editor on "Eighteen-Year Follow-Up Study of 2 Alternative Bearing Surfaces Used in Total Hip Arthroplasty in the Same Young Patients". *J. Arthroplast.* **2020**, *35*, 1956–1957. [CrossRef]
34.   Learmonth, I.; Hussell, J.; Smith, E. Inadequate polyethylene thickness and osteolysis in cementless hip arthroplasty. *J. Arthroplast.* **1997**, *12*, 305–309. [CrossRef]

35. Takahashi, Y.; Tateiwa, T.; Shishido, T.; Masaoka, T.; Kubo, K.; Yamamoto, K. Size and thickness effect on creep behavior in conventional and vitamin E-diffused highly crosslinked polyethylene for total hip arthroplasty. *J. Mech. Behav. Biomed. Mater.* **2016**, *62*, 399–406. [CrossRef]
36. Goebel, P.; Kluess, D.; Wieding, J.; Souffrant, R.; Bader, R.; Heyer, H.; Sander, M. The influence of head diameter and wall thickness on deformations of metallic acetabular press-fit cups and UHMWPE liners: A finite element analysis. *J. Orthop. Sci.* **2013**, *18*, 264–270. [CrossRef]
37. Li, G.; Peng, Y.; Zhou, C.; Jin, Z.; Bedair, H. The effect of structural parameters of total hip arthroplasty on polyethylene liner wear behavior: A theoretical model analysis. *J. Orthop. Res.* **2019**, *38*, 1587–1595. [CrossRef]
38. Solarino, G.; Piazzolla, A.; Mori, C.M.; Moretti, L.; Patella, S.; Notarnicola, A. Alumina-on-alumina total hip replacement for femoral neck fracture in healthy patients. *BMC Musculoskelet. Disord.* **2011**, *12*, 32. [CrossRef] [PubMed]
39. Esposito, C.I.; Carroll, K.M.; Sculco, P.; Padgett, D.E.; Jerabek, S.A.; Mayman, D.J. Total Hip Arthroplasty Patients With Fixed Spinopelvic Alignment Are at Higher Risk of Hip Dislocation. *J. Arthroplast.* **2018**, *33*, 1449–1454. [CrossRef] [PubMed]
40. DelSole, E.M.; Vigdorchik, J.; Schwarzkopf, R.; Errico, T.; Buckland, A. Total Hip Arthroplasty in the Spinal Deformity Population: Does Degree of Sagittal Deformity Affect Rates of Safe Zone Placement, Instability, or Revision? *J. Arthroplast.* **2017**, *32*, 1910–1917. [CrossRef]
41. Hannouche, D.; Zingg, M.; Miozzari, H.; Nizard, R.; Lübbeke, A. Third-generation pure alumina and alumina matrix composites in total hip arthroplasty: What is the evidence? *EFORT Open Rev.* **2018**, *3*, 7–14. [CrossRef] [PubMed]
42. Solarino, G.; Vicenti, G.; Piazzolla, A.; Maruccia, F.; Notarnicola, A.; Moretti, B. Total hip arthroplasty for dysplastic coxarthrosis using a cementless Wagner Cone stem. *J. Orthop. Traumatol.* **2021**, *22*, 1–7. [CrossRef] [PubMed]
43. Rogmark, C.; Johnell, O. Orthopaedic treatment of displaced femoral neck fractures in elderly patients. *Disabil. Rehabil.* **2005**, *27*, 1143–1149. [CrossRef] [PubMed]
44. Bystrom, S.; Espehaug, B.; Furnes, O.; Havelin, L.I. Norwegian Arthroplasty Register Femoral head size is a risk factor for total hip luxation: A study of 42,987 primary hip arthroplasties from the Norwegian Arthroplasty Register. *Acta Orthop. Scand.* **2003**, *74*, 514–524. [CrossRef] [PubMed]
45. Berry, D.J.; Von Knoch, M.; Schleck, C.D.; Harmsen, W.S. Effect of femoral head diameter and operative approach on risk of dislocation after primary total hip arthroplasty. *J. Bone Jt. Surg.* **2005**, *87*, 2456–2463.
46. Santavirta, S.; Bohler, M.; Harris, W.H.; Konttinen, Y.T.; Lappalainen, R.; Muratoglu, O.; Rieker, C.; Salzer, M. Alternative materials to improve total hip replacement tribology. *Acta Orthop. Scand.* **2003**, *74*, 380–388.
47. Bierbaum, B.E.; Nairus, J.; Kuesis, D.; Morrison, J.C.; Ward, D. Ceramic-on-Ceramic Bearings in Total Hip Arthroplasty. *Clin. Orthop. Relat. Res.* **2002**, *405*, 158–163. [CrossRef] [PubMed]
48. Yoo, J.J.; Kim, Y.M.; Yoon, K.S.; Koo, K.H.; Song, W.S.; Kim, H.J. Alumina-on-alumina total hip arthroplasty. A five-year minimum fol-low-up study. *J. Bone Jt. Surg.* **2005**, *87*, 530–535. [CrossRef]
49. Zagra, L.; Ceroni, R.G.; Corbella, M. THA Ceramic-Ceramic Coupling: The Evaluation of the Dislocation Rate with Bigger Heads. In *Bioceramics in Joint Arthroplasty. Ceramics in Orthopaedics*; Steinkopff: Heidelberg, Germany, 2004; pp. 163–169. [CrossRef]

Journal of
Composites Science

Review

Bioactive Calcium Phosphate-Based Composites for Bone Regeneration

Marta Tavoni †, Massimiliano Dapporto *,†, Anna Tampieri and Simone Sprio *

Institute of Science and Technology for Ceramics—National Research Council of Italy (ISTEC-CNR), 48018 Faenza, Italy; marta.tavoni@istec.cnr.it (M.T.); anna.tampieri@istec.cnr.it (A.T.)
* Correspondence: massimiliano.dapporto@istec.cnr.it (M.D.); simone.sprio@istec.cnr.it (S.S.); Tel.: +39-0546699760 (M.D.)
† Co-first author, these authors contributed equally to this work.

Abstract: Calcium phosphates (CaPs) are widely accepted biomaterials able to promote the regeneration of bone tissue. However, the regeneration of critical-sized bone defects has been considered challenging, and the development of bioceramics exhibiting enhanced bioactivity, bioresorbability and mechanical performance is highly demanded. In this respect, the tuning of their chemical composition, crystal size and morphology have been the matter of intense research in the last decades, including the preparation of composites. The development of effective bioceramic composite scaffolds relies on effective manufacturing techniques able to control the final multi-scale porosity of the devices, relevant to ensure osteointegration and bio-competent mechanical performance. In this context, the present work provides an overview about the reported strategies to develop and optimize bioceramics, while also highlighting future perspectives in the development of bioactive ceramic composites for bone tissue regeneration.

Keywords: calcium phosphates; hydroxyapatite; scaffolds; bone cements; bioactive composites; bone regeneration

Citation: Tavoni, M.; Dapporto, M.; Tampieri, A.; Sprio, S. Bioactive Calcium Phosphate-Based Composites for Bone Regeneration. *J. Compos. Sci.* **2021**, *5*, 227. https://doi.org/10.3390/jcs5090227

Academic Editor: Francesco Tornabene

Received: 28 July 2021
Accepted: 18 August 2021
Published: 27 August 2021

Publisher's Note: MDPI stays neutral with regard to jurisdictional claims in published maps and institutional affiliations.

1. Introduction

Musculoskeletal diseases are a worldwide cause of disability and pain, as they involve bones, teeth and joints, which are anatomical districts relevant for structural support, handling, protection, locomotion, mastication and many other physiological functions [1–3].

Bones are complex structures continuously undergoing dynamic remodeling due to a complex interaction of multiple biochemical processes, primarily ascribable on two different cell lines, osteoblasts and osteoclasts, as actors of bone deposition and resorption, respectively. Such processes can occur spontaneously in the case of minimal bone damage; however, if massive bone defects occur, as a result of a metabolic or traumatic cause, the physiological bone healing process has to be supported by a solid 3D scaffold, acting as a physical and instructive guide for cells [4–8].

Some properties requested for ideal bone scaffolds include *biocompatibility*, which is the ability of a biomaterial to function in vivo without eliciting any adverse side effects; *bioactivity*, which is the additional ability of a biomaterial to chemically bond with the surrounding tissue and to participate in specific biologically relevant phenomena (e.g., ion exchange); and *bioresorbability*, which is the ultimate ability of the implanted material to be resorbed over time, by active participation in physiological turnover reactions, favoring the formation of new tissue [9–12]. More specifically, scaffolds should exhibit *osteoinductivity* and *osteoconductivity*, both stimulating the osteointegration of the scaffold, which consists of a direct bone–scaffold interaction without fibrous tissue at the interface, essential to ensure mechanical stability and also the in-growth of blood vessels. In this respect, a leading concept guiding scaffold development is the achievement of high

95

mimicry with targeted bony tissues, aiming to achieve a physiological cell response while preventing adverse foreign body reactions [13].

The design and development of biomimetic bone scaffolds have to be inspired by the complex physiological bone composition and structure. The bone microstructure is the result of the biomineralization of type I collagen, secreted by fibroblasts and osteoblasts cells, as a major component of the extracellular matrix of skin, tendon and bone [14]. Osteoblasts create the nano-composite structure of bone by secreting the ions responsible for the formation of apatite crystals. In turn, the ECM influences the adhesion, proliferation and differentiation of osteoblast, osteoclast and osteocyte [5,12]. The ECM is composed of inorganic and organic phases and water: the organic component consists of collagen and non-collagenous proteins, and the inorganic component contains calcium phosphate (mainly plate-like nanocrystalline hydroxyapatite, HA), calcium carbonate, magnesium phosphate and magnesium fluoride doped with various anionic (HPO_4^{2-}, CO_3^{2-} and Cl^-) and cationic species (Na^+, K^+, Mg^{2+}, Sr^{2+}, Zn^{2+}, Ba^{2+}, Cu^{2+}, Al^{3+}, Fe^{2+} and Si^{2+}) trapped in the crystal structure. Carbonate ions are found in extent up to 8 wt%, while Na^+, Mg^{2+}, K^+, Sr^{2+}, Zn^{2+}, Ba^{2+}, Cu^{2+}, Al^{3+}, $Fe^{2+/3+}$, F^-, Cl^- and Si^{4+} ions occur at trace (<1 wt%) [15]. Biogenic HA in bony tissue is non-stoichiometric with a Ca/P ratio between 1.5 and 1.67, where the inclusion of foreign ions in the crystal structure influences solubility, bioactivity, surface chemistry and morphology [16,17]. The general chemical formula for biogenic apatite is $Ca_{10-x}(PO_4)_{6-x}(HPO_4$ or $CO_3)_x(OH$ or $\frac{1}{2}CO_3)_{2-x}$ with $0 \leq x \leq 2$. One of the most common doping ions in biogenic HA is CO_3^{2-} ions, which can replace both phosphate and hydroxyl ions (leading to type B and type A carbonated apatite, respectively). For example, in B-type carbonated HA, the presence of CO_3^{2-} ions in the phosphate site inhibits the crystal growth and decreases the crystallinity; this structural disorder increases the chemical reactivity and enhances the solubility without changing the affinity of the osteoblast cells. Other possible anionic substitutions are with fluoride and chloride ions [17,18]. Cationic substitutions generally involve monovalent and bivalent cationic in the calcium sites of HA crystal lattice as reported in Table 1 [18–20].

Table 1. Relevant cation substitutions in natural HA crystal structure.

Cations	Biological Effects
Magnesium	Enhancing skeletal metabolism and bone growth
Strontium	Increasing bone mass: stimulating bone formation and reducing bone resorption (anti-osteoporotic agent)
Silicon	Stimulating extracellular matrix formation and mineralization
Zinc	Stimulating osteoblastic activity in vitro and inhibiting bone resorption in vivo

The bone structure exhibits a complex hierarchical architecture resulting from complex interactions of multilevel components, from micrometric osteons to apatite nanocrystals [21] (Table 2).

Table 2. Main components of bone structure, from macroscale to nanoscale.

Macrostructure	Cortical bone Spongy bone
Microstructure	Osteons (100 µm) Haversian canals (10 µm) Collagen fibrils (25–500 nm)
Nanostructure	Tropocollagen triple helix Collagen molecule Hydroxyapatite nanocrystals (30 nm)

In particular, it is possible to classify the levels and structures of components as follows:

1. Macrostructure: cancellous and cortical bone;
2. Microstructure: (10–500 mm): Haversian channel, osteons and single trabeculae;
3. Sub-microstructure (1–10 mm): lamellae;
4. Nanostructure (100 nm–1 mm): fibrillar collagen and embedded mineral;
5. Sub-nanostructures (<100 nm): molecular structure of constituent elements, such as minerals, collagen and non-collagenous proteins [22].

Such a complexity is the main responsible factor for the outstanding mechanical performance of bone and its self-repair ability [23].

The ideal bone scaffolds should be endowed by several physico-chemical features, including chemical composition mimicking both the natural bone ECM and mineral phase, open and interconnected porosity capable of promoting neo-vascularization, tissue ingrowth, nutrient and oxygen supply, nano-structured surface topography positively driving adhesion, proliferation and differentiation of cells, that are adequate mechanical properties able to sustain the biomechanical loads toward the effective regeneration of the tissue.

Several studies have been carried out on the research of biomaterials, such as metals, natural or synthetic polymers, ceramics and composites, which can match all these characteristics, but no one fully satisfies all these requirements [24–30]. In particular, bioceramic-based scaffolds are widely used in numerous biomedical applications, including maxillofacial reconstruction, the stabilization of jaw bones, periodontal disease, as space fillers, self-hardening bone pastes/cements and as a coating on implants, due to their positive interaction with human tissue. Bioceramic-based materials can be classified as bioactive and bioinert materials. Ceramics considered as bioinert include alumina and zirconia; they show high chemical stability in vivo as well as high mechanical strength. However, they do not have osteogenic properties [31]. Bioactive ceramics, such as calcium phosphates (CaPs), silicates, bioactive glass, and titanium oxide, are capable of interacting with cells and thus able to promote and stimulate bone regeneration [28–33].

CaP bioceramics are widely used as bone substitutes since the 1920s and are considered as the golden standard in bone regeneration due to their similarity to the inorganic bone [34–37]. The chemical composition of CaPs relies on multiple ions, including calcium (Ca^{2+}), orthophosphate (PO_4^{3-}), metaphosphate (PO_3^-), pyrophosphate ($P_2O_7^{4-}$) and hydroxide (OH^-) [9,37] (Table 3).

Table 3. Some CaP materials: name, abbreviation, chemical formula, Ca/P ratio and solubility.

Name	Abbreviation	Chemical Formula	Ca/P Ratio	Solubility at 25 °C, mg/L
Hydroxyapatite	HA	$Ca_{10}(PO_4)_6(OH)_2$	1.67	~0.3
Calcium-deficient hydroxyapatite	CDHA	$Ca_{10-x}(PO_4)_{6-x}(HPO_4 \text{ or } CO_3)_x(OH \text{ or } \frac{1}{2}CO_3)_{2-x}$	1.5–1.67	~9.4
Dicalcium phosphate dihydrate	DCPD	$CaHPO_4 \cdot 2H_2O$	1	~88
α-Tricalcium phosphate	α-TCP	$\alpha\text{-}Ca_3(PO_4)_2$	1.5	~2.5
β-Tricalcium phosphate	β-TCP	$\beta\text{-}Ca_3(PO_4)_2$	1.5	~0.5

The solubility of CaP compounds strongly influences their behavior in vivo [37].

Among CaPs, HA is particularly promising for bone tissue regeneration due to its very close composition with natural apatite. In the last decades, the synthesis of HA has been investigated for different applications, including scaffolds, injectable pastes/cements, coatings for metallic implants and in nanomedicine as drug delivery platforms [38,39].

HA can be produced by several methods: high-temperature solid-state reactions or low-temperature precipitation [38]. Stoichiometric HA exhibits high stability at physiological pH, limiting its long-term resorption. Therefore, various recent studies have been focused on increasing the solubility and osteogenic activity of HA by ionic doping [39,40].

The notable interest in TCP comes from the combination of its solubility and low Ca/P ratio, particularly interesting when obtaining apatite crystals in an aqueous environment [16].

There are two polymorphs of TCP: the high-temperature α-TCP and the low-temperature β-TCP[41]. The β-TCP polymorph is stable at room temperature, while a transformation into α-TCP occurs at temperatures higher than 1125 °C. Besides a similar chemical composition, the TCP polymorphs have different crystalline structures, density and solubility, thus also resulting in different biological performance. The α-TCP phase is more soluble than β-TCP and can be easily hydrolyzed in calcium-deficient hydroxyapatite (1).

$$3\,Ca_3\,(PO_4)_2 + H_2O \rightarrow Ca_9(HPO_4)(PO_4)_5OH \tag{1}$$

In addition, several ions can be introduced in the structure of TCP (Mg^{2+}, Sr^{2+}, Zn^{2+}, Si^{2+}, etc.), opening different thermodynamic scenarios in terms of polymorph stabilization; e.g., silicon was reported to stabilize α-TCP, while magnesium ions stabilize β-TCP.

Due to its high solubility, TCP has been used for the preparation of biphasic CaP scaffolds, able to conjugate the osteogenic properties of HA and the resorption behavior of TCP [42,43].

DCPD is biocompatible, biodegradable and osteoconductive [9]. DCPD can be prepared by the neutralization of phosphoric acid with calcium hydroxide at pH 3–4 at room temperature. DCPD can be obtained by double decomposition between calcium- and phosphate-containing solutions in slightly acidic media. It can also be formed by the conversion of calcium phosphate salts, in acidic media, or by the reaction of calcium salts, such as calcium carbonate in acidic orthophosphate solutions. In vivo studies showed that DCPD converts into HA or it degrades and is replaced by bone [44–46]. Brushite, in medicine, is used in CaP paste/cement and as an intermediate for tooth remineralization [44,47].

Other silica-based bioceramics have also been studied as bone scaffolds, including wollastonite ($CaSiO_3$), larnite (Ca_2SiO_4), hatrurite (Ca_3SiO_5), monticellite ($CaMgSiO_4$), diopside ($CaMgSi_2O_6$), akermanite ($Ca_2MgSi_2O_7$), merwinite ($Ca_3MgSi_2O_8$), silicocarnotite ($Ca_5(PO_4)_2SiO_4$), nagelschmidtite ($Ca_7(SiO_4)_3(PO_4)$) and bioglass [48]. Silicon ions participate in bone metabolism, and silica-based materials exhibit good biological response in vitro, resulting in bioactive, biocompatible, bioresorbable, osteoinductive and osteoconductive behavior. The favored formation of apatite in physiological fluid was reported, thus facilitating the chemical interaction into the living bone structure following implantation [29,32].

The following steps explain the formation of apatite on the surface of silica-based bioceramics:

- The rapid exchange of Ca^{2+} with H^+ or H_3O^+ from a body fluid solution results in the hydrolysis of silica groups, which creates silanol, according to Si-O-Ca$^+$ +H$^+$ $\rightarrow$ Si-OH$^+$Ca^{2+}(aq).
- The loss of soluble silica in the form of $Si(OH)_4$ to the body fluid, resulting from the breaking of Si-O-Si bonds and the formation of silanol (Si-OH) at the glass solution interface: Si-O-Si + H_2O $\rightarrow$ 2Si-OH.
- The condensation and polymerization of a SiO_2-rich layer on the surface short in alkalis and alkaline earth cations: Si-OH+HO-Si $\rightarrow$ Si-O-Si+H_2O.
- The migration of Ca_2^+ and PO_4^{3-} groups to the surface via the SiO_2-rich layer forming a CaO-P_2O_5-rich film by the incorporation of soluble calcium and phosphates from the solution.
- The crystallization of the amorphous CaO-P_2O_5-rich film by the addition of OH^- and CO_3^{2-} anions from body fluid forms a mixed hydroxyl, carbonated apatite layer.
- The adsorption and desorption of biological growth factors on the carbonated apatite layer to activate stem cells.
- The action of macrophages to remove debris from the site allowing cells to occupy their space.
- The attachment of stem cells to the bioactive surface and its differentiation to form osteoblasts.

- The generation of ECM by the osteoblast to form new bone and its crystallization in the living composite structure.

Bioglasses are also a class of bioactive, osteoconductive and osteoinductive materials essentially composed of silicate, calcium, sodium and phosphate (e.g., composition of Bioglass 45S5® (wt%) 45 SiO_2, 24.5 CaO, 24.5 Na_2O, 6 P_2O_5). Upon implantation, bioglasses are not surrounded by fibrous tissue but form a strong, integrated bond to bone. In fact, when immersed in body fluids, the formation of a silica-rich layer on its surface takes place, which converts to a silica-CaO/P_2O_5-rich gel layer as a precursor of HA layer formation [24,33]. In addition, they are able to release ions, which enhance gene up-regulation and favor bio-degradation, in turn favoring bone regeneration [49]. Major drawbacks are related to the difficult consolidation of bioglasses into 3D porous scaffolds, as the required thermal treatment easily provokes the crystallization of oxides, thus losing the bioactive properties related to the material in its amorphous state. Therefore, alternative consolidation methods are currently under investigation; however, a major issue remains regarding the achievement of substantial mechanical properties associated with open porosity [50].

The mechanical properties of scaffolds play an important role in bone tissue engineering. The relevant mechanical properties of bone include Young's modulus, toughness, shear modulus, tensile strength, fatigue and compressive strength. Several approaches have been reported to increase the mechanical performance and load transfer efficiency between the scaffold and the surrounding bone tissue, mainly related to stronger interfacial bonding of the coating layer to the substrate [51].

The mechanical strength of ceramics mainly relies on their chemical composition, grain size, porosity extent and internal structural defects [37] (Table 4).

Table 4. Ideal features of scaffolds for bone regeneration, with respective proposed strategies to improve them.

Properties	Proposed Improving Strategies
Open and interconnected porosity	<ul><li>Traditional techniques for the fabrication of a 3D porous device (sacrificial template, direct foaming)</li><li>Low-temperature self-hardening methods</li><li>Biomimetic and biomorphic synthesis</li><li>3D printing technology</li></ul>
Mechanical properties	<ul><li>Reinforced scaffold by compression, using fibers (polymeric or ceramic) or a dual setting system</li></ul>
Biofunctionality	<ul><li>Biomimetic and biomorphic synthesis</li><li>Surface topography modifications</li></ul>
Bioactivity	<ul><li>Biomimetic and biomorphic synthesis</li><li>Ion-doping ceramic-based scaffold</li><li>Ceramic-based composites</li></ul>

Bioceramics typically exhibit higher compressive than tensile strength, but they are also intrinsically brittle, leading to sudden failure during handling and fixation [52]. In this respect, a critical challenge is related to the optimization of toughening mechanisms for ceramics [53,54].

The enhancement of the performance of bioceramic scaffolds has been widely explored by the combination of different calcium phosphate phases into bioceramic composites. The present work aims to provide the reader with an overview about the recently reported strategies to enhance the biofunctionality and mechanical properties of bioceramic scaffolds. In particular, various manufacturing techniques are explored, including the replica method, the sacrificial template, direct foaming, the low-temperature self-hardening method and biomorphic and biomimetic synthesis, as well as 3D printing, while also highlighting future perspectives for the development of bioactive ceramic composites and devices with enhanced biofunctional properties (Figure 1).

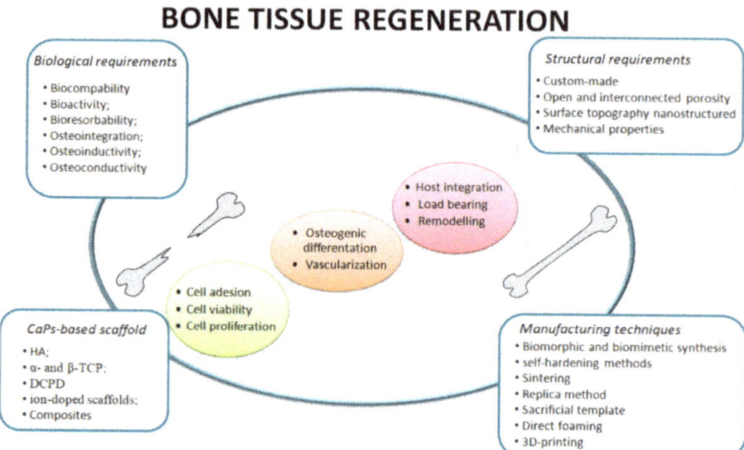

Figure 1. Flow chart of biological and structural requirements underlying the ideal scaffold for bone tissue regeneration.

2. Fabrication of Bioceramic Composites

The biological events occurring upon implantation of a scaffold for bone regeneration are strongly influenced by pore size distribution. The scaffold porosity affects the capability of the surrounding tissue to promote cell infiltration, migration, vascularization and nutrient and oxygen flows [18,55]. The morphological properties of scaffolds in terms of pore volume and size are important at both the macroscopic and the microscopic level.

It was reported that osteointegration and angiogenesis can be favored by interconnected macroporosity (100–600 µm) with channel-like microporosity [18]. A pore size increase is generally associated with an increase in permeability and the new bone ingrowth, while small pores are more suitable for soft tissue in-growth.

Over the past two decades, several technologies have been developed for the manufacturing of highly porous bioceramic-based scaffold for bone tissue regeneration [15,17,31,40–47]. In the next paragraphs, we explore the main fabrication techniques of porous scaffolds: traditional methods (partial sintering, replica method, sacrificial template and direct foaming), low-temperature self-hardening methods, biomorphic and biomimetic synthesis and 3D printing technology.

2.1. Macroporous Compositescaffolds

The development of materials with tailored porosity has been a matter of intense research in the last decades, particularly in the case of composite scaffolds for bone tissue regeneration, because of the crucial role of voids in the structure to guide and facilitate cell proliferation and neovascularization [56].

One of the first reported approaches to tune the porosity of ceramics was the partial sintering process: the pore size distribution is mainly affected by powder particle size and sintering temperature, as higher sintering temperatures induce a significant decrease in intergranular porosity [57,58].

A great research effort has also been devoted to the preparation of macroporous bioceramic scaffolds, leading to the establishment of various techniques, including template-assisted (replica and sacrificial template) and template-free techniques (direct foaming) [56,59,60] (Table 5) [56,60].

Table 5. Main processing steps involved in the fabrication of porous bioceramics.

Template-Assisted Techniques	Processing
Replica	• Preparation of stable ceramic suspension • Impregnation of synthetic/natural porous template into the ceramic suspension • Drying and template removal • Sintering
Sacrificial template	• Preparation of ceramic or ceramic precursor in solid or liquid form • Addition of sacrificial phase • Drying, pyrolysis and evaporation • Sintering
Template-Free Technique	
Direct foaming	• Preparation of stable ceramic suspension • Addition of surfactants • Incorporation of gas • Drying of the foamed suspension • Sintering

These methods generally involve the preparation of slurries, intended as aqueous suspensions of dispersed powders; then, the slurries are properly manipulated, dried and thermally consolidated.

The replica method is a template-assisted technique based on the impregnation of a polymeric sponge with a defined porous structure and pore size into the ceramic slurry in order to produce microporous structures exhibiting the original sponge morphology [56]. The templates used in this technique can be either synthetic or natural polymers (e.g., polyurethane and cellulose, respectively). The macroporous scaffolds obtained with this method can reach an anisotropic porosity ranging from 40 to 95% and are characterized by a cross-linked structure with highly interconnected pores ranging in size from 200 μm to 3 mm [56].

The sacrificial template method involves the homogeneous dispersion of sacrificial phases into a continuous matrix of ceramic particles or ceramic precursors, followed by drying and sintering. A wide variety of sacrificial materials can be used as pore-forming agents, including natural polymers (e.g., gelatin, potato starch, cotton), synthetic polymers (e.g., polymer beads, organic fibers, polyethylene) and inorganic polymers (e.g., NaCl, K_2SO_4). The removal of sacrificial materials from the matrix can be achieved by thermal treatments or chemical processes. This method leads to porosity ranging from 20 to 90%, with an average pore diameter of 1–700 μm [18,56].

Template-free foaming techniques are particularly promising due to the absence of massive amounts of organic phases to be eliminated during thermal consolidation. Direct foaming represents an easy, cheap and fast way to prepare macroporous bioceramics with open porosity from 40 to 97% and pore size 10 μm–1 mm by incorporating gas bubbles into ceramic slurries, followed by drying and sintering [18,56,61]. The total porosity volume is related to the amount of gas bubbles incorporated during the foaming process, whereas the pore size depends on the stability of the poured foam before drying [18,56,61].

The sacrificial template approach also includes the freeze-casting method, which is based on the controlled freezing of liquid-based ceramic slurries [18]. The freezing of the liquid, generally water, induces the formation of anisotropic ice structures, intended as fugitive materials, during the subsequent freeze-drying process [62]. The efficacy of the process is affected by several parameters, including the viscosity of the slurry, the solvent and the freezing control in space and time. Typical structures obtained by freeze-casting methods showed well-defined pore connectivity along with directional and completely open porosity, such as a lamellar morphology after sintering [63]. The channel-like anisotropic porosity obtained by the freeze-casting method may lead to scaffolds with channels similar to cortical bone, particularly useful for long bone applications [18].

2.2. Self-Hardening Bioceramic Composites

The possibility to obtain bioactive ceramics through low-temperature self-hardening processes has been widely explored in the form of bone cements for injectable orthopedic applications, including spinal fusion, vertebroplasty and kyphoplasty [30,64–66]. Bone cements refer to pastes able to self-harden under physiological conditions and can be injected in vivo through minimally invasive surgery [64]. The first bone cement used in orthopedics was based on polymers, in particular polymethylmethacrylate (PMMA) in 1958, and, in the 1970s, the FDA approved bone cement for use in hip and knee prosthetic fixation [67]. Despite PMMA-based cements exhibiting good handling, setting times and mechanical performance, they are not osteogenic nor bioresorbable. Calcium phosphate cements (CPC) were discovered by Brown and Chow in the 1980s [68–70], overcoming the drawbacks of PMMA cements in terms of exothermic polymerization hardening and chemical composition. In this respect, CPCs exhibit bioactivity, bioresorbability and a physiological hardening at 37 °C, also allowing the incorporation of biomolecules [68]. The main drawback of CPCs hampering their clinical applications is related to their poor mechanical performance, which limits their applicability to a moderate- or non-load-bearing situation [71].

CPCs can be classified by several parameters, including the number of components in the solid phase, the type of setting reaction and the type of end product (Table 6) [38,68].

Table 6. Classification of CPC.

	Apatitic CPC		Brushitic CPC
	Single Component	**Multiple Components**	
Reactives	α-TCP	TTCP + DCPA/DCPD	B-TCP + MCPM/MCPA
Reaction type	Hydrolysis	Acid-Base	
Reaction	$3\alpha - Ca_3(PO_4)_2 + H_2O \rightarrow Ca_9(HPO_4)(PO_4)_5(OH)$	$Ca_4(PO_4)_2O + 2CaHPO_4 \rightarrow Ca_{10}(PO_4)_6(OH)$	$\beta - Ca_3(PO_4)_2 + Ca(H_2PO_4)_2 \cdot H_2O + 7H_2O \rightarrow 4CaHPO_4 \cdot 2H_2O$

Many different formulations of CPCs have been developed, and they can be divided into two groups based on the type of end product: brushite (DCPD) and apatite (HA or CDHA) cements. Both brushite and apatite CPCs are produced upon mixing one or more CaP powders with aqueous solutions, which induces the dissolution of the initial CaPs; this is followed by precipitation into crystals of DCPD, HA or CDHA depending on the compositions of the powders and the setting reactions that take place [38,72]. During precipitation, new apatitic crystals grow and their physical entanglement causes the hardening or setting at body temperature.

Apatitic CPCs can be obtained by mixing single or multi-components with aqueous solutions that undergo hydrolysis or acid–base reactions, respectively. In the first case, the end product is calcium-deficient hydroxyapatite (CDHA), and in the latter, it is stoichiometric HA [64,68]. Some examples are as follows:

- Hydrolysis of metastable α-TCP:

$$3\ \alpha\text{-}Ca_3(PO_4)_2 + H_2O \rightarrow Ca_9(HPO_4)(PO_4)_5(OH)$$

- Acid–base reaction between tetra calcium phosphate, TTCP (basic), and di calcium phosphate anhydrous, DCPA (acidic):

$$Ca_4(PO_4)_2O + CaHPO_4 \rightarrow Ca_5(PO_4)_3(OH)$$

Brushite CPC obtained by an acid–base reaction between TCP (almost neutral) and monocalcium phosphate monohydrate, MCPM (acidic):

$$\beta\text{-}Ca_3(PO_4)_2 + Ca(H_2PO_4)_2 \cdot H_2O + 7H_2O \rightarrow 4\, CaHPO_4 \cdot 2H_2O$$

Two of the most important parameters that play a key role in the final CPC features are the liquid-to-powder ratio (LPR) and the particle size of the starting powder [37,68]. The LPR influences setting time, injectability, cohesion, mechanical properties and the porosity of harder CPC [73]. The setting time is the "time required from the start of powdered agent and liquid agent blending until hardening of the cement", according to ISO/DIS 18531 for CaPs [30,74], and influences the clinical applicability of both apatite and brushite cements as well as their injectability [30,74].

Both particle size and the LPR influence the final surface morphology of the brushite or apatite crystals and the total porosity of the final scaffolds, which affects the mechanical performance and the resorbability of scaffolds and therefore the overall bioactivity (Table 7) [37,68]. The reduction in the particle size of CaPs increases the surface area, thus affecting the reaction kinetics and yielding small needle-like crystals rather than large plate-like crystals as observed when larger CaP precursor particles are used [38,75]. Moreover, porosity is also attributed to the amount of liquid phase used; thus, by increasing the LPR, the amount of liquid phase decreases, and the porosity increases. This effect of the LPR explains the difference between brushite and apatite cement in terms of microstructure porosity: the water consumption during the setting reaction of brushite cement is larger than that of the apatite, which leads to the formation of a larger crystal size and makes the total porosity smaller and average pore size greater than those of the apatitic cements [37,73]. The typical porosity of CPC ranges between nano- and sub-micrometer size, allowing the flow of physiological fluids within the microstructure of the cement, but the pores are too small to facilitate the growth of bone tissue; in this regard, porogens are often used [69].

Table 7. Effect of particle size and liquid-to-powder ratio on the crystals' morphology and pore distribution.

	Particle Size		Liquid-to-Powder Ratio	
	Fine Particles	**Coarse Particles**	**Low L/P**	**High L/P**
Final crystal morphology	Needle-like crystals	Plate-like crystals	Low inter-aggregate distance	High inter-aggregate distance
Pore size distribution	Fine	Coarse	Fine	Coarse

As mentioned above, increasing porosity leads to decreasing mechanical strength; thus, a compromise must be sought between mechanical performance and porosity degree.

One of the advantages of CPC is the room-temperature self-hardening mechanism, which, combined with the intrinsic porosity, allows the incorporation of drugs, biologically active molecules and cells, obtaining drug delivery materials [76,77]. The incorporation of active molecules in CPCs can be achieved by dissolving it in the liquid phase or by a combination with the powder phase of the CPC mixing setting [68,78]. Another possible approach is the superficial adsorption of drugs on the CPC surface by incubation of the scaffold in the drug solution: the kinetic release of drugs depends on the functionalization, microstructure and resorbability of the CPC matrix [68,78].

2.3. Biomorphic Transformations

A valuable approach to obtain bioceramic composite scaffolds with a complex structural hierarchy relies on biomorphic transformations of natural structures mimicking the morphology and microstructure of the target tissue [20,77,79].

Since the 1970s, biomorphic transformations from natural sources have been proposed for the fabrication of bioceramic scaffolds due to their 3D highly interconnected porous architecture, including the replica of the porous microstructure of $CaCO_3$-based corals, which are impossible to create artificially, and the replica of marine sponges, soft vegetal structures and fruit- and wood-template bioceramics [75,80,81].

The approach of wood biotransformation is particularly interesting, as many ligneous species exhibit a porous and hierarchically organized structure very close to that of cortical and cancellous bone. The transformation of wood generally involves pyrolysis followed by a hydrothermal treatment; in particular, a complex multi-step strategy to convert rattan wood structures into biomimetic HA scaffolds was proposed [76,77,82]. In particular, several subsequent and strictly controlled reactions are required, including (i) the *pyrolysis* of wood to produce a carbon template; (ii) *carburization*, calcium infiltration to transform carbon in CaC_2; (iii) an *oxidation* process that leads to CaO formation; (iv) *carbonation* by hydrothermal processes or by heterogeneous processes carried out at supercritical conditions and high pressure; and finally (v) *phosphorylation* through the hydrothermal process generating biomimetic, hierarchically organized scaffolds made of ion-doped HA.

2.4. 3D Printing

Three-dimensional (3D) printing represents an additive manufacturing (AM) technique (also known as rapid prototyping) to produce complex-shaped devices with complex geometry and design flexibility from 3D model schemes [83–86]. A wide range of materials have been employed with 3D printing techniques, including metals, polymers, ceramics and composites [85,86].

Different 3D printing methods have been proposed [85,87–90]. Extrusion-based techniques consist of the deposition of ink to create designed structures by forcing the ink through a nozzle as a melt, in fused deposition modeling (FDM), or viscous suspensions, in direct ink writing (DIW), to form lines that solidify onto a build plate [90].

DIW represents an easy manufacturing technique that allows the creation of a wide range of structures, from solid monolithic parts to highly complex porous scaffolds and composite materials. The use of pastes also allows shape retention due to the high solid loading and visco-elastic properties. The use of high viscous inks requires larger diameter nozzles compared to the conventional inkjet printing ink; it can therefore be used successfully to print extremely viscous pastes that are HA based [88].

Three-dimensional printing technology finds a wide range of biomedical applications: craniofacial implants, dental models, prosthetic parts, scaffold for tissue regenerations (bone and skin), organ printing, tumor therapy and tissue modeling for drug discovery [90–92]. In these kinds of applications, printable materials are formulated from biomaterials and bio-inspired materials to achieve patient-specific scaffolds with high structural complexity [93]. Moreover, printable biomaterials should be biocompatible and bioactive and should have good degradation kinetics, appropriate mechanical properties, give desirable cellular responses and exhibit tissue biomimicry [94,95].

Bioceramic powders, natural or synthetic hydrogels, polymers and their composites have been used as raw materials to formulate inks for 3D printing; in this review, we focused on ceramic-based scaffolds and bioceramic/polymer composites. Bioceramics commonly printed are calcium phosphate-based bioceramics (HA, TCP and biphasic CaP), calcium silicate-based bioceramics and bioactive glasses [91,93].

Moreover, the precise tuning of the macro- and micro-porosity permitted by 3D printing technology not only allows the fabrication of scaffolds with hierarchical porosity but also leads to the controlled release of biomolecules or drug loaded in the scaffold matrix or adsorbed on the scaffold surface [96,97].

Three-dimensional-printed bioceramics include sintered 3D-printed bioceramics, non-*sintered 3D-printed bioceramics* and *composites with polymers.*In the first case, bioceramic scaffolds are printed and sintered, removing the organic phases and improving the mechanical properties of the structure [93]. In the presence of biologically active ions, such as magnesium or strontium, in addition to an improvement of mechanical properties, an increase in biological performance in vivo was also reported [98]. Another study described biphasic CaP scaffolds (HA:β-TCP with a weight ratio 60:40) coated with calcium peroxide and polycaprolactone in order to promote bone growth with greater proliferation of osteoblasts under hypoxic conditions, following the release of oxygen dependent on the concentration of calcium peroxide in the PCL coating [99].

In non-sintered 3D-printed bioceramics, a small amount of organic solvent is used as a binder for bioceramic powders and is not removed after printing. Sun et al. developed a porous 3D scaffold of biodegradable CaP loaded with antibiotics for the regeneration of the bone tissue of the jaw, achieving a controlled drug release. This scaffold was based on an HA or biphasic mixture of CaP (β-TCP and HA with a weight ratio of 1:1) cross-linked with sodium alginate in the presence of the drug, and the paste was then extruded by the 3D printer. By modulating the degree and the time of cross-linking, it is possible to control the drug release kinetics. In vitro studies show low cytotoxicity and good cell adhesion and proliferation on the scaffold surface [100].

Bioceramic and polymer composite are synthesized to combine the bioactivity and osteoconductivity of bioceramics with the handling performance of polymers [87]. For example, the presence of strontium-doped HA nanoparticles in 3D-printed PCL scaffolds leads to a significant increase in cell proliferation and bone regeneration, due to the simultaneous release of calcium and strontium ions, associated with an improvement in mechanical properties as related to the inorganic phase content [101]. HA nanoparticles were also used as an external coating for 3D-printed polymer scaffolds in order to enhance cell proliferation and differentiation while also strengthening the scaffold [94].

Recently, 4D printing approaches have been developed, which, in addition to three-dimensional spatial control, introduces the concept of temporal control, i.e., active smart materials responsive and mechanically converted into other shapes via external stimuli. This technique enables the production of smart 3D scaffolds responding to external stimuli, such as changes in pH and temperature or when subjected to magnetism or light radiation of adequate energy [95,102,103].

3. Enhancing the Biological Performance of Bioceramic Composites

3.1. Biofunctionalization

Biofunctionalization is the modification of a material to achieve improved biological function and/or stimulus, whether permanent or temporary. The biofunctionality of scaffolds for regenerative medicine has been considered to play a key role for effective tissue regeneration [92,95].

Several parameters can be tuned, including surface energy and roughness, Ca/P ratio, solubility, particle size and crystallinity, in order to improve the biological events beyond the interaction with the biological environment, e.g., protein adsorption, cell attachment, cell proliferation and cell differentiation [93,104] (Figure 2).

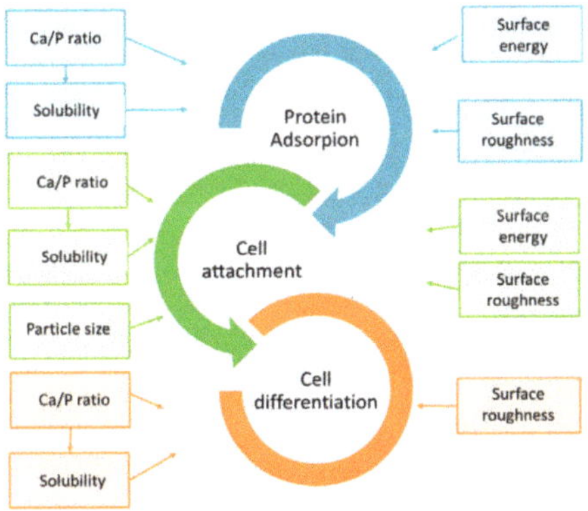

Figure 2. Key properties of CaP-based bioceramics that have an impact on biological events.

The architecture of biomimetic scaffolds greatly affects the chance to obtain a suitable microenvironment for bone regeneration. The presence of a diffuse macroporosity favors cell adhesion, cell proliferation and vascular growth. In turn, the surface micro-architecture enhances protein adsorption, and specific nano-topography could directly modulate the osteogenic differentiation, producing a favorable osteoimmune microenvironment [97]. Among the various microstructures, microgrooves have strong effects in the regulation of cell orientation and adhesion [96,105]. The width of the micro-channels controls the orientation, while the depth regulates the adhesion strength of the cells, which decreases as the depth of the groove increases. Micro–nano hybrid structures (micropattern–nanorod hybrid structure) showed higher cell adhesion, proliferation and ALP (alkaline phosphatase protein) activity than a single-scale structure (including nanorods and micropatterns) [96,106].

The roughness of the surface plays a crucial role in cellular behavior [107–109] (Table 8).

Table 8. Effects of structural size, morphology and roughness surface of CaP biomaterials on cellular behavior.

Surface Structure	Parameters (Size, Morphology, Roughness)	Biological Function	
		Enhance	Decrease
Micro/nano size (CaP)	Microgroove width: From 20–40 um to 60–100 um	Cell number inside the pattern	Cell alignment/orientation
	Microgroove depth From 3 um to 5.5 um		Cell adhesion force
	Microgroove depth pattern: From nano-hybrid to micro-hybrid	Cell adhesion, proliferation, osteogenesis	
Micro-/nano-morphology (CaP)	Micro-morphology: Plate-like and net-like	Cell attachment expansion	
	Nano-morphology: Plate-like and wire-like	Osteogenesis	
Micro-/nano-roughness (CaP)	Micro-roughness: Ra from 1 um to 2 um	Cell attachment osteogenesis	
	Nano-roughness: Ra from 5.3 nm to 9.8 nm	Focal adhesion osteogenesis	

It was demonstrated that specifically designed roughness can enhance osteogenesis due to the modulated concentration of calcium ions and osteocalcin in the grooves [110].

The surface chemistry also plays a key role in cell behavior. The crystallinity of nanometric bioceramics, i.e., ACP and HA, was observed to affect cell attachment efficiency, proliferation and differentiation of bone marrow-derived mesenchymal stem/stromal cells (BMSCs) [111]. In particular, nano-HA allows a better adhesion, proliferation and differentiation of BMSCs into osteoblasts than ACP.

The chemical approach of creating functional groups on the surface of the scaffolds is also promising for the improvement of cell adhesion and osteogenic differentiation. For example, functional groups, such as –COOH and –NH$_2$, improve protein adsorption due to the formation of hydrogen bonds linking proteins, finally resulting in improved cell adhesion [97].

3.2. Enhancing the Mechanical Performance of Bioceramics Composites

CaP-based scaffolds generally exhibit poor mechanical properties compared to teeth and bone, especially due to their intrinsic brittleness, limiting their load-bearing bone applications [82,112]. Brittle materials are more likely to fail under tension or shearing rather than compression, essentially due to the crack propagation in preexisting flaws, such as micro-cracks or macro-pores [70,113].

Common approaches to improve mechanical performance and reduce the brittleness of ceramic materials are classified as intrinsic or extrinsic modifications (Figure 3).

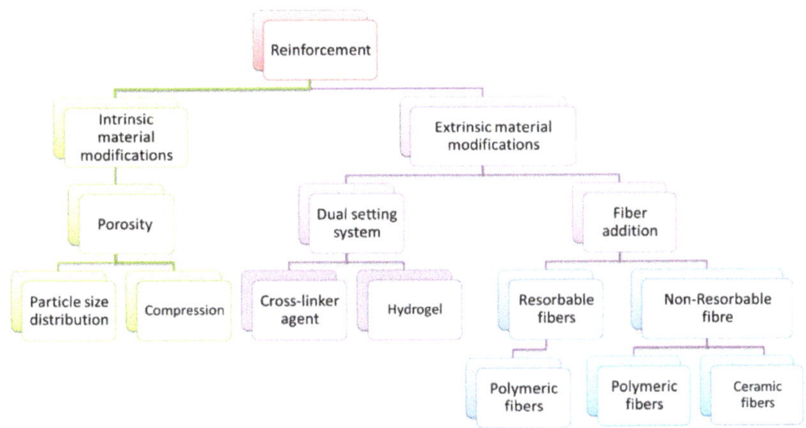

Figure 3. Mechanical reinforcement strategies for CaP-based biomaterials in a load-bearing application.

The intrinsic approach involves changes in the inherent properties of the scaffold, such as the composition, porosity and microstructure, whereas extrinsic modifications involve the use of reinforcing fibers or, in the case of CaP-based cements and pastes, the use of a cross-linker agent or hydrogel for the optimization of the dual setting system [72].

3.2.1. Intrinsic Material Modifications

The mechanical strength of scaffolds closely depends on their microstructure. Several factors, such as composition, crystal size and porosity, greatly affect the microstructure of scaffolds and its final strength [102]. One of the major factors affecting the mechanical performance is porosity, as the strength exponentially decreases with increasing voids [56,114,115].

A possible strategy to increase the mechanical strength is the reduction of intergranular voids by favoring the packing of the particles or using multimodal particle size distributions, leading to a decrease in the microporosity extent, especially in the struts [116].

The pore size distribution influences the degradation performance of the scaffold, and, therefore, the biodegradation kinetics can be modulated by varying the pore architecture [91]. Triangular, rectangular and elliptic pores were reported to support angiogenesis and faster cell migration due to their greater curvature [55]. Nevertheless, the increase in scaffold porosity is inversely related to mechanical strength; this is a key problem, difficult to solve and strongly limiting to load-bearing applications. In this respect, an exponential decrease in the compressive strength with increasing porosity was observed [115].

$$\sigma = [(E_0 R)/(\pi c)]^{0.5} \exp(-KP) \tag{2}$$

where E_0 is the Young's modulus at zero porosity; c is the average pore size; R is the fracture surface energy; K is an empirical constant, which can be extracted from the slope of a semi-logarithmic plot of the strength–porosity curve; and P is the porosity extent (in volume) [117,118].

Higher compressive moduli are associated with smaller pore sizes, porosity gradients and oriented pores [114,115]. The capability to modulate the porosity extent and distribution is helpful in limiting the concentration of mechanical stresses toward damage-tolerant structures; that is, micro-fractures occur until the scaffold's failure.

3.2.2. Extrinsic Material Modifications

The approaches proposed to increase the mechanical strength while limiting the brittleness of bioceramics include the combination with polymers, fibers or a dual setting system, especially for cements [104,107,108,119,120].

The dual setting system refers to the addition of reactive monomers to the liquid phase, together with an initiator into the inorganic component of the cement or eventually polymeric component that can be cross-linked [121,122]. In the first case, during the setting, there are simultaneous gelation/polymerization and dissolution–precipitation reactions, thus obtaining cement with a porous microstructure reinforced with a hydrogel-based matrix. As a consequence, an increase in compressive strength and hardness with stable rheological properties was achieved [102]. In turn, the cross-linking agent permits the binding of Ca^{2+} ions with carboxylic acid or organic phosphate fractions in the polymer chain, thus resulting in a reduction in brittleness and an increase in compressive strength [82,122].

The addition of fibers is one of the most effective approaches to increase the strength and toughness of bioceramics [123,124]. The mechanical behavior of fiber bioceramic composites is based on the interaction between the composite components and is time dependent due to the potential degradation of both fiber and CaP-based materials after implantation to allow bone regeneration. The reinforcements are related to several parameters, including (i) composition, mechanical properties and degradation of the matrix; (ii) fiber–matrix interface properties; and (iii) type, length, diameter, volume fraction, orientation and mechanical properties of fibers [124,125]. It was observed that the long-term strengthening effect of fibers was related to the type of fibers: the addition of non-resorbable fibers led to a stable increase in mechanical performance over time, while resorbable/biodegradable fibers provided only an initial reinforcement, followed by the creation of a macroporosity in the ceramic matrix after degradation of the fibers, favoring osteointegration [102].

The application of critical loads to brittle materials induces catastrophic fractures without any reversible deformation. The incorporation of fibers provides intergranular bridges increasing the tensile strength, flexural strength and fracture toughness.

There are three main fiber-reinforcing mechanisms [108,119,120]:

- Fiber bridging: the fibers bridge the existing crack, limiting its opening and propagation;
- Crack deflection: the fibers increase the length of the crack propagation, requiring more energy in newly formed surfaces;
- Frictional sliding: the presence of intergranular fibers in the matrix increases the fracture resistance of the composite.

Fibers can be classified as natural and man-made fibers, further divided into resorbable and non-resorbable [71] (Table 9).

Table 9. Fiber classification and some examples of fibers used in bioceramic reinforcement.

Natural Fibers	Man-Made Fibers			
	Resorbable		Non-Resorbable	
	Natural Polymer	Synthetic Polymer	Polymeric	Ceramics
Silk fibroin [107]	Polylactide [109] Cellulose [112]	Poly-caprolactone [109]	Polyamide [103,113]	Carbon [104,115,116,118] Silicate based [117,121,122,126] HA whiskers [122,127]

The introduction of carbon fibers (CF) in bioceramic scaffolds, including bone cements, has been explored in the past decades [123–125]. In particular, previous works showed that the addition of fibers led to an increase in compressive strength without interfering with HA formation during the setting of CPC [127]. The presence of CF induced a significant reinforcement also in calcinated HA-based scaffolds, while preserving biocompatibility and bioactivity; the mechanisms underlying the increase in mechanical properties were attributed to crack deflection, interlocking of the fibers, pullout and crack bridging [118]. Basically, the interaction between fibers and the surrounding ceramic matrix is based on several properties of fibers, including chemical composition, wettability and surface modifications. HA bioceramics reinforced with silicon-coated CF with controllable alignment were prepared via hot pressing and pressureless sintering, leading to the formation of a SiO_2 protective layer upon thermal decomposition of HA [118].

Various oxidation treatments were also implemented to improve the performance of CF as a strengthening agent of CPCs, including a preliminary treatment with aqua regia followed by immersion in $CaCl_2$ [104]. This treatment favored the heterogeneous nucleation of apatite nanocrystals on the surface of fibers, thus reducing the setting times; the addition of 1 wt% of fibers led to a significant increase in both bending strength and the work of fracture, essentially due to the deflection of crack propagation, while the in vitro biocompatibility was preserved.

Moreover, silicate-based fibers, calcium silicate, glass and basalt fibers have been used to reinforce bioceramics [117,121,122,126]. In particular, wollastonite ($CaSiO_3$) fibers were introduced into CPCs, showing that Si could favor the crystallization of needle-like apatite during cement setting, associated with a significant increase in compressive strength [126]. Furthermore, the presence of $CaSiO_3$ fibers was a promoter of cell viability and ALP activity [121].

Glass fibers (GF), such as E-glass and bioactive glass fibers (BGF), have been proposed as CPC-reinforcing agents [128,129]. E-glass fibers are composed of alumino-borosilicate with about 1 wt% alkali oxides, while BGF is described by the ternary system SiO_2-CaO-P_2O_5 [122]. Xu and co-workers had incorporated short and long E-glass fibers into CPC, obtaining an increase in elastic modulus, flexural strength and the work of fracture [115]. The addition of 15 wt% of BGF also determined an improvement of compressive strength, toughness and elastic modulus of CPCs [122].

In addition to fibers, apatite whiskers were proposed to improve the mechanical properties of CPCs; the enhancement of 120% of the work of fracture and 60% of flexural strength was obtained by adding 30 vol% of HA whiskers [127].

4. Ion-Doped Bioceramics and Composite Scaffolds

4.1. Ion Doping

Calcium phosphates, especially HA, are capable of hosting a variety of foreign (i.e., different from Ca and P) ions, involving the formation of atomic defects but with a limited modification of the overall crystal structure [39]. As biological apatites forming the inorganic part of bone are characterized by nanocrystallinity, poor crystal ordering and multiple ion doping, in the last few decades, various approaches were proposed to tune the biological properties of ceramics [129–136] in order to obtain novel biomaterials with multifunctional abilities, including antibacterial [137–140] and magnetic properties [128].

Some of the most studied substituting ions in bioceramics, with related biological roles, are listed in Table 10.

Table 10. Doping ions in calcium phosphate bioceramics, with related biofunctional ability.

Ion	Biological Effects	References
Si^{4+}	- Induction of the biomimetic precipitation of HA	[39,78,114]
Sr^{2+}	- Osteogenic activity - Anti-osteoporotic agent - Enhancement of mechanical properties	[39,132,141]
Mg^{2+}	- Enhancement of bone growth - Induction of angiogenesis - Antibacterial agent	[39,132,139,141]

Table 10. *Cont.*

Ion	Biological Effects	References
Zn^{2+}	- Stimulation of osteoblastic activity - Inhibition of bone resorption - Antibacterial agent	[142–144]
Ag^+	- Antibacterial agent	[137,138,140]
Mn^{2+}	- Regulation of osteoblastic differentiation - Control of bone resorption - Promotion of cell adhesion - Promotion the synthesis of extracellular matrix proteins	[131,136]
Cu^{2+}	- Antibacterial agent	[131,136]
Co^{2+}	- Neo-vascularization promotion - High cell proliferation - Osteogenic activity	[131,136]
$Fe^{2+/3+}$	- Antibacterial agent - Super-paramagnetism - Promotion of bone formation - Osteoinductivity	[39,128,131,145]
F^-	- Shrinkage of HA crystal lattice - Decreasing solubilization and increasing stability of HA - Induction of biomineralization - Osteogenic activity - Antibacterial agent	[136,142,144,146,147]

4.1.1. Magnesium

Magnesium is considered as the main ion capable of replacing Ca in biological apatite, in an amount close to 1 wt% [132]. Mg^{2+} ions play a key role in bone metabolism, taking part of the biochemical reaction beyond bone formation, while also controlling bone growth and metabolism [47,142,148].

Magnesium phosphates are also associated with a higher dissolution rate than calcium phosphates [149]. Mg has been shown to inhibit the formation of crystalline minerals, such as hydroxyapatite, whereas more soluble phases, such as brushite, are minimally affected by the presence of Mg [150–152]. Specifically, it was observed in basic solutions that HAP precipitation is inhibited by Mg substitution for Ca higher than 10%, and amorphous calcium phosphate (ACP) or whitlockite, the Mg polymorph of β-tricalcium phosphate, forms [153,154].

The incorporation of magnesium was also associated with increased protein adsorption and cell adhesion on the surface of bioceramics [17,18]. Furthermore, an intrinsic antibacterial activity was described for Mg-HA [136,139].

4.1.2. Strontium

Strontium (Sr^{2+}) is a natural component of bones and teeth and have affinity with Ca^{2+} ions, thus representing a calcium-like entity within cells, acting along similar biochemical and cellular pathways [136,141]. At a low concentration, strontium inhibits osteoclast activity, reduces bone resorption, enhances osteoblast proliferation and promotes bone formation. In this context, the addition of strontium in bioceramics is promising for the local treatment of bone affected by metabolic diseases, such as osteoporosis [45,155–159]. Several approaches can be implemented to obtain Sr-doped bioceramics, including the addition of strontium salts in wet synthesis processes [160] or of Sr-doped inorganic reactants involved in solid-state reactions at high temperatures [73,159]. The incorporation of strontium ions replacing Ca^{2+} in the crystal lattice of calcium phosphates generally induces deformations in the crystal lattice due to its higher ionic radius in respect to calcium. This affects the physicochemical properties of CaPs; for instance, it was observed that Sr^{2+} ions stabilize the β-TCP polymorph during thermal synthesis reactions. Furthermore, various previous studies reported a mechanical reinforcement ascribed to strontium doping, possibly due to enhancement of the interatomic bond strength in the CaP crystal in respect to calcium [161–163].

4.1.3. Silicon

Silicon plays a key role in the physiological formation of bone and cartilage tissues, especially due to its intrinsic capacity to act both as a cross-linker in ECM and to favor the precipitation of HA and bone mineralization [29,135]. When used in the synthesis of bioceramics, such as tricalcium phosphate (TCP), normally obtained with high-temperature treatments, silicon has the capacity to favor the formation of α-TCP polymorph against β-TCP [164,165]. Silicon-containing bioceramics exhibit high bioactivity, including bio-glasses (Na-Ca-P-Si), wollastonite ($CaSiO_3$) and Si-doped bioceramics (e.g., Si-HA and Si-TCP) [24,27,29,32,48].

The pivotal role of Si-containing bioceramics, such as silicon-doped HA, in bone tissue regeneration was confirmed by in vivo studies revealing the enhanced formation of collagen fibrils after 6 weeks at the bone/Si-HA interface and after 12 weeks with the bone/HA interface [134,135]. In addition, the enhanced formation of mature osteoclasts from mononuclear precursor cells was observed, thus showing the potential of silicon to favor the complex bone regeneration cascade by stimulating the various cell lines involved in new bone formation and remodeling. Long-term in vivo studies also reported the significantly higher bioresorbability of Si-doped HA scaffolds compared to pure HA scaffolds, as only few residues of the doped scaffold were observed at one year upon implantation, while non-doped HA scaffolds remained unchanged even after five years from implantation [129].

4.1.4. Silver

The incorporation of silver ions into bioceramics, as a replacing element for calcium, is possible due to their similar ionic radius [133].

Silver doping has been proposed as a valuable antibacterial strategy due to its ability to interfere with the electron transfer process on bacterial membranes and to promote the production of reactive oxygen species (ROS), finally causing cell death [148].

4.1.5. Iron

The incorporation of iron ions into bioceramics has been widely studied in recent decades, together with its neighboring transition elements from the fourth period of the periodic table (Mn, Co, Ni, Zn) [39], with the purpose of generating new bioceramics with magnetic proper-ties. Indeed, super-paramagnetic iron oxide nanoparticles (SPIONs) are widely approved mag-netic biomaterials (usually composed of magnetite Fe_3O_4 or maghemite γ-Fe_2O_3) as a contrast agent in magnetic resonance imaging applications for cancer diagnosis or hyperthermia-based cancer treatments. Nevertheless, their iron oxide core also causes long-term cytotoxicity; there-fore, intensive effort is today dedicated to develop iron-doped magnetic ceramics preserving good biocompatibility and bioactivity [166].

In this respect, iron-doped CPCs for magnetic hyperthermia were obtained, exhibiting improved osteoconductive and antibacterial properties [167–169]. A new concept of mag-netic CaP was obtained by synthesizing apatite nanocrystals doped with $Fe^{2+/3+}$ ions, so as to generate intrinsic superparamagnetic ability, generated by the specific positioning of Fe^{2+} and Fe^{3+} ions in the crystal lattice and in the outer hydrated layer of the apatite [128,145]. Such a new phase exhibited a magnetization ability similar to that of iron oxides but with excellent biocompatibility and enhanced osteogenic ability [170].

4.1.6. Fluorine

Fluorine ions take part in several biochemical processes, becoming particularly impor-tant for oral care applications, neuromodulation and bone structure [136]. Fluorine promotes osteoblast proliferation and inhibits osteoclast activation and differentiation; moreover, when present in a low concentration, it can enhance in vivo bone formation [147,171].

The substitution of OH^- groups of apatite with F^- ions accelerated the crystallization process, increasing the stability of the crystals while decreasing their solubility [147,171]; the incorporation of fluorine also affected the crystal morphology toward flattened hexagonal rod-like shapes [147,171].

Fluorine-doped HA also exhibited antibacterial behavior, inhibiting the adhesion and proliferation of bacteria typically detected in an oral environment [147,171].

4.2. Composites with Silicates

The preparation of composites containing both calcium phosphates and silicates has been explored with several approaches for the purpose of enhancing the mechanical properties of scaffolds. In fact, various studies showed that calcium phosphate composites with calcium silicates exhibited enhanced compressive and flexural strength [172,173].

Si-containing bioceramics include colloidal silica nanoparticles [174,175], silicates (i.e., calcium silicates and zinc silicates) [176–178], glasses [179,180] and silicate-phosphates (i.e., silicocarnotite and nagelschmidtite) [181,182]. Regarding the preparation of bone cements, previous studies showed that the addition of silica nanoparticles led to a decrease in the setting times and led to improved mechanical properties, especially due to the formation of Si-O-Si bonds among the particles [174,175]. Calcium phosphate cements containing zinc silicate and PLGA microspheres were also prepared [162]: the role of Si and Zn in improving setting times, injectability and compression strength was observed, while the addition of the microspheres did not affect the porosity.

The incorporation of silicates becomes particularly interesting in bioglass-reinforced cements, e.g., single-phase crystalline or amorphous calcium silicate phosphates (CaO-SiO_2-P_2O_5, CaSiP) or Bioglass A5S4, resulting in increased setting times and injectability [179,180]. The incorporation of bioglasses also significantly improved the bioactivity of the scaffold, promoting osteoblast attachment, proliferation and differentiation in vivo. The effect of CaSiP (silicocarnotite, $Ca_5(PO_4)_2SiO_4$) in brushite cements was also investigated, showing the role of Si in favoring the formation of HA, osteoblast proliferation and the formation of novel bone tissue [181].

The application of single-phase calcium phosphate silicate bioceramics (CaSiP) is not limited to bone regeneration but also to periodontal repair. In this respect, various works showed the preparation of 3D-printed silicate bioceramics, such as nagelschmidtite ($Ca_7(SiO_4)_3(PO_4)_2$, CSP) and silicocarnotite ($Ca_5(PO_4)_2SiO_4$, S_{ss}) [48]. CaSiP showed good biological performance with the formation of flake-like apatite layers (in the cases of S_{ss} and CSP, respectively). The ion release positively induced cell proliferation and differentiation as well as the formation of the extracellular matrix and the mineralization of periodontal tissue [183,184].

4.3. Composites with Carbon

The interest in the synthesis of composites with calcium phosphates and carbon-derived structures rapidly rose in recent years, especially considering graphene, a 2D material made of nanosheets of hexagonally bonded carbon atoms characterized by a high surface area, high conductivity, excellent mechanical properties and good biocompatibility, particularly interesting for tissue engineering applications [185–187].

The synthesis strategies to obtain graphene/carbon nanotubes–hydroxyapatite composites have been reported, evidencing hemocompatibility, antibacterial properties and the ability of graphene–hydroxyapatite composites to increase osteogenic activity [161,163,188–194].

A hybrid composite made of graphene oxide (GO), chitosan (CS) and HA (GO-CS-HA) was developed as a coating for titanium implants, exhibiting an increased formation of biomimetic apatite and also antibacterial properties, possibly ascribed to the increased production of reactive oxygen species [177].

Furthermore, 3D-printed composite scaffolds made of β-TCP, reduced graphene oxide (RGO), magnesium nanoparticles and arginine were prepared [176]. The combination of amino groups of arginine, released Mg ions and the nanotopography of GO resulted in increased mechanical performance.

The effect of RGO and carbon nanotubes (CNT) in α-TCP-based cements was evaluated. The setting times decreased when increasing the concentration of RGO, while negligible variations were observed with the addition of CNT; the mechanical performance was also valuable for load-bearing applications [178,195].

The use of microwaves resulted in a reduction in setting time and an increase in mechanical properties, ascribed to the evaporation of gas from the surface of RGO and CNT, strengthening the final composite [195]. The formation of an external HA layer was observed, favoring cell adhesion and proliferation.

4.4. Composites with Titanates

Titanium and its alloys have been used in combination with calcium phosphates for bone tissue engineering due to their excellent mechanical properties [196]. Metallic prostheses and implants are widely used to replace damaged bones and teeth, and their interaction with the surrounding tissue depends on the chemistry and microstructure of the surface [197], but their main drawback is related to their poor bioresorbability. In this respect, the preparation of bioceramic composites containing titanium oxides was considered as a valuable approach, exhibiting good biocompatibility and enhancing in vivo osteointegration [198,199].

Titanium oxide nanomaterials can also be added to injectable cements and pastes, leading to higher injectability and improved mechanical performance [175].

Some titanates, such as barium titanate ($BaTiO_3$, BT) and strontium titanate ($SrTiO_3$), are also characterized by piezoelectric properties, potentially providing microstructural accumulation of charges mimicking the mechanotransduction of bone cells [183,184,200,201]. BT-HA composites were investigated to combine the bioactivity of HA with the piezoelectricity of BT [184,200]. Three-dimensional-printed highly porous piezoelectric scaffolds based on BT and HA were obtained, with good cytocompatibility and cell attachment [184]. An aligned porous BT-HA piezoelectric composite was obtained by the ice-template method, exhibiting high porosity, cell proliferation, differentiation and adhesion of osteoblastic cells [200].

5. Conclusions

Calcium phosphates are widely accepted biomaterials and the gold standard to promote the regeneration of bone tissue. CaP scaffolds with biomimetic composition can exhibit osteogenic ability, bioresorbability and antibacterial properties. However, appropriate mechanical properties are required if the target is the regeneration of critical-sized bone defects, particularly when load bearing.

The co-existence of various factors, such as bioactive chemical composition, nanostructure and bone-like mechanical performance, is a major problem with ceramics due to the need of sintering and the difficulty of achieving complex bone-mimicking 3D structures. In fact, several technologies developed in the last decades for the manufacturing of a highly porous bioceramic-based scaffold from traditional methods (partial sintering, replica method, sacrificial template and direct foaming, as well as various 3D printing technologies) usually fail in generating bioactive and effective bone scaffolds. Hence, future perspectives are strongly related to the development of new approaches that can generate bone scaffolds endowed with bone-mimicking features yielding effective regenerative ability. To this end, recently developed innovative approaches targeting low-temperature processes, including chemically induced consolidation of CaP pastes or biomorphic transformation processes, are examples of radically new methods enabling the possibility to create scaffolds retaining nanocrystallinity and bioactive, ion-doped composition or even multi-scale hierarchically organized architectures inherited from natural sources. These results open new perspectives in ceramic science and are encouraging for further research in the field, targeting the decisive resolution of many still unmet clinical problems related to bone regeneration.

Author Contributions: Conceptualization, S.S., A.T. and M.D.; methodology, S.S., A.T. and M.D.; writing—original draft preparation, M.T.; writing—review and editing, M.D.; supervision, M.D. and S.S.; All authors have read and agreed to the published version of the manuscript.

Funding: This research received no external funding.

Conflicts of Interest: The authors declare no conflict of interest.

References

1. Bean, A.C. Basic Science Concepts in Musculoskeletal Regenerative Medicine. In *Regenerative Medicine for Spine and Joint Pain*; Springer: Berlin/Heidelberg, Germany, 2020; pp. 5–27. [CrossRef]
2. Li, J.J.; Ebied, M.; Xu, J.; Zreiqat, H. Current Approaches to Bone Tissue Engineering: The Interface between Biology and Engineering. *Adv. Healthc. Mater.* **2018**, *7*, 1–8. [CrossRef] [PubMed]
3. Sommerfeldt, D.; Rubin, C. Biology of bone and how it orchestrates the form and function of the skeleton. *Eur. Spine J.* **2001**, *10*, 86–95. [CrossRef]
4. Armiento, A.R.; Hatt, L.P.; Sanchez Rosenberg, G.; Thompson, K.; Stoddart, M.J. Functional Biomaterials for Bone Regeneration: A Lesson in Complex Biology. *Adv. Funct. Mater.* **2020**, *1909874*, 1–41. [CrossRef]
5. Berthiaume, F.; Maguire, T.J.; Yarmush, M.L. Tissue Engineering and Regenerative Medicine: History, Progress, and Challenges. *Annu. Rev. Chem. Biomol. Eng.* **2011**, *2*, 403–430. [CrossRef] [PubMed]
6. Cooper, G.; Herrera, J.; Kirkbride, J.; Perlman, Z. Introduction to Regenerative Medicine. In *Regenerative Medicine for Spine and Joint Pain*; Springer: Berlin/Heidelberg, Germany, 2020; pp. 1–4. [CrossRef]
7. Oryan, A.; Alidadi, S.; Moshiri, A.; Maffulli, N. Bone regenerative medicine: Classic options, novel strategies, and future directions. *J. Orthop. Surg. Res.* **2014**, *9*, 1–27. [CrossRef]
8. Vaish, A.; Murrell, W.; Vaishya, R. History of regenerative medicine in the field of orthopedics. *J. Arthrosc. Surg. Sports Med.* **2020**, *1*, 154–158. [CrossRef]
9. Eliaz, N.; Metoki, N. Calcium phosphate bioceramics: A review of their history, structure, properties, coating technologies and biomedical applications. *Materials* **2017**, *10*, 334. [CrossRef] [PubMed]
10. Jodati, H.; Bengi, Y.; Evis, Z. A review of bioceramic porous scaffolds for hard tissue applications: Effects of structural features. *Ceram. Int.* **2020**, *46*, 15725–15739. [CrossRef]
11. Pereira, H.F.; Cengiz, I.F.; Samuel, F.; Rui, S.; Reis, L.; Oliveira, J.M. Scaffolds and coatings for bone regeneration. *J. Mater. Sci. Mater. Med.* **2020**, *31*, 27. [CrossRef]
12. Williams, D.F. On the mechanisms of biocompatibility. *Biomaterials* **2008**, *29*, 2941–2953. [CrossRef]
13. Polo-Corrales, L.; Latorre-Esteves, M.; Ramirez-Vick, J.E. Scaffold design for bone regeneration. *J. Nanosci. Nanotechnol.* **2014**, *14*, 15–56. [CrossRef] [PubMed]
14. Kielty, C.M.; Grant, M.E. The Collagen Family: Structure, Assembly, and Organization in the Extracellular Matrix. In *Connective Tissue and Its Heritable Disorders*; Wiley: Hoboken, NJ, USA, 2003; pp. 159–221. [CrossRef]
15. No, Y.J.; Holzmeister, I.; Lu, Z.; Prajapati, S.; Shi, J.; Gbureck, U.; Zreiqat, H. Effect of baghdadite substitution on the physicochemical properties of brushite cements. *Materials* **2019**, *12*, 1719. [CrossRef] [PubMed]
16. Jeong, J.; Kim, J.H.; Shim, J.H.; Hwang, N.S.; Heo, C.Y. Bioactive calcium phosphate materials and applications in bone regeneration. *Biomater. Res.* **2019**, *23*, 1–11. [CrossRef] [PubMed]
17. Tampieri, A.; Iafisco, M.; Sprio, S.; Ruffini, A.; Panseri, S.; Montesi, M.; Adamiano, A.; Sandri, M. Hydroxyapatite: From Nanocrystals to Hybrid Nanocomposites for Regenerative Medicine. In *Handbook of Bioceramics and Biocomposites*; Springer: Berlin/Heidelberg, Germany, 2016; pp. 119–144. [CrossRef]
18. Sprio, S.; Sandri, M.; Iafisco, M.; Panseri, S.; Filardo, G.; Kon, E.; Marcacci, M.; Tampieri, A. Composite biomedical foams for engineering bone tissue. In *Biomedical Foams for Tissue Engineering Applications*; Woodhead Publishing: Sawston, UK, 2014; pp. 249–280. [CrossRef]
19. Barrère, F.; van Blitterswijk, C.A.; de Groot, K. Bone regeneration: Molecular and cellular interactions with calcium phosphate ceramics. *Int. J. Nanomed.* **2006**, *1*, 317–332.
20. Sprio, S.; Ruffini, A.; Valentini, F.; D'Alessandro, T.; Sandri, M.; Panseri, S.; Tampieri, A. Biomimesis and biomorphic transformations: New concepts applied to bone regeneration. *J. Biotechnol.* **2011**, *156*, 347–355. [CrossRef]
21. Wegst, U.G.K.; Bai, H.; Saiz, E.; Tomsia, A.P.; Ritchie, R.O. Bioinspired structural materials. *Nature* **2015**, *14*, 23–36. [CrossRef]
22. Rho, J.Y.; Kuhn-Spearing, L.; Zioupos, P. Mechanical properties and the hierarchical structure of bone. *Med. Eng. Phys.* **1998**, *20*, 92–102. [CrossRef]
23. Kang, J.; Dong, E.; Li, D.; Dong, S.; Zhang, C.; Wang, L. Anisotropy characteristics of microstructures for bone substitutes and porous implants with application of additive manufacturing in orthopaedic. *Mater. Des.* **2020**, *191*, 108608. [CrossRef]

24. Arcos, D.; Vallet-Regí, M. Sol-gel silica-based biomaterials and bone tissue regeneration. *Acta Biomater.* **2010**, *6*, 2874–2888. [CrossRef]
25. Burduşel, A.-C. Bioactive composites for bone regeneration. *Biomed. Eng. Int.* **2019**, *1*, 9–15. [CrossRef]
26. Fernandez de Grado, G.; Keller, L.; Idoux-Gillet, Y.; Wagner, Q.; Musset, A.M.; Benkirane-Jessel, N.; Bornert, F.; Offner, D. Bone substitutes: A review of their characteristics, clinical use, and perspectives for large bone defects management. *J. Tissue Eng.* **2018**, *9*, 2041731418776819. [CrossRef] [PubMed]
27. Lei, Q.; Guo, J.; Noureddine, A.; Wang, A.; Wuttke, S.; Brinker, C.J.; Zhu, W. Sol–Gel-Based Advanced Porous Silica Materials for Biomedical Applications. *Adv. Funct. Mater.* **2020**, *1909539*, 1–28. [CrossRef]
28. Shue, L.; Yufeng, Z.; Mony, U. Biomaterials for periodontal regeneration A review of ceramics and polymers. *BioMatter* **2012**, *2*, 271–277. [CrossRef]
29. Venkatraman, S.K.; Swamiappan, S. Review on calcium- and magnesium-based silicates for bone tissue engineering applications. *J. Biomed. Mater. Res. Part A* **2020**, *108*, 1546–1562. [CrossRef]
30. Yousefi, A.-M. A review of calcium phosphate cements and acrylic bone cements as injectable materials for bone repair and implant fixation. *J. Appl. Biomater. Funct. Mater.* **2019**, *17*, 228080001987259. [CrossRef] [PubMed]
31. Yamamuro, T. Bioceramics. In *Biomechanics and Biomaterials in Orthopedics*; Springer: Berlin/Heidelberg, Germany, 2004; pp. 22–33. [CrossRef]
32. Gul, H.; Khan, M.; Khan, A.S. 3—Bioceramics: Types and clinical applications. In *Handbook of Ionic Substituted Hydroxyapatites*; Elsevier: Amsterdam, The Netherlands, 2020; pp. 53–83. [CrossRef]
33. Li, X.; Wang, J.; Joiner, A.; Chang, J. The remineralisation of enamel: A review of the literature. *J. Dent.* **2014**, *42*, S12–S20. [CrossRef]
34. Doremus, R.H. Review Bioceramics. *J. Mater. Sci.* **1992**, *27*, 285–297. [CrossRef]
35. Low, K.L.; Tan, S.H.; Zein, S.H.S.; Roether, J.A.; Mouriño, V.; Boccaccini, A.R. Calcium phosphate-based composites as injectable bone substitute materials. *J. Biomed. Mater. Res. Part B Appl. Biomater.* **2010**, *94*, 273–286. [CrossRef]
36. Surmenev, R.A.; Surmeneva, M.A.; Ivanova, A.A. Significance of calcium phosphate coatings for the enhancement of new bone osteogenesis—A review. *Acta Biomater.* **2014**, *10*, 557–579. [CrossRef]
37. Kucko, N.W.; Herber, R.-P.; Leeuwenburgh, S.C.G.; Jansen, J.A. Calcium Phosphate Bioceramics and Cements. In *Principles of Regenerative Medicine*; Elsevier: Amsterdam, The Netherlands, 2019; pp. 591–611. [CrossRef]
38. Ginebra, M.P.; Espanol, M.; Maazouz, Y.; Bergez, V.; Pastorino, D. Bioceramics and bone healing. *EFORT Open Rev.* **2018**, *3*, 173–183. [CrossRef]
39. Uskoković, V. Ion-doped hydroxyapatite: An impasse or the road to follow? *Ceram. Int.* **2020**, *46*, 11443–11465. [CrossRef]
40. Graziani, G.; Boi, M.; Bianchi, M. A review on ionic substitutions in hydroxyapatite thin films: Towards complete biomimetism. *Coatings* **2018**, *8*, 269. [CrossRef]
41. Carrodeguas, R.G.; De Aza, S. α-Tricalcium phosphate: Synthesis, properties and biomedical applications. *Acta Biomater.* **2011**, *7*, 3536–3546. [CrossRef] [PubMed]
42. Lobo, S.E.; Livingston Arinzeh, T. Biphasic Calcium Phosphate Ceramics for Bone Regeneration and Tissue Engineering Applications. *Materials* **2010**, *3*, 815–826. [CrossRef]
43. Maji, K.; Mondal, S. Calcium Phosphate Biomaterials for Bone Tissue Engineering: Properties and Relevance in Bone Repair. In *Racing for the Surface: Antimicrobial and Interface Tissue Engineering*; Springer: Berlin/Heidelberg, Germany, 2020; pp. 535–555. [CrossRef]
44. Cabrejos-Azama, J.; Alkhraisat, M.H.; Rueda, C.; Torres, J.; Blanco, L.; López-Cabarcos, E. Magnesium substitution in brushite cements for enhanced bone tissue regeneration. *Mater. Sci. Eng. C* **2014**, *43*, 403–410. [CrossRef]
45. Pina, S.; Ferreira, J.M.F. Brushite-forming Mg-, Zn- and Sr-substituted bone cements for clinical applications. *Materials* **2010**, *3*, 519–535. [CrossRef]
46. Theiss, F.; Apelt, D.; Brand, B.; Kutter, A.; Zlinszky, K.; Bohner, M.; Matter, S.; Frei, C.; Auer, J.A.; Von Rechenberg, B. Biocompatibility and resorption of a brushite calcium phosphate cement. *Biomaterials* **2005**, *26*, 4383–4394. [CrossRef]
47. Huan, Z.; Chang, J. Novel bioactive composite bone cements based on the β-tricalcium phosphate-monocalcium phosphate monohydrate composite cement system. *Acta Biomater.* **2009**, *5*, 1253–1264. [CrossRef] [PubMed]
48. Sepantafar, M.; Mohammadi, H.; Maheronnaghsh, R.; Tayebi, L.; Baharvand, H. Single phased silicate-containing calcium phosphate bioceramics: Promising biomaterials for periodontal repair. *Ceram. Int.* **2018**, *44*, 11003–11012. [CrossRef]
49. Jablonská, E.; Horkavcová, D.; Rohanová, D.; Brauer, D.S. A review of: In vitro cell culture testing methods for bioactive glasses and other biomaterials for hard tissue regeneration. *J. Mater. Chem. B* **2020**, *8*, 10941–10953. [CrossRef]
50. Hench, L.L.; Jones, J.R. Bioactive glasses: Frontiers and Challenges. *Front. Bioeng. Biotechnol.* **2015**, *3*, 1–12. [CrossRef]
51. Prasadh, S.; Wong, R.C.W. Unraveling the mechanical strength of biomaterials used as a bone scaffold in oral and maxillofacial defects. *Oral Sci. Int.* **2018**, *15*, 48–55. [CrossRef]
52. Roseti, L.; Parisi, V.; Petretta, M.; Cavallo, C.; Desando, G.; Bartolotti, I.; Grigolo, B. Scaffolds for Bone Tissue Engineering: State of the art and new perspectives. *Mater. Sci. Eng. C* **2017**, *78*, 1246–1262. [CrossRef] [PubMed]
53. Peroglio, M.; Gremillard, L.; Chevalier, J.; Chazeau, L.; Gauthier, C.; Hamaide, T. Toughening of bio-ceramics scaffolds by polymer coating. *J. Eur. Ceram. Soc.* **2007**, *27*, 2679–2685. [CrossRef]
54. Steinbrech, R.W. Toughening mechanisms for ceramic materials. *J. Eur. Ceram. Soc.* **1992**, *10*, 131–142. [CrossRef]

55. Abbasi, N.; Hamlet, S.; Love, R.M.; Nguyen, N.T. Porous scaffolds for bone regeneration. *J. Sci. Adv. Mater. Devices* **2020**, *5*, 1–9. [CrossRef]

56. Studart, A.R.; Gonzenbach, U.T.; Tervoort, E.; Gauckler, L.J. Processing routes to macroporous ceramics: A review. *J. Am. Ceram. Soc.* **2006**, *89*, 1771–1789. [CrossRef]

57. Champion, E. Sintering of calcium phosphate bioceramics. *Acta Biomater.* **2013**, *9*, 5855–5875. [CrossRef]

58. Eom, J.H.; Kim, Y.W.; Raju, S. Processing and properties of macroporous silicon carbide ceramics: A review. *J. Asian Ceram. Soc.* **2013**, *1*, 220–242. [CrossRef]

59. Kim, I.J.; Park, J.G.; Han, Y.H.; Kim, S.Y.; Shackelford, J.F. Wet foam stability from colloidal suspension to porous ceramics: A review. *J. Korean Ceram. Soc.* **2019**, *56*, 211–232. [CrossRef]

60. Ohji, T.; Fukushima, M. Macro-porous ceramics: Processing and properties. *Int. Mater. Rev.* **2012**, *57*, 115–131. [CrossRef]

61. Dapporto, M.; Sprio, S.; Fabbi, C.; Figallo, E.; Tampieri, A. A novel route for the synthesis of macroporous bioceramics for bone regeneration. *J. Eur. Ceram. Soc.* **2016**, *36*, 2383–2388. [CrossRef]

62. Deville, S. Freeze-casting of porous ceramics: A review of current achievements and issues. *Adv. Eng. Mater.* **2008**, *10*, 155–169. [CrossRef]

63. Babaie, E.; Bhaduri, S.B. Fabrication Aspects of Porous Biomaterials in Orthopedic Applications: A Review. *ACS Biomater. Sci. Eng.* **2018**, *4*, 1–39. [CrossRef] [PubMed]

64. Ginebra, M.P. Cements as bone repair materials. In *Bone Repair Biomaterials*; Elsevier: Amsterdam, The Netherlands, 2009; pp. 271–308. [CrossRef]

65. Lewis, G. Injectable bone cements for use in vertebroplasty and kyphoplasty: State-of-the-art review. *J. Biomed. Mater. Res. Part B Appl. Biomater.* **2006**, *76*, 456–468. [CrossRef]

66. Şahin, E.; Kalyon, D.M. The rheological behavior of a fast-setting calcium phosphate bone cement and its dependence on deformation conditions. *J. Mech. Behav. Biomed. Mater.* **2017**, *72*, 252–260. [CrossRef] [PubMed]

67. Vaishya, R.; Chauhan, M.; Vaish, A. Bone cement. *J. Clin. Orthop. Trauma* **2013**, *4*, 157–163. [CrossRef]

68. Ginebra, M.P.; Canal, C.; Espanol, M.; Pastorino, D.; Montufar, E.B. Calcium phosphate cements as drug delivery materials. *Adv. Drug Deliv. Rev.* **2012**, *64*, 1090–1110. [CrossRef]

69. O'Neill, R.; McCarthy, H.O.; Montufar, E.B.; Ginebra, M.P.; Wilson, D.I.; Lennon, A.; Dunne, N. Critical review: Injectability of calcium phosphate pastes and cements. *Acta Biomater.* **2017**, *50*, 1–19. [CrossRef]

70. Zhang, J.; Liu, W.; Schnitzler, V.; Tancret, F.; Bouler, J.M. Calcium phosphate cements for bone substitution: Chemistry, handling and mechanical properties. *Acta Biomater.* **2014**, *10*, 1035–1049. [CrossRef]

71. Canal, C.; Ginebra, M.P. Fibre-reinforced calcium phosphate cements: A review. *J. Mech. Behav. Biomed. Mater.* **2011**, *4*, 1658–1671. [CrossRef]

72. Bigi, A.; Boanini, E. Functionalized biomimetic calcium phosphates for bone tissue repair. *J. Appl. Biomater. Funct. Mater.* **2017**, *15*, e313–e325. [CrossRef] [PubMed]

73. Dapporto, M.; Gardini, D.; Tampieri, A.; Sprio, S. Nanostructured Strontium-Doped Calcium Phosphate Cements: A Multifactorial Design. *Appl. Sci.* **2021**, *11*, 2075. [CrossRef]

74. International Organization for Standardization. *Implants for Surgery—Calcium Phosphate Bioceramics—Characterization of Hardening Bone Paste Materials—ISO/DIS 18531(en)*; ISO: Geneva, Switzerland, 2015.

75. Baino, F.; Ferraris, M. Learning from Nature: Using bioinspired approaches and natural materials to make porous bioceramics. *Int. J. Appl. Ceram. Technol.* **2017**, *14*, 507–520. [CrossRef]

76. Sprio, S.; Panseri, S.; Montesi, M.; Dapporto, M.; Ruffini, A.; Dozio, S.M.; Cavuoto, R.; Misseroni, D.; Paggi, M.; Bigoni, D.; et al. Hierarchical porosity inherited by natural sources affects the mechanical and biological behaviour of bone scaffolds. *J. Eur. Ceram. Soc.* **2020**, *40*, 1717–1727. [CrossRef]

77. Tampieri, A.; Sprio, S.; Ruffini, A.; Celotti, G.; Lesci, I.G.; Roveri, N. From wood to bone: Multi-step process to convert wood hierarchical structures into biomimetic hydroxyapatite scaffolds for bone tissue engineering. *J. Mater. Chem.* **2009**, *19*, 4973–4980. [CrossRef]

78. Lucas-Aparicio, J.; Manchón, Á.; Rueda, C.; Pintado, C.; Torres, J.; Alkhraisat, M.H.; López-Cabarcos, E. Silicon-calcium phosphate ceramics and silicon-calcium phosphate cements: Substrates to customize the release of antibiotics according to the idiosyncrasies of the patient. *Mater. Sci. Eng. C* **2020**, *106*, 110173. [CrossRef]

79. Sprio, S.; Sandri, M.; Iafisco, M.; Panseri, S.; Adamiano, A.; Montesi, M.; Campodoni, E.; Tampieri, A. Bio-inspired assembling/mineralization process as a flexible approach to develop new smart scaffolds for the regeneration of complex anatomical regions. *J. Eur. Ceram. Soc.* **2016**, *36*, 2857–2867. [CrossRef]

80. Vincent, J.F.V.; Bogatyreva, O.A.; Bogatyrev, N.R.; Bowyer, A.; Pahl, A.K. Biomimetics: Its practice and theory. *J. R. Soc. Interface* **2006**, *3*, 471–482. [CrossRef]

81. White, R.A.; Weber, J.N.; White, E.W. Replamineform: A new process for preparing porous ceramic, metal, and polymer prosthetic materials. *Science* **1972**, *176*, 922–924. [CrossRef] [PubMed]

82. Tampieri, A.; Ruffini, A.; Ballardini, A.; Montesi, M.; Panseri, S.; Salamanna, F.; Fini, M.; Sprio, S. Heterogeneous chemistry in the 3-D state: An original approach to generate bioactive, mechanically-competent bone scaffolds. *Biomater. Sci.* **2019**, *7*, 307–321. [CrossRef]

83. Guvendiren, M.; Molde, J.; Soares, R.M.D.; Kohn, J. Designing Biomaterials for 3D Printing. *ACS Biomater. Sci. Eng.* **2016**, *2*, 1679–1693. [CrossRef]
84. Hart, L.R.; He, Y.; Ruiz-Cantu, L.; Zhou, Z.; Irvine, D.; Wildman, R.; Hayes, W. 3D and 4D printing of biomaterials and biocomposites, bioinspired composites, and related transformers. In *3D and 4D Printing of Polymer Nanocomposite Materials: Processes, Applications, and Challenges*; Elsevier: Amsterdam, The Netherlands, 2019; pp. 467–504. [CrossRef]
85. Ngo, T.D.; Kashani, A.; Imbalzano, G.; Nguyen, K.T.Q.; Hui, D. Additive manufacturing (3D printing): A review of materials, methods, applications and challenges. *Compos. Part B Eng.* **2018**, *143*, 172–196. [CrossRef]
86. Wang, X.; Peng, X.; Yue, P.; Qi, H.; Liu, J.; Li, L.; Guo, C.; Xie, H.; Zhou, X.; Yu, X. A novel CPC composite cement reinforced by dopamine coated SCPP fibers with improved physicochemical and biological properties. *Mater. Sci. Eng. C* **2020**, *109*. [CrossRef]
87. Bose, S.; Robertson, S.F.; Bandyopadhyay, A. Surface modification of biomaterials and biomedical devices using additive manufacturing. *Acta Biomater.* **2018**, *66*, 6–22. [CrossRef]
88. Chen, Z.; Li, Z.; Li, J.; Liu, C.; Lao, C.; Fu, Y.; Liu, C.; Li, Y.; Wang, P.; He, Y. 3D printing of ceramics: A review. *J. Eur. Ceram. Soc.* **2019**, *39*, 661–687. [CrossRef]
89. Ghorbani, F.; Li, D.; Ni, S.; Zhou, Y.; Yu, B. 3D printing of acellular scaffolds for bone defect regeneration: A review. *Mater. Today Commun.* **2020**, *22*, 100979. [CrossRef]
90. Sinha, S.K. Additive manufacturing (AM) of medical devices and scaffolds for tissue engineering based on 3D and 4D printing. In *3D and 4D Printing of Polymer Nanocomposite Materials: Processes, Applications, and Challenges*; Elsevier: Amsterdam, The Netherlands, 2019; pp. 119–160. [CrossRef]
91. Ma, H.; Feng, C.; Chang, J.; Wu, C. 3D-printed bioceramic scaffolds: From bone tissue engineering to tumor therapy. *Acta Biomater.* **2018**, *79*, 37–59. [CrossRef] [PubMed]
92. Stevens, M.M. Biomaterials for bone tissue engineering. *Mater. Today* **2008**, *11*, 18–25. [CrossRef]
93. Wang, C.; Huang, W.; Zhou, Y.; He, L.; He, Z.; Chen, Z.; He, X.; Tian, S.; Liao, J.; Lu, B.; et al. 3D printing of bone tissue engineering scaffolds. *Bioact. Mater.* **2020**, *5*, 82–91. [CrossRef]
94. Mondal, S.; Nguyen, T.P.; Pham, V.H.; Hoang, G.; Manivasagan, P.; Kim, M.H.; Nam, S.Y.; Oh, J. Hydroxyapatite nano bioceramics optimized 3D printed poly lactic acid scaffold for bone tissue engineering application. *Ceram. Int.* **2020**, *46*, 3443–3455. [CrossRef]
95. Pina, S.; Ribeiro, V.P.; Marques, C.F.; Maia, F.R.; Silva, T.H.; Reis, R.L.; Oliveira, J.M. Scaffolding Strategies for Tissue Engineering and Regenerative Medicine Applications. *Materials* **2019**, *12*, 1824. [CrossRef] [PubMed]
96. Xiao, D.; Zhang, J.; Zhang, C.; Barbieri, D.; Yuan, H.; Moroni, L.; Feng, G. The role of calcium phosphate surface structure in osteogenesis and the mechanisms involved. *Acta Biomater.* **2020**, *106*, 22–33. [CrossRef] [PubMed]
97. Zhu, L.; Luo, D.; Liu, Y. Effect of the nano/microscale structure of biomaterial scaffolds on bone regeneration. *International J. Oral Sci.* **2020**, *12*, 1–15. [CrossRef]
98. Tarafder, S.; Dernell, W.S.; Bandyopadhyay, A.; Bose, S. SrO- and MgO-doped microwave sintered 3D printed tricalcium phosphate scaffolds: Mechanical properties and in vivo osteogenesis in a rabbit model. *J. Biomed. Mater. Res. Part B Appl. Biomater.* **2015**, *103*, 679–690. [CrossRef]
99. Touri, M.; Moztarzadeh, F.; Abu Osman, N.A.; Dehghan, M.M.; Brouki Milan, P.; Farzad-Mohajeri, S.; Mozafari, M. Oxygen-Releasing Scaffolds for Accelerated Bone Regeneration. *ACS Biomater. Sci. Eng.* **2020**, *6*, 2985–2994. [CrossRef]
100. Sun, H.; Hu, C.; Zhou, C.; Wu, L.; Sun, J.; Zhou, X.; Xing, F.; Long, C.; Kong, Q.; Liang, J.; et al. 3D printing of calcium phosphate scaffolds with controlled release of antibacterial functions for jaw bone repair. *Mater. Des.* **2020**, *189*, 108540. [CrossRef]
101. Liu, D.; Nie, W.; Li, D.; Wang, W.; Zheng, L.; Zhang, J.; Zhang, J.; Peng, C.; Mo, X.; He, C. 3D printed PCL/SrHA scaffold for enhanced bone regeneration. *Chem. Eng. J.* **2019**, *362*, 269–279. [CrossRef]
102. Geffers, M.; Groll, J.; Gbureck, U. Reinforcement strategies for load-bearing calcium phosphate biocements. *Materials* **2015**, *8*, 2700–2717. [CrossRef]
103. Xu, H.H.K.; Quinn, J.B.; Takagi, S.; Chow, L.C.; Eichmiller, F.C. Strong and macroporous calcium phosphate cement: Effects of porosity and fiber reinforcement on mechanical properties. *J. Biomed. Mater. Res.* **2001**, *57*, 457–466. [CrossRef]
104. Boehm, A.V.; Meininger, S.; Tesch, A.; Gbureck, U.; Müller, F.A. The mechanical properties of biocompatible apatite bone cement reinforced with chemically activated carbon fibers. *Materials* **2018**, *11*, 192. [CrossRef]
105. Holthaus, M.G.; Stolle, J.; Treccani, L.; Rezwan, K. Orientation of human osteoblasts on hydroxyapatite-based microchannels. *Acta Biomater.* **2012**, *8*, 394–403. [CrossRef]
106. Zhao, C.; Wang, X.; Gao, L.; Jing, L.; Zhou, Q.; Chang, J. The role of the micro-pattern and nano-topography of hydroxyapatite bioceramics on stimulating osteogenic differentiation of mesenchymal stem cells. *Acta Biomater.* **2018**, *73*, 509–521. [CrossRef] [PubMed]
107. Hu, M.; He, Z.; Han, F.; Shi, C.; Zhou, P.; Ling, F.; Zhu, X.; Yang, H.; Li, B. Reinforcement of calcium phosphate cement using alkaline-treated silk fibroin. *Int. J. Nanomed.* **2018**, *13*, 7183–7193. [CrossRef] [PubMed]
108. Ritchie, R.O. The conflicts between strength and toughness. *Nat. Mater.* **2011**, *10*, 817–822. [CrossRef] [PubMed]
109. Zuo, Y.; Yang, F.; Wolke, J.G.C.; Li, Y.; Jansen, J.A. Incorporation of biodegradable electrospun fibers into calcium phosphate cement for bone regeneration. *Acta Biomater.* **2010**, *6*, 1238–1247. [CrossRef] [PubMed]
110. Zan, X.; Sitasuwan, P.; Feng, S.; Wang, Q. Effect of Roughness on in Situ Biomineralized CaP-Collagen Coating on the Osteogenesis of Mesenchymal Stem Cells. *Langmuir* **2016**, *32*, 1808–1817. [CrossRef] [PubMed]

111. Hu, Q.; Tan, Z.; Liu, Y.; Tao, J.; Cai, Y.; Zhang, M.; Pan, H.; Xu, X.; Tang, R. Effect of crystallinity of calcium phosphate nanoparticles on adhesion, proliferation, and differentiation of bone marrow mesenchymal stem cells. *J. Mater. Chem.* **2007**, *17*, 4690–4698. [CrossRef]

112. Mansoorianfar, M.; Shahin, K.; Mirström, M.M.; Li, D. Cellulose-reinforced bioglass composite as flexible bioactive bandage to enhance bone healing. *Ceram. Int.* **2021**, *47*, 416–423. [CrossRef]

113. Dos Santos, L.A.; De Oliveira, L.C.; da Silva Rigo, E.C.; Carrodéguas, R.G.; Boschi, A.O.; Fonseca de Arruda, A.C. Fiber Reinforced Calcium Phosphate Cement. *J. Mech. Behav. Biomed. Mater.* **2000**, *4*, 1658–1671. [CrossRef]

114. Sprio, S.; Ruffini, A.; Dapporto, M.; Tampieri, A. New biomimetic strategies for regeneration of load-bearing bones. In *Bio-Inspired Regenerative Medicine: Materials, Processes and Clinical Applications*; Pan Stanford Publishing: Singapore, 2016; Volume 6, pp. 85–117.

115. Xu, H.H.K.; Eichmiller, F.C.; Giuseppetti, A.A. Reinforcement of a self-setting calcium phosphate cement with different fibers. *J. Biomed. Mater. Res.* **2000**, *52*, 107–114. [CrossRef]

116. Zhang, Y.; Tan, S.; Yin, Y. C-fibre reinforced hydroxyapatite bioceramics. *Ceram. Int.* **2003**, *29*, 113–116. [CrossRef]

117. Li, G.; Zhang, K.; Pei, Z.; Liu, P.; Chang, J.; Zhang, K.; Zhao, S.; Zhang, K.; Jiang, N.; Liang, G. Basalt fibre reinforced calcium phosphate cement with enhanced toughness. *Mater. Technol.* **2020**, *35*, 152–158. [CrossRef]

118. Zhao, X.; Zheng, J.; Zhang, W.; Chen, X.; Gui, Z. Preparation of silicon coated-carbon fiber reinforced HA bio-ceramics for application of load-bearing bone. *Ceram. Int.* **2020**, *46*, 7903–7911. [CrossRef]

119. Dorner-Reisel, A.; Müller, E.; Tomandl, G. Short fiber reinforced hydroxyapatite-based bioceramics. *Adv. Eng. Mater.* **2004**, *6*, 572–577. [CrossRef]

120. Krüger, R.; Groll, J. Fiber reinforced calcium phosphate cements e On the way to degradable load bearing bone substitutes? *Biomaterials* **2012**, *33*, 5887–5900. [CrossRef] [PubMed]

121. Domingues, J.A.; Motisuke, M.; Bertran, C.A.; Hausen, M.A.; De Rezende Duek, E.A.; Camilli, J.A. Addition of Wollastonite Fibers to Calcium Phosphate Cement Increases Cell Viability and Stimulates Differentiation of Osteoblast-Like Cells. *Sci. World J.* **2017**, *2017*, 5260106. [CrossRef] [PubMed]

122. Nezafati, N.; Moztarzadeh, F.; Hesaraki, S.; Mozafari, M. Synergistically reinforcement of a self-setting calcium phosphate cement with bioactive glass fibers. *Ceram. Int.* **2011**, *37*, 927–934. [CrossRef]

123. Petersen, R. Carbon fiber biocompatibility for implants. *Fibers* **2016**, *4*, 1. [CrossRef]

124. Saito, N.; Aoki, K.; Usui, Y.; Shimizu, M.; Hara, K.; Narita, N.; Ogihara, N.; Nakamura, K.; Ishigaki, N.; Kato, H.; et al. Application of carbon fibers to biomaterials: A new era of nano-level control of carbon fibers after 30-years of development. *Chem. Soc. Rev.* **2011**, *40*, 3824–3834. [CrossRef]

125. Siddiqui, H.A.; Pickering, K.L.; Mucalo, M.R. A review on the use of hydroxyapatite- carbonaceous structure composites in bone replacement materials for strengthening purposes. *Materials* **2018**, *11*, 1813. [CrossRef]

126. Motisuke, M.; Santos, V.R.; Bazanini, N.C.; Bertran, C.A. Apatite bone cement reinforced with calcium silicate fibers. *J. Mater. Sci. Mater. Med.* **2014**, *25*, 2357–2363. [CrossRef]

127. Müller, F.A.; Gbureck, U.; Kasuga, T.; Mizutani, Y.; Barralet, J.E.; Lohbauer, U. Whisker-reinforced calcium phosphate cements. *J. Am. Ceram. Soc.* **2007**, *90*, 3694–3697. [CrossRef]

128. Iannotti, V.; Adamiano, A.; Ausanio, G.; Lanotte, L.; Aquilanti, G.; Coey, J.M.D.; Lantieri, M.; Spina, G.; Fittipaldi, M.; Margaris, G.; et al. Fe-Doping-Induced Magnetism in Nano-Hydroxyapatites. *Inorg. Chem.* **2017**, *56*, 4446–4458. [CrossRef] [PubMed]

129. Mastrogiacomo, M.; Corsi, A.; Francioso, E.; Di Comite, M.; Monetti, F.; Scaglione, S.; Favia, A.; Crovace, A.; Bianco, P.; Cancedda, R. Reconstruction of extensive long bone defects in sheep using resorbable bioceramics based on silicon stabilized tricalcium phosphate. *Tissue Eng.* **2006**, *12*, 1261–1273. [CrossRef]

130. Gopi, D.; Shinyjoy, E.; Kavitha, L. Synthesis and spectral characterization of silver/magnesium co-substituted hydroxyapatite for biomedical applications. *Spectrochim. Acta Part A Mol. Biomol. Spectrosc.* **2014**, *127*, 286–291. [CrossRef] [PubMed]

131. Hurle, K.; Oliveira, J.M.; Reis, R.L.; Pina, S.; Goetz-Neunhoeffer, F. Ion-doped Brushite Cements for Bone Regeneration. *Acta Biomater.* **2021**, *123*, 51–71. [CrossRef] [PubMed]

132. Iafisco, M.; Ruffini, A.; Adamiano, A.; Sprio, S.; Tampieri, A. Biomimetic magnesium-carbonate-apatite nanocrystals endowed with strontium ions as anti-osteoporotic trigger. *Mater. Sci. Eng. C* **2014**, *35*, 212–219. [CrossRef] [PubMed]

133. Lim, P.N.; Teo, E.Y.; Ho, B.; Tay, B.Y.; Thian, E.S. Effect of silver content on the antibacterial and bioactive properties of silver-substituted hydroxyapatite. *J. Biomed. Mater. Res. Part A* **2013**, *101 A*, 2456–2464. [CrossRef]

134. Patel, N.; Best, S.M.; Bonfield, W.; Gibson, I.R.; Hing, K.A.; Damien, E.; Revell, P.A. A comparative study on the in vivo behavior of hydroxyapatite and silicon substituted hydroxyapatite granules. *J. Mater. Sci. Mater. Med.* **2002**, *13*, 1199–1206. [CrossRef]

135. Pietak, A.M.; Reid, J.W.; Stott, M.J.; Sayer, M. Silicon substitution in the calcium phosphate bioceramics. *Biomaterials* **2007**, *28*, 4023–4032. [CrossRef]

136. Shi, H.; Zhou, Z.; Li, W.; Fan, Y.; Li, Z.; Wei, J. Hydroxyapatite based materials for bone tissue engineering: A brief and comprehensive introduction. *Crystals* **2021**, *11*, 149. [CrossRef]

137. Gokcekaya, O.; Ueda, K.; Narushima, T.; Ergun, C. Synthesis and characterization of Ag-containing calcium phosphates with various Ca/P ratios. *Mater. Sci. Eng. C* **2015**, *53*, 111–119. [CrossRef] [PubMed]

138. Jadalannagari, S.; Deshmukh, K.; Ramanan, S.R.; Kowshik, M. Antimicrobial activity of hemocompatible silver doped hydroxyapatite nanoparticles synthesized by modified sol–gel technique. *Appl. Nanosci.* **2014**, *4*, 133–141. [CrossRef]

139. Sprio, S.; Preti, L.; Montesi, M.; Panseri, S.; Adamiano, A.; Vandini, A.; Pugno, N.M.; Tampieri, A. Surface Phenomena Enhancing the Antibacterial and Osteogenic Ability of Nanocrystalline Hydroxyapatite, Activated by Multiple-Ion Doping. *ACS Biomater. Sci. Eng.* **2019**, *5*, 5947–5959. [CrossRef] [PubMed]
140. Stanić, V.; Janaćković, D.; Dimitrijević, S.; Tanasković, S.B.; Mitrić, M.; Pavlović, M.S.; Krstić, A.; Jovanović, D.; Raičević, S. Synthesis of antimicrobial monophase silver-doped hydroxyapatite nanopowders for bone tissue engineering. *Appl. Surf. Science* **2011**, *257*, 4510–4518. [CrossRef]
141. Pors Nielsen, S. The biological role of strontium. *Bone* **2004**, *35*, 583–588. [CrossRef]
142. Ge, X.; Leng, Y.; Bao, C.; Xu, S.L.; Wang, R.; Ren, F. Antibacterial coatings of fluoridated hydroxyapatite for percutaneous implants. *J. Biomed. Mater. Res. Part A* **2010**, *95 A*, 588–599. [CrossRef]
143. Pina, S.; Torres, P.M.; Goetz-Neunhoeffer, F.; Neubauer, J.; Ferreira, J.M.F. Newly developed Sr-substituted α-TCP bone cements. *Acta Biomater.* **2010**, *6*, 928–935. [CrossRef]
144. Wang, L.; He, S.; Wu, X.; Liang, S.; Mu, Z.; Wei, J.; Deng, F.; Deng, Y.; Wei, S. Polyetheretherketone/nano-fluorohydroxyapatite composite with antimicrobial activity and osseointegration properties. *Biomaterials* **2014**, *35*, 6758–6775. [CrossRef]
145. Tampieri, A.; D'Alessandro, T.; Sandri, M.; Sprio, S.; Landi, E.; Bertinetti, L.; Panseri, S.; Pepponi, G.; Goettlicher, J.; Bañobre-López, M.; et al. Intrinsic magnetism and hyperthermia in bioactive Fe-doped hydroxyapatite. *Acta Biomater.* **2012**, *8*, 843–851. [CrossRef]
146. Degli Esposti, L.; Markovic, S.; Ignjatovic, N. Thermal crystallization of amorphous calcium phosphate combined with citrate and fluoride doping: A novel route to produce hydroxyapatite bioceramics. *J. Mater. Chem. B* **2021**, *10*. [CrossRef]
147. Yin, X.; Bai, Y.; Zhou, S.j.; Ma, W.; Bai, X.; Chen, W.d. Solubility, Mechanical and Biological Properties of Fluoridated Hydroxyapatite/Calcium Silicate Gradient Coatings for Orthopedic and Dental Applications. *J. Therm. Spray Technol.* **2020**, *29*, 471–488. [CrossRef]
148. He, W.; Zhou, Y.T.; Wamer, W.G.; Boudreau, M.D.; Yin, J.J. Mechanisms of the pH dependent generation of hydroxyl radicals and oxygen induced by Ag nanoparticles. *Biomaterials* **2012**, *33*, 7547–7555. [CrossRef]
149. Klammert, U.; Ignatius, A.; Wolfram, U.; Reuther, T.; Gbureck, U. In vivo degradation of low temperature calcium and magnesium phosphate ceramics in a heterotopic model. *Acta Biomater.* **2011**, *7*, 3469–3475. [CrossRef] [PubMed]
150. Cao, X.; Harris, W. Carbonate and magnesium interactive effect on calcium phosphate precipitation. *Environ. Sci. Technol.* **2008**, *42*, 436–442. [CrossRef]
151. Salimi, M.H.; Heughebaert, J.C.; Nancollas, G.H. Crystal Growth of Calcium Phosphates in the Presence of Magnesium Ions. *Langmuir* **1985**, *1*, 119–122. [CrossRef]
152. Wang, L.; Nancollas, G.H. Calcium orthophosphates: Crystallization and Dissolution. *Chem. Rev.* **2008**, *108*, 4628–4669. [CrossRef]
153. Cao, X.; Harris, W.G.; Josan, M.S.; Nair, V.D. Inhibition of calcium phosphate precipitation under environmentally-relevant conditions. *Sci. Total Environ.* **2007**, *383*, 205–215. [CrossRef] [PubMed]
154. Diallo-Garcia, S.; Laurencin, D.; Krafft, J.M.; Casale, S.; Smith, M.E.; Lauron-Pernot, H.; Costentin, G. Influence of magnesium substitution on the basic properties of hydroxyapatites. *J. Phys. Chem. C* **2011**, *115*, 24317–24327. [CrossRef]
155. Bianchi, M.; Degli Esposti, L.; Ballardini, A.; Liscio, F.; Berni, M.; Gambardella, A.; Leeuwenburgh, S.C.G.; Sprio, S.; Tampieri, A.; Iafisco, M. Strontium doped calcium phosphate coatings on poly(etheretherketone) (PEEK) by pulsed electron deposition. *Surf. Coat. Technol.* **2017**, *319*, 191–199. [CrossRef]
156. Guo, D.; Xu, K.; Zhao, X.; Han, Y. Development of a strontium-containing hydroxyapatite bone cement. *Biomaterials* **2005**, *26*, 4073–4083. [CrossRef]
157. Marie, P.J. Strontium as therapy for osteoporosis. *Curr. Opin. Pharmacol.* **2005**, *5*, 633–636. [CrossRef]
158. Montesi, M.; Panseri, S.; Dapporto, M.; Tampieri, A.; Sprio, S. Sr-substituted bone cements direct mesenchymal stem cells, osteoblasts and osteoclasts fate. *PLoS ONE* **2017**, *12*, e0172100. [CrossRef]
159. Sprio, S.; Dapporto, M.; Montesi, M.; Panseri, S.; Lattanzi, W.; Pola, E.; Logroscino, G.; Tampieri, A. Novel Osteointegrative Sr-Substituted Apatitic Cements Enriched with Alginate. *Materials* **2016**, *9*, 763. [CrossRef]
160. Landi, E.; Tampieri, A.; Celotti, G.; Sprio, S.; Sandri, M.; Logroscino, G. Sr-substituted hydroxyapatites for osteoporotic bone replacement. *Acta Biomater.* **2007**, *3*, 961–969. [CrossRef]
161. Biris, A.R.; Mahmood, M.; Lazar, M.D.; Dervishi, E.; Watanabe, F.; Mustafa, T.; Baciut, G.; Baciut, M.; Bran, S.; Ali, S.; et al. Novel multicomponent and biocompatible nanocomposite materials based on few-layer graphenes synthesized on a gold/hydroxyapatite catalytic system with applications in bone regeneration. *J. Phys. Chem. C* **2011**, *115*, 18967–18976. [CrossRef]
162. Liang, W.; Gao, M.; Lou, J.; Bai, Y.; Zhang, J.; Lu, T.; Sun, X.; Ye, J.; Li, B.; Sun, L.; et al. Integrating silicon/zinc dual elements with PLGA microspheres in calcium phosphate cement scaffolds synergistically enhances bone regeneration. *J. Mater. Chem. B* **2020**. [CrossRef] [PubMed]
163. Zhu, J.; Wong, H.M.; Yeung, K.W.K.; Tjong, S.C. Spark plasma sintered hydroxyapatite/graphite nanosheet and hydroxyapatite/multiwalled carbon nanotube composites: Mechanical and in vitro cellular properties. *Adv. Eng. Mater.* **2011**, *13*, 336–341. [CrossRef]
164. Dong, G.; Zheng, Y.; He, L.; Wu, G.; Deng, C. The effect of silicon doping on the transformation of amorphous calcium phosphate to silicon-substituted α-tricalcium phosphate by heat treatment. *Ceram. Int.* **2016**, *42*, 883–890. [CrossRef]

165. Mestres, G.; Le Van, C.; Ginebra, M.P. Silicon-stabilized α-tricalcium phosphate and its use in a calcium phosphate cement: Characterization and cell response. *Acta Biomater.* **2012**, *8*, 1169–1179. [CrossRef] [PubMed]

166. Feng, Q.; Liu, Y.; Huang, J.; Chen, K.; Huang, J.; Xiao, K. Uptake, distribution, clearance, and toxicity of iron oxide nanoparticles with different sizes and coatings. *Sci. Rep.* **2018**, *8*, 2082. [CrossRef]

167. Manchón, A.; Hamdan Alkhraisat, M.; Rueda-Rodriguez, C.; Prados-Frutos, J.C.; Torres, J.; Lucas-Aparicio, J.; Ewald, A.; Gbureck, U.; López-Cabarcos, E. A new iron calcium phosphate material to improve the osteoconductive properties of a biodegradable ceramic: A study in rabbit calvaria. *Biomed. Mater.* **2015**, *10*, 055012. [CrossRef] [PubMed]

168. Perez, R.A.; Patel, K.D.; Kim, H.W. Novel magnetic nanocomposite injectables: Calcium phosphate cements impregnated with ultrafine magnetic nanoparticles for bone regeneration. *RSC Adv.* **2015**, *5*, 13411–13419. [CrossRef]

169. Xu, C.; Zheng, Y.; Gao, W.; Xu, J.; Zuo, G.; Chen, Y.; Zhao, M.; Li, J.; Song, J.; Zhang, N.; et al. Magnetic Hyperthermia Ablation of Tumors Using Injectable Fe3O4/Calcium Phosphate Cement. *ACS Appl. Mater. Interfaces* **2015**, *7*, 13866–13875. [CrossRef]

170. Panseri, S.; Cunha, C.; D'Alessandro, T.; Sandri, M.; Giavaresi, G.; Marcacci, M.; Hung, C.T.; Tampieri, A. Intrinsically superparamagnetic Fe-hydroxyapatite nanoparticles positively influence osteoblast-like cell behaviour. *J. Nanobiotechnol.* **2012**, *10*, 1. [CrossRef]

171. Degli Esposti, L.; Adamiano, A.; Tampieri, A.; Ramirez-Rodriguez, G.B.; Siliqi, D.; Giannini, C.; Ivanchenko, P.; Martra, G.; Lin, F.H.; Delgado-López, J.M.; et al. Combined Effect of Citrate and Fluoride Ions on Hydroxyapatite Nanoparticles. *Cryst. Growth Des.* **2020**, *20*, 3163–3172. [CrossRef]

172. Guo, H.; Wei, J.; Yuan, Y.; Liu, C. Development of calcium silicate/calcium phosphate cement for bone regeneration. *Biomed. Mater.* **2007**, *2*, S153–S159. [CrossRef]

173. Sprio, S.; Tampieri, A.; Celotti, G.; Landi, E. Development of hydroxyapatite/calcium silicate composites addressed to the design of load-bearing bone scaffolds. *J. Mech. Behav. Biomed. Mater.* **2009**, *2*, 147–155. [CrossRef] [PubMed]

174. Hesaraki, S.; Alizadeh, M.; Borhan, S.; Pourbaghi-Masouleh, M. Polymerizable nanoparticulate silica-reinforced calcium phosphate bone cement. *J. Biomed. Mater. Res. Part B Appl. Biomater.* **2012**, *100 B*, 1627–1635. [CrossRef]

175. Mohammadi, M.; Hesaraki, S.; Hafezi-Ardakani, M. Investigation of biocompatible nanosized materials for development of strong calcium phosphate bone cement: Comparison of nano-titania, nano-silicon carbide and amorphous nano-silica. *Ceram. Int.* **2014**, *40*, 8377–8387. [CrossRef]

176. Golzar, H.; Mohammadrezaei, D.; Yadegari, A.; Rasoulianboroujeni, M.; Hashemi, M.; Omidi, M.; Yazdian, F.; Shalbaf, M.; Tayebi, L. Incorporation of functionalized reduced graphene oxide/magnesium nanohybrid to enhance the osteoinductivity capability of 3D printed calcium phosphate-based scaffolds. *Compos. Part B Eng.* **2020**, *185*, 107749. [CrossRef]

177. Shi, Y.Y.; Li, M.; Liu, Q.; Jia, Z.J.; Xu, X.C.; Cheng, Y.; Zheng, Y.F. Electrophoretic deposition of graphene oxide reinforced chitosan–hydroxyapatite nanocomposite coatings on Ti substrate. *J. Mater. Sci. Mater. Med.* **2016**, *27*, 1–13. [CrossRef] [PubMed]

178. Wang, S.; Zhang, S.; Wang, Y.; Sun, X.; Sun, K. Reduced graphene oxide/carbon nanotubes reinforced calcium phosphate cement. *Ceram. Int.* **2017**, *43*, 13083–13088. [CrossRef]

179. Medvecky, L.; Stulajterova, R.; Giretova, M.; Sopcak, T.; Faberova, M. Properties of CaO-SiO$_2$-P$_2$O$_5$ reinforced calcium phosphate cements and in vitro osteoblast response. *Biomed. Mater.* **2017**, *12*, aa5b3b. [CrossRef]

180. Yu, L.; Li, Y.; Zhao, K.; Tang, Y.; Cheng, Z.; Chen, J.; Zang, Y.; Wu, J.; Kong, L.; Liu, S.; et al. A Novel Injectable Calcium Phosphate Cement-Bioactive Glass Composite for Bone Regeneration. *PLoS ONE* **2013**, *8*, e62570. [CrossRef] [PubMed]

181. Aparicio, J.L.; Rueda, C.; Manchón, Á.; Ewald, A.; Gbureck, U.; Alkhraisat, M.H.; Jerez, L.B.; Cabarcos, E.L. Effect of physicochemical properties of a cement based on silicocarnotite/calcium silicate on in vitro cell adhesion and in vivo cement degradation. *Biomed. Mater.* **2016**, *11*, 045005. [CrossRef]

182. Sopcak, T.; Medvecky, L.; Giretova, M.; Stulajterova, R.; Molcanova, Z.; Podobova, M.; Girman, V. Physical, mechanical and in vitro evaluation of a novel cement based on akermantite and dicalcium phosphate dihydrate phase. *Biomed. Mater.* **2019**, *14*, ab216d. [CrossRef]

183. Kapat, K.; Shubhra, Q.T.H.; Zhou, M.; Leeuwenburgh, S. Piezoelectric Nano-Biomaterials for Biomedicine and Tissue Regeneration. *Adv. Funct. Mater.* **2020**, *30*, 201909045. [CrossRef]

184. Polley, C.; Distler, T.; Detsch, R.; Lund, H.; Springer, A.; Boccaccini, A.R.; Seitz, H. 3D printing of piezoelectric barium titanate-hydroxyapatite scaffolds with interconnected porosity for bone tissue engineering. *Materials* **2020**, *13*, 1773. [CrossRef]

185. Li, M.; Xiong, P.; Yan, F.; Li, S.; Ren, C.; Yin, Z.; Li, A.; Li, H.; Ji, X.; Zheng, Y.; et al. An overview of graphene-based hydroxyapatite composites for orthopedic applications. *Bioact. Mater.* **2018**, *3*, 1–18. [CrossRef]

186. Tanaka, M.; Aoki, K.; Haniu, H.; Kamanaka, T.; Takizawa, T.; Sobajima, A.; Yoshida, K.; Okamoto, M.; Kato, H.; Saito, N. Applications of carbon nanotubes in bone regenerative medicine. *Nanomaterials* **2020**, *10*, 659. [CrossRef] [PubMed]

187. Zhang, Y.; Nayak, T.R.; Hong, H.; Cai, W. Graphene: A versatile nanoplatform for biomedical applications. *Nanoscale* **2012**, *4*, 3833–3842. [CrossRef] [PubMed]

188. Barabás, R.; de Souza Ávila, E.; Ladeira, L.O.; Antônio, L.M.; Tötös, R.; Simedru, D.; Bizo, L.; Cadar, O. Graphene Oxides/Carbon Nanotubes–Hydroxyapatite Nanocomposites for Biomedical Applications. *Arab. J. Sci. Eng.* **2020**, *45*, 219–227. [CrossRef]

189. Gonçalves, E.M.; Oliveira, F.J.; Silva, R.F.; Neto, M.A.; Fernandes, M.H.; Amaral, M.; Vallet-Regí, M.; Vila, M. Three-dimensional printed PCL-hydroxyapatite scaffolds filled with CNTs for bone cell growth stimulation. *J. Biomed. Mater. Res. Part B Appl. Biomater.* **2016**, *104*, 1210–1219. [CrossRef] [PubMed]

190. Han, Y.H.; Gao, R.; Bajpai, I.; Kim, B.N.; Yoshida, H.; Nieto, A.; Son, H.W.; Yun, J.; Jang, B.K.; Jhung, S.; et al. Spark plasma sintered bioceramics–from transparent hydroxyapatite to graphene nanocomposites: A review. *Adv. Appl. Ceram.* **2020**, *119*, 57–74. [CrossRef]
191. Jakus, A.E.; Shah, R.N. Multi and mixed 3D-printing of graphene-hydroxyapatite hybrid materials for complex tissue engineering. *J. Biomed. Mater. Res. Part A* **2017**, *105*, 274–283. [CrossRef]
192. Liu, Y.; Dang, Z.; Wang, Y.; Huang, J.; Li, H. Hydroxyapatite/graphene-nanosheet composite coatings deposited by vacuum cold spraying for biomedical applications: Inherited nanostructures and enhanced properties. *Carbon* **2014**, *67*, 250–259. [CrossRef]
193. Mata, D.; Oliveira, F.J.; Neto, M.A.; Belmonte, M.; Bastos, A.C.; Lopes, M.A.; Gomes, P.S.; Fernandes, M.H.; Silva, R.F. Smart electroconductive bioactive ceramics to promote in situ electrostimulation of bone. *J. Mater. Chem. B* **2015**, *3*, 1831–1845. [CrossRef]
194. Zhang, L.; Liu, W.; Yue, C.; Zhang, T.; Li, P.; Xing, Z.; Chen, Y. A tough graphene nanosheet/hydroxyapatite composite with improved in vitro biocompatibility. *Carbon* **2013**, *61*, 105–115. [CrossRef]
195. Wang, S.; Sun, X.; Wang, Y.; Sun, K.; Bi, J. Properties of reduced graphene/carbon nanotubes reinforced calcium phosphate bone cement in a microwave environment. *J. Mater. Sci. Mater. Med.* **2019**, *30*, 37. [CrossRef]
196. Dapporto, M.; Tampieri, A.; Sprio, S. Composite Calcium Phosphate/Titania Scaffolds in Bone Tissue Engineering. In *Application of Titanium Dioxide*; InTechOpen: London, UK, 2017.
197. Roguska, A.; Pisarek, M.; Andrzejczuk, M.; Dolata, M.; Lewandowska, M.; Janik-Czachor, M. Characterization of a calcium phosphate-TiO2 nanotube composite layer for biomedical applications. *Mater. Sci. Eng. C* **2011**, *31*, 906–914. [CrossRef]
198. Cunha, C.; Sprio, S.; Panseri, S.; Dapporto, M.; Marcacci, M.; Tampieri, A. High biocompatibility and improved osteogenic potential of novel Ca-P/titania composite scaffolds designed for regeneration of load-bearing segmental bone defects. *J. Biomed. Mater. Res. Part A* **2013**, *101 A*, 1612–1619. [CrossRef]
199. Sprio, S.; Guicciardi, S.; Dapporto, M.; Melandri, C.; Tampieri, A. Synthesis and mechanical behavior of β-tricalcium phosphate/titania composites addressed to regeneration of long bone segments. *J. Mech. Behav. Biomed. Mater.* 2013 17, 1–10. [CrossRef]
200. Liu, H.; Guan, Y.; Wei, D.; Gao, C.; Yang, H.; Yang, L. Reinforcement of injectable calcium phosphate cement by gelatinized starches. *J. Biomed. Mater. Res. Part B Appl. Biomater.* **2016**, *104*, 615–625. [CrossRef] [PubMed]
201. Tandon, B.; Blaker, J.J.; Cartmell, S.H. Piezoelectric materials as stimulatory biomedical materials and scaffolds for bone repair. *Acta Biomater.* **2018**, *73*, 1–20. [CrossRef] [PubMed]

Journal of
Composites Science

Review

Toughening of Bioceramic Composites for Bone Regeneration

Zahid Abbas †, Massimiliano Dapporto *,†, Anna Tampieri and Simone Sprio *

Institute of Science and Technology for Ceramics-National Research Council of Italy (ISTEC-CNR),
48018 Faenza, Italy; zahid.abbas@istec.cnr.it (Z.A.); anna.tampieri@istec.cnr.it (A.T.)
* Correspondence: massimiliano.dapporto@istec.cnr.it (M.D.); simone.sprio@istec.cnr.it (S.S.);
   Tel.: +39-054-669-9760 (M.D.)
† Co-first author, these authors contributed equally to this work.

Abstract: Bioceramics are widely considered as elective materials for the regeneration of bone tissue, due to their compositional mimicry with bone inorganic components. However, they are intrinsically brittle, which limits their capability to sustain multiple biomechanical loads, especially in the case of load-bearing bone districts. In the last decades, intense research has been dedicated to combining processes to enhance both the strength and toughness of bioceramics, leading to bioceramic composite scaffolds. This review summarizes the recent approaches to this purpose, particularly those addressed to limiting the propagation of cracks to prevent the sudden mechanical failure of bioceramic composites.

Keywords: bioceramics; mechanical properties; calcium phosphate; carbon fibers; mineralization

Citation: Abbas, Z.; Dapporto, M.; Tampieri, A.; Sprio, S. Toughening of Bioceramic Composites for Bone Regeneration. *J. Compos. Sci.* **2021**, *5*, 259. https://doi.org/10.3390/jcs5100259

Academic Editor: Francesco Tornabene

Received: 29 July 2021
Accepted: 24 September 2021
Published: 29 September 2021

Publisher's Note: MDPI stays neutral with regard to jurisdictional claims in published maps and institutional affiliations.

1. Introduction

Bone tissue is classified as a calcified connective tissue with several important roles in the human body, including storing minerals, protecting vital organs, enabling movement, providing internal support, and providing the sites of attachment for muscles and tendons [1,2]. Bone can be considered as a natural composite made of inorganic components (naturally doped calcium phosphates, ~70 wt %), organics (Collagen Type I, non-collagenous proteins, proteoglycans, cells, ~22 wt %), and water (~8 wt %) [2–4].

The complex metabolism and 3D hierarchic structure of bone tissue give it an innate ability to heal from minor defects. However, the natural healing process of bone is limited when major injuries due to traumas or metabolic or neoplastic bone pathologies occur [1,2]. In such instances, the orthopedic surgeon is challenged to find out adequate regenerative approaches [3]. The use of natural bone grafting (i.e., autologous or heterologous bone) can be pursued to replace the bone defect. Autografts are considered as the gold standard for bone grafting, as they closely resemble the natural bone structure, without immunogenic response. Despite these benefits, some limitations are evident, including the morbidity of the donor site, increased operation time, increased blood loss, and risk of immunogenicity and pathogenicity [4–7]. In addition, the sterilization and irradiation processes of natural bone grafts have been reported as critical steps that limit their bioactivity [8–11].

In this context, a great deal of research effort has been devoted in the last decades towards the synthesis of synthetic scaffolds [12–15]. The naturally occurring mineral phase in bone tissue is represented by poorly crystalline calcium phosphates with the crystal structure of hydroxyapatite (HA). HA can be synthesized in laboratory, and it is currently under study for the development of bone grafts, due to its excellent biocompatibility, osteoconductivity, and osteoinductivity [16–22].

In the last decades, bioceramics have been considered as ideal candidates for bone grafting due to their ability to locally deliver biomolecules in vivo. Calcium phosphates are a major member of bioceramics, covering a wide range of biomedical applications in tissue engineering, including orthopedic and dental surgeries [23–26].

Bioceramics must meet strict criteria to be approved for biomedical applications, such as biocompatibility, bioactivity, and an absence of proinflammatory features [27]. The classification of bioceramics is generally based on their chemical composition, as well as on the basis of their interaction with natural tissues; thus, bioceramics can be considered as bioinert or bioactive, considering biodegradability as an added value that enables the replacement of damaged bone parts with new ones during the scaffold bioresorption [28–32]. In this respect, recent studies demonstrated that the modulation of composition and textural properties can be considered as a valuable strategy to control material resorption and bone formation [33,34].

Bioinert ceramics, including alumina, zirconia, and silicon nitride, are not able to undergo any modification upon implantation, and thus maintain their chemical structure and represent a foreign body within the biological environment [35,36]. In contrast, bioactive ceramics have the capability to form chemical bonds with the surrounding tissues, and actively interact with the surrounding environment [37,38]. Among them, calcium phosphates (CaPs), bioglasses, and calcium silicate (Ca-Si) bioceramics are intensively studied for skeletal bone regeneration applications [22,39–43].

The osteogenic capability of bioceramic scaffolds is significantly correlated to their intrinsic pore size distribution and interconnection, enabling cell infiltration, migration, and neo-vascularization. The pore distribution and geometry of the scaffold strongly influence the ability of cells to penetrate, proliferate, and differentiate as well as the rate of scaffold degradation [44–48]. In spite of their great potential, a main drawback associated with bioceramics is their intrinsic brittleness, i.e., incapacity to withstand deformation without rupture, which is a major problem that can potentially cause a sudden failure of the scaffold structure under physiological mechanical loading. This is particularly relevant for porous calcium phosphates that associate brittleness to limited fracture strength, in comparison with inert ceramics such as zirconia or alumina [38,49,50]. In the last century, an intense research effort has been devoted to the reinforcement of bioceramics for different applications. In this respect, various approaches have been proposed, including modified sintering treatments [51–53], combination with polymeric phases to produce composites [49,54,55], the addition of fibers or the development of additive manufacturing as a 3D technique to prepare complex-shaped bioceramic structures [34,56–60]. A major approach to this purpose is the addition of ceramic particles, whiskers, and fibers to the ceramic matrix to improve the fracture toughness [61–65]. Ceramic fibers selected for their lightweight, adequate strength and modulus, and biocompatibility have been tested in the last decade for improving the mechanical properties of bioceramics [66–69]. The key factor influencing the performance of the final material is represented by the interfacial adhesion between fibers and the surrounding matrix [70]. The main factors affecting the fabrication of fiber-reinforced scaffolds include the chemical composition of fibers and matrix, the physical interaction between them, and the amount and alignment of fibers [59,71,72]. These factors affect the mechanical strength and degradation properties of the scaffold, leading to changes in the cell response. In this respect, many studies have reported the biocompatibility of fiber-reinforced ceramics both in vitro and in vivo [73–77].

It was observed that smooth fibers with a chemically inert surface are provided with less reactive functional groups, resulting in poor adhesion with the matrix [78]. Some studies reported chemical approaches to activate the fibers' surfaces, in order to strengthen this interaction [59,79–85].

The present review summarizes the relevant progress made on the mechanical reinforcement of bioceramic composites. The fabrication techniques for these scaffolds, along with the current strategies for toughening mechanisms, are described. Furthermore, the concerns related to porosity along with the mechanical and biological properties of fibrous ceramics are reported. As the advances in bone tissue engineering move toward application in the clinical setting, achieving adequate bioceramic toughness for clinical applications is particularly critical. In this context, recent computational approaches have been proposed in order to predict the crack propagation pathways, while increasing the toughness of ceramic-based bioinspired materials [86].

2. Bioceramic Composites in Bone Regeneration

2.1. Bone Tissue Formation and Remodeling

The adult human skeleton is made up of 206 bones. Each bone is a very complex hierarchical structure consisting of osteon, lamellar, fibrils, and mineral and collagen fibers (Figure 1). The bone is a dynamic tissue that continuously remodels during the life span of an individual: the term "bone remodeling" refers to a complex biochemical process involving the degradation of the mineralized bone via osteoclasts followed by the deposition of newly formed bone matrix by osteoblasts [87]. Due to this remodeling, the timing for complete bone tissue renewal is about 5 to 10 years [88–90]. This helps it in adapting to ever changing biomechanical forces by replacing the old or micro-damaged bone with a new and mechanically stronger bone, thereby preserving bone strength [91]. The bone has a unique ability to shift the intricate balance between osteoclastic and osteoblastic activities depending upon the external stimuli [92–94]. Such mechano-transduction signals can amplify the osteoblastic activity, resulting in an enhanced deposition of bone matrix [44]. In other circumstances, this equilibrium can be triggered by a chemical rather than a mechanical signal [44,95,96]. After receiving the mechanical signal, osteoclasts are deployed to the bone to initiate resorption. This process results in the release of calcium or phosphate to the body fluid as it is crucial for specific metabolic reactions. It is also assumed that chemokines are responsible for the differentiation-fusion of monocytes into osteoclasts and for carrying out the subsequent osteoclastic activity [97–99].

In this context, the regulation of osteoblasts-osteoclasts mediated processes plays a key role in achieving effective bone tissue regeneration [100]. Bioceramic composites can be engineered for better resorption and bone remodeling by mixing different ceramic materials [101]. The incorporation of strontium (Sr) into bioceramic composites can improve the bone tissue density via increasing osteoblast function and inhibiting osteoclast activity [102–105]. In addition, the surface topography also affects the resorption capacity of osteoclasts [104,105]. It was observed that human peripheral blood monocyte derived osteoclasts were more actively resorbed onto sub-micro structured β-TCP compared to microscale topography [106].

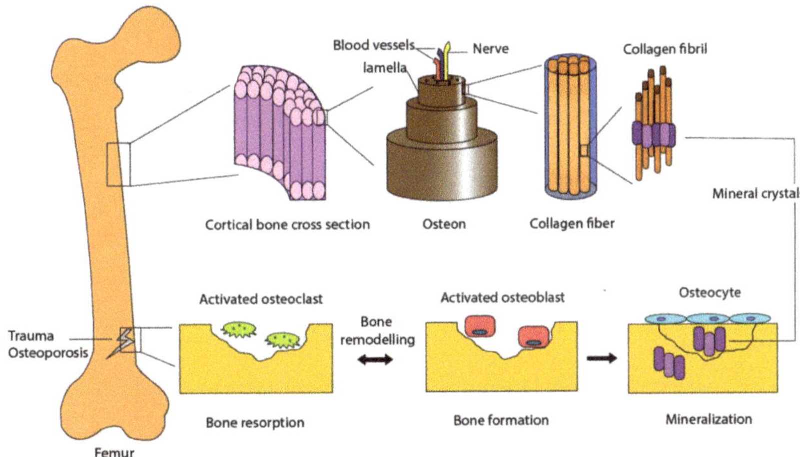

Figure 1. Hierarchical structure and mechanism of the formation and remodeling of long bone.

2.2. Classification of Bioceramic Composites

One major classification of bioceramics relies on the biochemical reactions occurring between the implanted scaffold and the surrounding tissue, particularly on the increasing capacity to be resorbed upon implantation in vivo. In this way, bioceramics can be considered as bioinert, bioactive, or bioresorbable materials [107]. Zirconia (ZrO_2) and alumina (Al_2O_3) are examples of inert bioceramics with minimal adverse reactions on tissue and body organs after coming into contact with the human physiology [108]. They have inherently low levels of reactivity compared with other materials, such as polymers and metals, as well as surface reactive or resorbable ceramics. Strategies for the improvement of the biocompatibility of inert ceramics were proposed, such as surface modifications, coatings, or ion doping [109–112].

In contrast, bioactive ceramics, such as bioglasses, show an ability to superficially bond with the surrounding bone, thus improving their interfacial strength [103,113]. The ability to bond to bone tissue is a unique property of bioactive ceramics. Analyses of the bone implant interface revealed that the presence of hydroxyapatite is one of the key features in the bonding zone [103].

Bioresorbable bioceramics represent a further improvement in their long-term interaction with surrounding tissues, because in addition to their chemical similarity to the mineral component of bone, they are able to be gradually resorbed and replaced by new bone tissue over time. The in vivo behavior of ceramic bone substitutes includes three main steps: (i) solubility: if the compound is soluble in physiological conditions, dissolution and removal can occur; (ii) the dissolution kinetics, related to the speed at which the particular ceramic is removed from the body; and (iii) conversion into another compound via a dissolution–precipitation mechanism [114]. Bioresorbable bioceramics are represented by calcium sulphates (Plaster of Paris) and calcium phosphates, especially with ion doping with Sr, Mg, Si, and Zn, which can improve their biological activity [19,115–120]. However, there are some drawbacks of calcium phosphates, such as their poor mechanical strength, differences between the bone regeneration and degradation rate, inflammatory reaction of synthetic bioceramics, and limited ingrowth due to pore size [121–123].

In this context, the possibility to introduce additional inorganic phases in bioceramics opened a wider choice of materials for their use as implants. Some of these materials include ceramic/ceramic, ceramic/metal, and ceramic/polymer composites. However, ceramic/polymer composites have been observed to release toxic components in the surrounding tissues, while metals undergo corrosion-related issues as the ceramic coating on the metallic implant degrades over time [124–126]. Ceramic/ceramic composites are thought to have a better performance because they resemble bone minerals and exhibit high biocompatibility [127–129]. Nevertheless, the biological activity of bioceramic composites has to be defined, especially considering the specific implantation site [130]. Bioceramic composites have exceptional biocompatibility and are non-toxic [131–133]. Some additional features of bioceramics composites include their hydrophilicity and antibacterial properties [134–136].

2.3. Surface Chemistry

The effective chemical interaction between the surfaces of the implanted scaffolds and the surrounding tissues plays a crucial role in the regeneration of bone tissue [137,138]. The three different types of bioceramics (bioinert, bioresorbable, and bioactive) have significantly different superficial interactions [139]. It is worth mentioning here that these fundamental differences in surface chemistry result in different interacting conditions at the biomolecular interface with cells and proteins [23,140,141].

The implantation of a scaffold into a biological environment is followed by the hydration and hydrolysis of the surfaces, typically leading to the formation of chemical bonds, containing either hydrogen or hydroxyl groups, with a rate dependent on the pH environment [142].

The superficial properties essentially modulate the interaction with water molecules and the mechanisms of adsorption of biological macromolecules (e.g., proteins). This interaction ultimately determines the interplay between the implanted bioceramics and bone cells.

It was reported that an electrostatic attraction primarily affects the protein adsorption on bioceramic surfaces, and effective surface charge modulation can be achieved by the immobilization of biomolecules such as bisphosphonates (BPs), amino acids, or carboxylic acid on the bioceramic surface [140]. Additional factors affecting the protein adsorption and cell adhesion include surface wettability and surface energy. The tuning of the chemical and morphological features of bioceramics can be performed by chemical or physical surface modifications, including atomic layer deposition, chemical vapor deposition, plasma vapor deposition, and electrochemical deposition [141,143,144].

Chemical treatments generally result in the formation of coating layers or the induction of specific chemical functional groups (e.g., carbidization, nitration, oxidation), while physical modifications result in micro- to nanoscale morphological or topographical alterations via a multitude of processes (e.g., machining, grit-blasting, and etching) [145–147].

2.4. Mechanical Properties

The mechanical properties of bioceramics, including compressive strength, stiffness, fracture toughness, and fatigue resistance, represent the key factors for effective bone regeneration [148]. These criteria include "static" mechanical properties (e.g., stiffness, hardness, strength), as well as "dynamic" mechanical properties (e.g., fatigue cycle resistance, crack propagation stability, and fracture toughness) [149].

A major concept in defining mechanical properties of ceramics is the difference between strength and toughness. They are frequently considered to be overlapped, despite the fact that they are mutually exclusive—strength is a stress representing the intrinsic capability of a material to resist to irreversible deformations, while toughness refers to the energy required to induce a fracture [150].

Toughness can also be determined using fracture mechanics methods, which determine the critical value of a crack-driving force, such as the stress intensity K, strain-energy release rate G, or nonlinear elastic J-integral, required to initiate and/or propagate a previously formed crack.

However, the intrinsic brittleness of ceramics basically limit the capability to improve the toughness, primarily because they cannot be toughened by promoting plasticity [151].

The compressive strength of the human cortical bone is reportedly in the range 90–209 MPa, while the reported flexural strength is 135–193 MPa [152,153]. The mechanical strength of bioceramics is reported to be in the range of cortical and cancellous bones [154]. The ideal scaffolds for bone regeneration should be designed considering this feature, but also considering that extensive bone penetration in a porous scaffold will increase the mechanical properties of the bone-scaffold construct until reaching physiological levels [44]. In particular, fracture toughness is important because it refers to the ability of the scaffold to contrast the propagation of a crack defect [155]. Hence, compressive strength and fracture toughness are relevant properties to be considered for effective bone regeneration [156,157]. The particle size, composition, porosity, and crystallinity of bioceramics significantly affect their mechanical performance—an increase of porosity and particle size leads to the decrease of mechanical properties [158–160].

The fracture toughness of cortical bone ($K_{Ic} = 2$–12 MPa·m$^{0.5}$) is higher than that of ceramics or inorganic glass [160–162]. Numerous methods have been developed over time to measure the fracture toughness and hardness [163–165]. The low fracture toughness and poor mechanical strength of bioceramics limits their usage in load-bearing applications [166,167]. It was reported that the fracture toughness and flexural strength of bioceramics increase in wet environments [168].

The toughness and flexibility of bone tissue can be ascribed to the complex biomineralization of collagen fibers with apatitic crystals, associated to the multi-scale hierarchical architecture [168]. The toughness of bioceramics can be improved by including additional biocompatible phases [169,170], crack bridging, or phase transformation, in order to control the crack growth [171–173]. The dispersion of second phase such as fibers, whiskers, and particles for creating toughness in bioceramics was also reported [174–177].

The mechanisms for increasing the toughness of ceramics can be classified as either intrinsic or extrinsic. Intrinsic toughening is primarily related to plasticity, that is, enlarging the plastic zone, mainly against the initiation of a crack. Conversely, extrinsic toughening acts to limit an initiated crack, reduce the stress and strain fields at the crack tip, preventing further opening, including crack bridging by fibers or ductile phases in composites.

A significant increase of flexural strength, flexural modulus, and fracture toughness of ceramic dental composites was also reported through the addition of zirconia-silica (ZS) or zirconia-yttria-silica (ZYS) nanofibers (2.5 wt % or 5.0 wt %) [178,179].

Bioceramic composites made from HA and TZP (tetragonal zirconia polycrystal) powders coated with Al_2O_3 also exhibited significantly higher strength and fracture toughness, due to the integration of ZrO_2 (15 vol %) and Al_2O_3 (30 vol %) [180].

The microstructural and mechanical changes of Al_2O_3 matrixes, after the incorporation of Cr_2O_3, was also studied, resulting in improved hardness and elastic modulus, while fracture toughness deteriorated with the addition of 2 mol % Cr_2O_3 particles [181].

It was also reported that Zr–Ti–Nb–Cu–Be glasses containing 42–67 vol % dendrites exhibited 100–160 $MPa\sqrt{m}$ toughness at tensile yield strengths of 1.1–1.5 GPa [182]. A monolithic and amorphous Pd–Ag–P–Si–Ge glass alloy with 1.5 GPa tensile strength and 200 $MPa{\cdot}m^{0.5}$ toughness properties was also recently reported; its properties were a result of the generation of shear band after loading, which resembles large-scale plasticity [183]. Nevertheless, it has drawbacks related to critical processing and production costs [150].

Moreover, researchers produced novel dental restorative composites by using hydroxyapatite whiskers. They reported that the efficiency of reinforcement depends on the filler morphology. Hydroxyapatite has good wettability with polymer which leads to increased toughness in comparison to nano-size HA powder [184,185]. Two composite materials have been produced by using ZrO_2-Al_2O_3 system: zirconia toughened alumina (ZTA) and alumina toughened zirconia (ATZ) [186–191]. The ZTA ceramic composites with 0.5 wt % MgO content exhibited the best attributes, such as a fracture toughness value of 9.14 $MPa{\cdot}m^{0.5}$ and a hardness value of 1591 HV. Similarly, the effect of TiO_2 phase composition and mechanical properties of Ca-TZP (calcium stabilized tetragonal zirconia) ceramic have been observed, with fracture toughness values up to 9.1 $MPa\sqrt{m}$ after reinforcement with TiO_2 in the range of 0.5–0.65 mol % [192].

A great research effort for the reinforcement of bioceramics with carbon fibers (CF) has been established, due to their excellent biocompatibility and mechanical properties [193–195]. The addition of CF to HA matrix effectively improves the bending strength and fracture toughness of HA [177,196]. ZrO_2-HAp composites (40 and 60 vol % of ZrO_2) were fabricated and evaluated, demonstrating the reinforcing effect of ZrO_2 [174].

3. Toughening Strategies for Bioceramic Composites

The brittleness of bioceramics significantly limits their applications because in addition to strength, adequate toughness is required to sustain the biomechanical loads [86].

Any crystallographic defect or irregularity within the crystal structure represents the main cause for dislocations, the mechanisms of which are related to the Peierls–Nabarro (PN) barrier energy that defines the fracture toughness of a material [197].

Metals contain mobile dislocations, leading to local plasticity and desirable toughness [197,198], while ceramics are characterize by the immobility of dislocations and low fracture toughness, especially at room temperature [199]. In this respect, the high-strength ionic bond typical for ceramic structures plays a crucial role, limiting atomic slip systems.

The mechanical performance of bioceramics is closely related to several factors, including microstructure, chemical composition, ionic impurities, and structural defects.

The strategies to improve the toughness or fracture strength of ceramics refer to the capacity to control or limit the propagation of cracks along the powder particles and grains.

Several methods have been reported to improve the toughness of ceramics, including crack-bridging, crack-deflection, microcrack-induced toughening, generation of phase transformations, and reduction of the defect size (Figure 2).

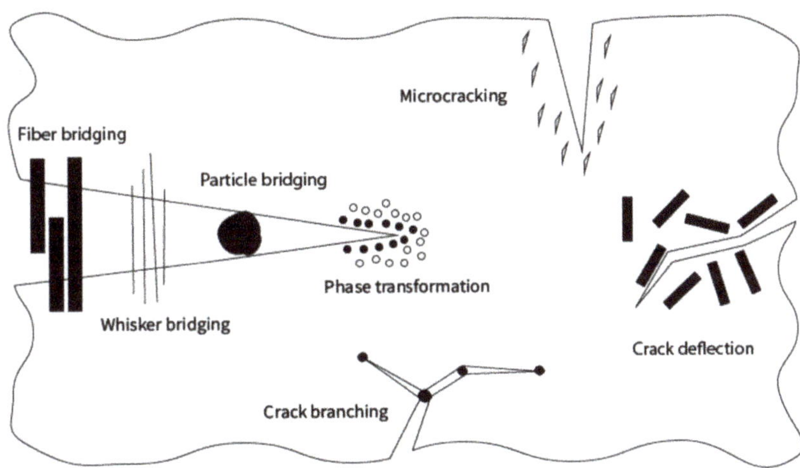

Figure 2. Different strategies to improve the toughness of bioceramics.

Moreover, the basic scaffold structure can be combined with polymer coatings, or interpenetrating polymer-bioactive ceramic microstructures can be formed to improve the toughness of the ceramic as a simple and effective approach [200].

3.1. Phase Transformation

Toughening phenomena related to phase transformation are well known for zirconia-containing composites, where the phase transition from the tetragonal to monoclinic phase is the essential mechanism behind the enhanced toughness of zirconia used for the development of dental and orthopedic implants. This approach is based on stress induced phase transformations, which are mainly responsible for microstructural additional compressive stresses during the propagation of cracks, which increase the crack growth resistance K_{Ic}.

Similar to precipitation hardening, the stabilization of particles in a metastable and thermodynamically unfavorable state requires overcoming an energy nucleation barrier. In this case, the modulation of particle volume can be achieved by the application of adequate tensile stress at the crack tip. Phase transformation is initiated by the presence of sufficiently large elastic energy. As the particle was metastable prior to the transformation, the decrease in stress due to an increase in volume does not hamper the process of transformation [201]. Moreover, compressive stress in radial direction and tensile stress in circumferential direction around a particle are superimposed to the external load during the transformation. These compressive residual stresses may result in the reduction of stress on the crack and hence may partially or completely close the crack. In addition, as the tensile stress is applied in circumferential direction around a particle it can generate microcracks, further leading to the dissipation of energy [202,203].

The stress induced transformation is also related to the free enthalpy reduction [204]. The addition of hydrostatic tensile stress strongly decreases the enthalpy of the phase with a larger volume. This, in turn, increases the driving force for the transformation, enabling the particle to overcome the nucleation barrier [205,206].

3.2. Defect Size Reduction

The main limitation of bioceramics' mechanical properties is related to their brittleness [207–209]. Ceramics generally exhibit higher compressive strength than tensile strength, essentially due to limitations in stress concentrations and crack propagation when micropores are flattened instead of dilated [210]. Bioceramics are characterized by very limited strain to failure and toughness, compared with ductile material counterparts (e.g., metals, polymers). Tensile stress could cause a fracture to propagate through the material, often causing failure in ceramic material.

Many defects can occur during the production, finishing, and application of ceramic because of the foreign particles, porous regions, or large grain sizes (Figure 3) [211].

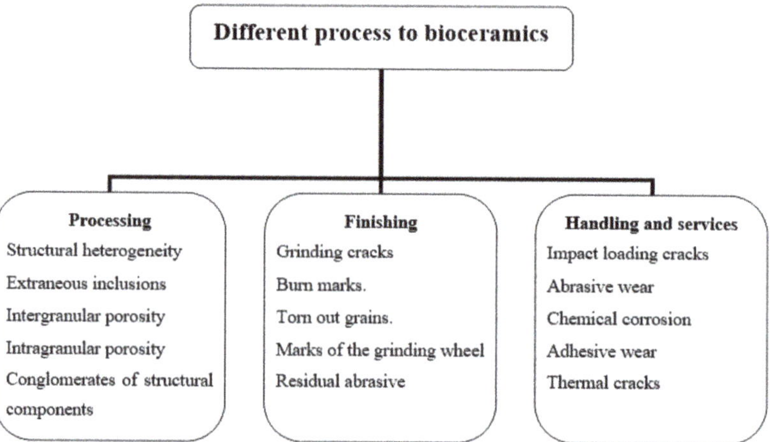

Figure 3. Different types of processing-derived defects in bioceramics.

A great research effort has been dedicated to the design of ceramic microstructures with increased toughness and damage tolerance. In this context, the reduction of defect sizes can also be obtained by the incorporation of various ions such as nickel, silver, tantalum [212], and strontium [212–214].

Some requirements for reducing the size of defects include an efficient, fast, and reliable fine grinding, a compact design, and the versatility of the process [215].

The fracture mechanics of bioceramics is mainly affected by the powder particle size distribution. Grain size is usually tuned towards monomodal or multimodal distributions, in order to increase the packing density of particles [216]. It was reported that the largest grain may control the size of largest flaw [217]. Alternatively, grain size can be measured at the fracture origin [218]. The microstructure is affected by multiple factors (e.g., powder impurities, thermal treatments of powders, sieving size), complicating the possibility to understand the role of each factor [24]. This methodology has been in use for the production of ceramics to obtain a more homogenous microstructure [219–221].

3.3. Crack Deflection

The propagation of cracks in bioceramics is a critical issue that can cause sudden failure of large structures. Crack deflection can be used as a strategy to increase the toughness of bioceramics. Some local areas in the bioceramics exhibit low resistance to crack propagation, resulting in crack deflection [222]. In particular, when a crack is deflected, the surface of the crack increases, leading to more energy required for crack propagation and an increase of fracture toughness [221,223]. The prediction of a crack path as the crack approaches a fiber can be based on an energy criterion or a stress criterion [217,219,220,224–227]. Young's modulus mismatches are also reported as mechanisms of crack deflection [211].

The microstructural paths for crack propagation in bioceramics generally reflect the grain boundaries [211,217,228–230].

The evaluation of crack deflection by disks, rods, and spheres [217] showed that (1) increased toughness as a result of crack deflection is dependent on particle shape and volume fraction, and is independent of particle size [231]; (2) a rod-shaped morphology is the most effective, followed by disk and sphere, for increasing the toughness [232]. An increase in volume fraction of up to 20% increases the toughness, however, very little increase was observed with a higher volume fraction.

The toughness significantly increases when using rods compared to toughness without deflecting particles [226,233]. Another cause of crack deflection is partial bridging by grains, occurring when a grain/whisker causes the deflection of a crack around it, hence leaving the grain/whisker to bridge the crack [234].

Interactions between crack bridging and crack deflection in silicon nitride containing rod-shaped grains and whiskers toughened alumina were observed, demonstrating that crack deflection is crucial for the development of crack bridging [235].

3.4. Microcrack Formation and Crack Branching

Stress-induced microcrack formation and crack branching represent irretrievable deformation phenomena associated with energy dissipation [151,236]. Microcracks appear to debond at a poorly bonded matrix particle interface [237,238]. The stress energy near the tip of the fracture can result in the formation of micro cracks at weak areas in the bioceramic, for example, due to the undesired orientation of grain boundaries.

It was observed that a microcrack can decreased crack resistance at the macrocrack tip, which encourages crack progression [239]. The microcrack's effect on fracture propagation can be examined in two ways: energy dissipation owing to microcrack generation [170] and change in local stress intensity factor by simulating the interaction of microcracks with the main cracks [240,241]. The crack shielding phenomena has a role in microcrack toughening because of two aspects: the material's lower elastic modulus as a result of microcracking and the microcracking-induced dilatation [242].

3.5. Crack Bridges Formation

A variety of reinforcing phases can improve the fracture toughness of bioceramic composites. The reinforcement in this case bridges the crack surfaces that effectively pins the crack and increases resistance for any further extension of the crack [243]. It was observed that these reinforcing phases bridge the crack in the region behind the crack tip [243]. When the two opposing crack surfaces interact during crack propagation, an increase in energy dissipation occurs during the propagation of a crack.

This behavior was observed in coarse-grained microstructures with intercrystallite crack propagation [243,244]. Different varieties of ceramic whiskers (high-strength crystals with length/diameter ratios of 10 or more), particulates, or fibers can be added to the matrix material of the host in order to generate a composite that can improve fracture toughness. This reinforcement strategy relies on two different mechanisms: (i) the presence of additional particles or fibers represents a deflection stimulus for opening cracks, against its propagation [238,245,246]; (ii) in case of weak bonds between the matrix and reinforcement phase, crack propagation energy can be absorbed by pulling out the fiber from its original location, thus preventing crack propagation by forming a bridge in a crack and holding the two face together [247].

3.5.1. Particles

Crack bridging is generally induced by the addition of particles that can confer ductile behavior (e.g., particles with lower Young's Modulus), as in this condition, additional work is required to achieve deformations and crack propagation [248]. Moreover, the addition of ductile particles in bioceramics can also significantly increase their fracture toughness by forming crack bridging behind the crack tip via a discontinuous but strong reinforcing second phase that imposes a closure force on the crack [249]. The mathematical description of non-linear fracture processes and stress transfer across cracks was proposed in the Dugdale–Barenblatt model, useful for estimating the effect of particles addition in increasing toughness [250]. This model encounters the behavior of crack extension when intersecting the particles: the primary crack propagation is impeded by particles, thus retarding its interaction with the surrounding cracks [246].

The doping of HA with strontium-doped particles can improve the mechanical properties [213,214]. The compressive strength was improved from 50 MPa to 66.57 MPa up to 5 mol % Sr/(Sr + Ca) doping [251]. The incorporation of Sr^{2+} in HA lattice replaces Ca^{2+} with Sr^{2+} and form a $Ca_{10-n}Sr_n(PO_4)_6(OH)_2$ (Sr-HA). This decreases the crystallization size and crystallization rate of HA and changes the lattice constant.

The addition of titanium particles was also proposed as a promising approach to improve the mechanical properties [252]. Titanium and its alloys are considered as some of the most attractive and important materials due to their unique properties, such as high tensile strength, resistance to body fluid effects, flexibility, and high corrosion resistance. They exhibit a unique combination of strength and biocompatibility, which makes them suitable for biomedical applications. Commercially pure titanium (c.p. Ti) is prominent in dental implants and Ti-6Al-4V is dominant in orthopedics applications [253].

3.5.2. Whiskers

A whisker is a single crystal in the form of a fiber. Whiskers can be considered as a sub-group of random fibers, possessing shorter lengths compared to conventional fibers. They are defect free and thus stronger and stiffer than fibers. Due to these properties, there is a more pronounced difference in the mechanical properties of a whisker when compared to bulk materials [254,255]. Materials are crystallized on a very small scale for the production of whiskers. Internal alignment within each whisker is observed to be extremely high. The processes that cause toughening in whisker-reinforced ceramics are considered to be fundamentally similar to those in ceramic matrix reinforced with aligned continuous fibers [256–258].

3.5.3. Nanosheets

Recently, regenerative medicine focused on the nanosheets applications owing to their excellent biocompatibility and unique mechanical and physicochemical properties. Two-dimensional (2D) structures of nanosheets (e.g., 1–100 nm thickness) are characterized by a large surface-to-volume ratio, ultrathin structure, and enhanced mechanical strength, which can be substituted with a large number of functional biomolecules [259–261]. They express a greater ability to interact with polymers through hydrophobic interaction, Van der Waals force, physical adsorption, and electrostatic attraction. The mechanical strength and biocompatibility of scaffolds can be improved by combinations of nanosheets with ceramics polymers [262].

Nanosheets are categorized into monolayered hydroxide nanosheets (MLDHs), polymeric nanosheets, metallic nanosheets, and nonmetallic nanosheets. Metallic and non-metallic nanosheets are used for tissue engineering. They have desirable features for tissue engineering, such as biocompatibility, mechanical strength, and photothermal and colloidal stability [78,145,263,264]. Molybdenum disulfide (MoS_2), manganese dioxide (MnO_2), and magnesium phosphate ($MgPO_4$) are frequently used metallic compounds, while the commonly used non-metallic components include graphene (GN), graphene oxide (GO), and black phosphorus (BP) [262,265].

Graphene oxide nanosheets (GOns) have improved mechanical ability, a large surface-to-volume ratio, protein adsorption, and biocompatibility, all of which are important properties required in tissue engineering [266]. Surface roughness, protein absorption, hydrophilicity, and cell adhesion can be improved by adding extracellular matrix (ECM) components such as Col and HAp to the above nanostructured composites [267].

Molybdenum disulfide (MoS_2) nanosheets exhibit excellent mechanical properties, including 300 GPa Young's modulus, a tensile strength of over 23 GPa, and excellent elasticity [266,268]. Graphene nanosheets are mostly used to strengthen HAp scaffolds. After nanosheets were incorporated into the scaffolds, the elastic modulus of the composite was increased by 40% to 141 ± 8.50 GPa and the fracture toughness was increased by 80% to 1.06 ± 0.03 MPa [269]. Plasma spray was used to create a graphene nanosheet (GNS) reinforced HA on Ti6Al4V substrate. The resulting GNS/HA composite coating has increased strength and toughness, with ~32.3% and ~54.7%, increases in fracture toughness and indentation yield strength, respectively [68]. The composite coating's improved strength and toughness were attributed to synergetic toughening and strengthening mechanisms such as load transfer, graphene nanosheet (GNS) pull-out, GNS inter-layer sliding, crack branching, and GNS bridging. Moreover, the frequent crack deflection when a crack comes into contact with GNS could tailor the trade-off between strength and toughness through crack branching and GNS bridging [68].

3.5.4. Fibers

Fibers in ceramic matrix composites (CMC) help increase the fracture toughness, due to their excellent mechanical properties [270]. Different types of fibers on the basis of length, i.e., particulates and fiber network, continuous fibers, and short fibers, can be used for the processing of ceramic matrix composites. For bridging by brittle short fibers, an increase in interfacial shear forces is observed until it either causes the particle to break or debond from the matrix. This interfacial debonding, when followed by the subsequent frictional pulling out process, has a great impact on the toughness of the material. Hull and Clyne (1996) expressed the fracture energy related to fibers pull-out with the following formula:

$$\Delta G_{PULL-OUT} = \int_0^l \frac{N\pi r x^2 \tau_i}{l} dx = \frac{V_f l^2 \tau_i}{3r} \tag{1}$$

where G represents the interfacial shear strength, r is fiber radius, l is pull-out length, and N is the number of fibers per unit area.

In bioceramics, the mechanism for fiber reinforcement involves fiber bridging the crack after its appearance due to stress, impeding its further propagation. Furthermore, the frictional sliding of fibers against the matrix during pullout further consumes the applied force that results in increased fracture toughness. The addition of different types of fibers (e.g., carbon, e-glass, aramid, and polyglactin) increased the strength of bioceramics and resulted in an increase of approximately two orders of magnitude in the fracture work [271].

Carbon fibers are the preferred choice among researchers compared to all other types of fibers due to their high strength-to-weight ratio, thermophysical properties, sorption, and high elastic modulus [272]. Carbon fibers are crystalline filaments of carbon that have a regular hexagonal pattern of carbon sheets. Moreover, due to their inherent biocompatibility (in vivo and in vitro), they are extensively used in the production of artificial heart valves, purulent wounds, in the treatment of bone fractures, and for making bio composites. Carbon fibers are produced by high temperature conversion during the pyrolysis of carbon-rich precursors.

The fracture toughness of bioceramic composites can be increased by adding carbon fibers. A 300% increase in fracture toughness of alumina-single-walled carbon nanotubes (SWCNTs) composites was reported [221]. In another report, a 69% improvement in fracture toughness for silica-CNT composites by loading only 0.05 wt % CNTs was obtained [273]. A significant increase in the fracture toughness of $BaTiO_3$-CNTs composites was described when loading 0.5, 1, and 3 wt %, respectively [274]. Wang also reported a moderate improvement of 15% for ZrB_2-SiC-multi-walled carbon nanotube (MWCNT) nanocomposite (2 wt %) [275]. He successfully manufactured composites comprising of micrometer-sized carbon fibers (CFs) and also made biocompatible nanocrystalline calcium hydroxyapatite that contained carbon fibers by 1.0, 2.0, and 5.0 wt %. Moreover, he reported the manufacturing of a HAp-carbon fiber composite via hot pressing by using high temperature, pressure, and argon atmosphere. The resulting bioceramic composite had improved fracture toughness and strength [276]. In another instance, the microabrasion resistance of carbon fiber based reinforced and non-reinforced hydroxyapatite was worked on. Commercial grade Hap and carbon fibers were used by hot pressing. The researchers used a temperature of 1000–1150 °C and 25 MPa pressure with 15 min pressing time in an argon atmosphere. Most researchers have used the microhardness indentation method to the measure fracture toughness (K_{Ic}) of carbon-based bioceramic composites due to the small sample sizes [277].

A chemical treatment performed to activate the fiber surface to improve the adhesion adhesion with surrounding matrix concerned the conditioning of the fibers surface, using molecules such as carboxylic acid, sulfuric acid, nitric acid, alkali, formaldehyde, and isocyanate [276,278–281].

The main drawbacks associated with the use of fiber-reinforced bioceramics include the tendency of fibers to agglomerate due to their high Van-der-Walls forces of interaction among carbon particles and light weight, and the low interfacial adhesion between the fibers and the matrix. This tendency to agglomerate has obstructed their application in various fields. In this context, surface functionalization/modification processes that can reduce this agglomeration tendency and increase the fiber–bioceramics interfacial adhesion through covalent or ionic bonding were proposed [276]. Several functionalization strategies were reported for fibers, including wet oxidation (oxidation using potassium permanganate, hydrogen peroxide, sulfuric acid, nitric acid, etc.), dry oxidation (oxidation by using plasma, air, ozone, etc.), surface adsorption, and encapsulation [282].

The oxidation of carbon fibers can be carried out in both wet and dry conditions [282]. Strong acids, such as H_2SO_4, HNO_3 or a mixture of the two with a strong oxidant, i.e., $KMnO_4$, is used for the wet oxidation of CFs and ozone or reactive plasma is used for the oxidation of CFs in dry conditions. Wet oxidation is the most cost-effective process for the surface modification of CFs. A few studies have indicated that the addition/activation of some functional groups on CF surface favors the bonding between bioceramics and carbon fibers; in particular, defects caused by oxidants on the surface of carbon fibers are stabilized by bonding with hydroxyl (-OH) or carboxylic acid (COOH) [282,283].

4. Processing Approaches towards Ceramic Toughening

Several approaches have been developed to improve the mechanical properties of bioceramics [26]. The accurate processing of toughened bioceramic composites involves many steps, from raw materials to the semi-finished processing, including the synthesis of powders, controlled drying, calcination, the debonding of organic components, the addition of second phases, and thermal sintering [284]. The intrinsic features of the ceramic powders significantly influence each physical (e.g., density, porosity), microstructural (e.g., shape of grains, grain size, grain boundaries), mechanical (e.g., strength, hardness, toughness, resistance to fatigue failure,), and chemical (e.g., dissolution, hydrolysis) property of the final bioceramic composite scaffold.

Essential criteria for the effective preparation and reinforcement of bioceramic composites are the homogeneous mixing of the matrix and reinforcement phase and a controlled particle size distribution to optimize the packing density of particles while avoiding agglomeration.

The preparation of powders involves several approaches, classified into dry and wet chemical methods. The formulation technique has a significant impact on surface characteristics, powder morphology, stoichiometry, and crystallinity. Dry methods involve three main types of chemical reactions: thermal decomposition, oxidation/reduction, and solid-state reactions. In contrast, various methods can be used for the liquid or wet reaction of bioceramic powders such as hydrothermal synthesis, precipitation, liquid drying, and sol–gel synthesis [24,285]. The preparation of bioceramic powders, in particular hydroxyapatite, mainly involves wet chemical methods, especially hydrothermal synthesis and solid-state reactions [286].

A promising approach for the toughening of bioceramics is the addition of carbon fibers into the matrix. The manipulation of carbon fibers can be performed by the solution powder mixing technique to prepare polymer–carbon composite materials [287] or biomimetic mineralization to improve the biocompatibility and bone inductivity [83,288].

The consolidation of bioceramic scaffolds is modulated by thermal treatments capable of improving the interactions among the particles. Sintering is a high-temperature treatment that can compact the ceramic particles of a pre-shaped *green body* or powder to consolidate a solid structure [289]. The major goal of sintering is the densification; fine and uniform microstructure and bioceramics are typically sintered at temperatures ranging from 500 to 1200 °C. The high temperature of sintering provides adequate energy to force material transport processes such as the migration of grain boundaries via the diffusion of atoms or evaporation–condensation phenomena, with the aim of reducing the superficial energy of ceramic particles and eliminating the pores [290]. The sintering can be performed in different atmospheres, including inert gas or air [291].

Semi-finished processing techniques for bioceramic composites involve a myriad of techniques, including hand layup, spray up, injection molding, resign transfer molding, compression molding, filament winding, and pultrusion, according to the type of filler (particles, whiskers, and fibers) [175,176].

A recently reported approach also explored the possibility to increase the toughness of bioceramics by introducing a large and controlled density of dislocations, thus leading to local plasticity [292]. It was observed that conventional sintering, the standard densification method for ceramics, actually yields ceramics virtually free of dislocations and dislocation sources. In other words, the brittleness of ceramics appears as merely a consequence of the established conventional production method.

5. Conclusions

The limited toughness of bioceramics highlights a relevant clinical need, especially when the regeneration of load-bearing bone portions is required. Despite the multitude of approaches that have been explored in the past decades, further research is still needed to improve the performance of sintered bioceramics for clinical use. In particular, fiber reinforcement is a promising approach, even though some critical issues still remain, mainly related to the achievement of a strong interface between fibers and the surrounding matrix and to the thermal fiber decomposition. In this respect, processes based on the activation of the fibers' surface or dislocation-toughening have been proposed and are promising for improving the reinforcement–matrix interface. Relevant research targets for material scientists in the future will be to focus on new forming processes that can generate reinforced ceramics with tailored porous architecture, thus enabling advanced applications in bone surgery.

Author Contributions: Conceptualization, M.D., S.S. and A.T.; writing—review and editing, Z.A., M.D.; supervision, S.S. and M.D. All authors have read and agreed to the published version of the manuscript.

Funding: This research received no external funding.

Conflicts of Interest: The authors declare no conflict of interest.

References

1. Annamalai, R.T.; Hong, X.; Schott, N.G.; Tiruchinapally, G.; Levi, B.; Stegemann, J.P. Injectable osteogenic microtissues containing mesenchymal stromal cells conformally fill and repair critical–size defects. *Biomaterials* **2019**, *208*, 32–44. [CrossRef]
2. Oryan, A.; Monazzah, S.; Bigham-Sadegh, A. Bone injury and fracture healing biology. *Biomed. Environ. Sci.* **2015**, *28*, 57–71.
3. Bose, S.; Sarkar, N. Natural medicinal compounds in bone tissue engineering. *Trends Biotechnol.* **2020**, *38*, 404–417. [CrossRef]
4. Shafiei, Z.; Bigham, A.; Dehghani, S.; Nezhad, S.T. Fresh cortical autograft versus fresh cortical allograft effects on experimental bone healing in rabbits: Radiological, histopathological and biomechanical evaluation. *Cell Tissue Bank.* **2009**, *10*, 19–26. [CrossRef] [PubMed]
5. Vaccaro, A.R. The role of the osteoconductive scaffold in synthetic bone graft. *Orthopedics* **2002**, *25*, s571–s578. [CrossRef] [PubMed]

6. Chiarello, E.; Cadossi, M.; Tedesco, G.; Capra, P.; Calamelli, C.; Shehu, A.; Giannini, S. Autograft, allograft and bone substitutes in reconstructive orthopedic surgery. *Aging Clin. Exp. Res.* **2013**, *25*, 101–103. [CrossRef] [PubMed]

7. Bal, Z.; Kaito, T.; Korkusuz, F.; Yoshikawa, H. Bone regeneration with hydroxyapatite-based biomaterials. *Emergent Mater.* **2020**, *3*, 521–544. [CrossRef]

8. Blokhuis, T.; Arts, J.C. Bioactive and osteoinductive bone graft substitutes: Definitions, facts and myths. *Injury* **2011**, *42*, S26–S29. [CrossRef]

9. Campana, V.; Milano, G.; Pagano, E.; Barba, M.; Cicione, C.; Salonna, G.; Lattanzi, W.; Logroscino, G. Bone substitutes in orthopaedic surgery: From basic science to clinical practice. *J. Mater. Sci. Mater. Med.* **2014**, *25*, 2445–2461. [CrossRef]

10. Haugen, H.J.; Lyngstadaas, S.P.; Rossi, F.; Perale, G. Bone grafts: Which is the ideal biomaterial? *J. Clin. Periodontol.* **2019**, *46*, 92–102. [CrossRef]

11. Jahangir, A.A.; Nunley, R.M.; Mehta, S.; Sharan, A. Bone-graft substitutes in orthopaedic surgery. *AAos Now* **2008**, *2*, 35–37.

12. Oryan, A.; Alidadi, S.; Moshiri, A.; Maffulli, N. Bone regenerative medicine: Classic options, novel strategies, and future directions. *J. Orthop. Surg. Res.* **2014**, *9*, 1–27. [CrossRef]

13. Matassi, F.; Nistri, L.; Paez, D.C.; Innocenti, M. New biomaterials for bone regeneration. *Clin. Cases Miner. Bone Metab.* **2011**, *8*, 21.

14. Basha, R.Y.; TS, S.K.; Doble, M. Design of biocomposite materials for bone tissue regeneration. *Mater. Sci. Eng. C* **2015**, *57*, 452–463. [CrossRef]

15. Ghanbari, H.; Vakili-Ghartavol, R. Bone regeneration: Current status and future prospects. *Adv. Tech. Bone Regen.* **2016**, 3–26. [CrossRef]

16. Eslami, H.; Tahriri, M.; Moztarzadeh, F.; Bader, R.; Tayebi, L. Nanostructured hydroxyapatite for biomedical applications: From powder to bioceramic. *J. Korean Ceram. Soc.* **2018**, *55*, 597–607. [CrossRef]

17. Boskey, A.; Camacho, N.P. FT-IR imaging of native and tissue-engineered bone and cartilage. *Biomaterials* **2007**, *28*, 2465–2478. [CrossRef]

18. Dorozhkin, S.V. *Calcium Orthophosphates: Applications in Nature, Biology, and Medicine*; CRC Press: Boca Raton, FL, USA, 2012.

19. Zhang, X.; Gubbels, G.M.; Terpstra, R.; Metselaar, R. Toughening of calcium hydroxyapatite with silver particles. *J. Mater. Sci.* **1997**, *32*, 235–243. [CrossRef]

20. Tsapikouni, T.S.; Missirlis, Y.F. Protein-material interactions: From micro-to-nano scale. *Mater. Sci. Eng. B* **2008**, *152*, 2–7. [CrossRef]

21. Dorozhkin, S.V. Amorphous calcium orthophosphates: Nature, chemistry and biomedical applications. *Int. J. Mater. Chem.* **2012**, *2*, 19–46. [CrossRef]

22. Jeong, J.; Kim, J.H.; Shim, J.H.; Hwang, N.S.; Heo, C.Y. Bioactive calcium phosphate materials and applications in bone regeneration. *Biomater. Res.* **2019**, *23*, 1–11. [CrossRef]

23. Brunello, G.; Panda, S.; Schiavon, L.; Sivolella, S.; Biasetto, L.; Del Fabbro, M. The impact of bioceramic scaffolds on bone regeneration in preclinical in vivo studies: A systematic review. *Materials* **2020**, *13*, 1500. [CrossRef]

24. Kokubo, T. *Bioceramics and Their Clinical Applications*; Elsevier: Amsterdam, The Netherlands, 2008.

25. Heydary, H.A.; Karamian, E.; Poorazizi, E.; Heydaripour, J.; Khandan, A. Electrospun of polymer/bioceramic nanocomposite as a new soft tissue for biomedical applications. *J. Asian Ceram. Soc.* **2015**, *3*, 417–425. [CrossRef]

26. Tavoni, M.; Dapporto, M.; Tampieri, A.; Sprio, S. Bioactive Calcium Phosphate-Based Composites for Bone Regeneration. *J. Compos. Sci.* **2021**, *5*, 227. [CrossRef]

27. Le Guéhennec, L.; Layrolle, P.; Daculsi, G. A review of bioceramics and fibrin sealant. *Eur. Cells Mater.* **2004**, *8*, 1e11. [CrossRef] [PubMed]

28. De Aza, P.; De Aza, A.; De Aza, S. Crystalline bioceramic materials. *Bol. Soc. Esp. Ceram.* **2005**, *44*, 135–145. [CrossRef]

29. Utneja, S.; Nawal, R.R.; Talwar, S.; Verma, M. Current perspectives of bio-ceramic technology in endodontics: Calcium enriched mixture cement—Review of its composition, properties and applications. *Restor. Dent. Endod.* **2015**, *40*, 1–13. [CrossRef]

30. Park, J.B.; Bronzino, J.D. (Eds.) *Biomaterials—Principles and Applications*; CRC Press: Boca Raton, FL, USA, 2003.

31. Gul, H.; Khan, M.; Khan, A.S. Bioceramics: Types and clinical applications. In *Handbook of Ionic Substituted Hydroxyapatites*; Elsevier: Amsterdam, The Netherlands, 2020; pp. 53–83.

32. Huang, J.; Best, S. Ceramic biomaterials for tissue engineering. In *Tissue Engineering Using Ceramic Polymymers*, 2nd ed.; Woodhead Publishing: Sawston, UK, 2014; pp. 3–34. [CrossRef]

33. Jiang, X.; Yao, Y.; Tang, W.; Han, D.; Zhang, L.; Zhao, K.; Wang, S.; Meng, Y. Design of dental implants at materials level: An overview. *J. Biomed. Mater. Res. Part A* **2020**, *108*, 1634–1661. [CrossRef]

34. Shao, H.; Ke, X.; Liu, A.; Sun, M.; He, Y.; Yang, X.; Fu, J.; Liu, Y.; Zhang, L.; Yang, G. Bone regeneration in 3D printing bioactive ceramic scaffolds with improved tissue/material interface pore architecture in thin-wall bone defect. *Biofabrication* **2017**, *9*, 025003. [CrossRef]

35. Udduttula, A.; Zhang, J.V.; Ren, P.-G. *Bioinert Ceramics for Biomedical Applications*; Biomedical Sci. and Tech. Series; Wiley–Scrivener: Salem, MA, USA, 2019.

36. Doremus, R. Bioceramics. *J. Mater. Sci.* **1992**, *27*, 285–297. [CrossRef]

37. Sainz, M.; Pena, P.; Serena, S.; Caballero, A. Influence of design on bioactivity of novel CaSiO3–CaMg (SiO3) 2 bioceramics: In vitro simulated body fluid test and thermodynamic simulation. *Acta Biomater.* **2010**, *6*, 2797–2807. [CrossRef] [PubMed]

38. Gao, C.; Yao, M.; Shuai, C.; Feng, P.; Peng, S. Advances in biocermets for bone implant applications. *Bio-Des. Manuf.* **2020**, *3*, 1–24. [CrossRef]

39. Wang, G.; Lu, Z.; Zreiqat, H. Bioceramics for skeletal bone regeneration. In *Bone Substitute Biomaterials*; Elsevier: Amsterdam, The Netherlands, 2014; pp. 180–216.

40. Zadehnajar, P.; Mirmusavi, M.; Soleymani Eil Bakhtiari, S.; Bakhsheshi-Rad, H.; Karbas, S.; RamaKrishna, S.; Berto, F. Recent advances on akermanite calcium-silicate ceramic for biomedical applications. *Int. J. Appl. Ceram. Technol.* **2021**, 1–20. [CrossRef]

41. Al-Harbi, N.; Mohammed, H.; Al-Hadeethi, Y.; Bakry, A.S.; Umar, A.; Hussein, M.A.; Abbassy, M.A.; Vaidya, K.G.; Berakdar, G.A.; Mkawi, E.M. Silica–based bioactive glasses and their applications in hard tissue regeneration: A review. *Pharmaceuticals* **2021**, *14*, 75. [CrossRef] [PubMed]

42. Gerhardt, L.-C.; Boccaccini, A.R. Bioactive glass and glass-ceramic scaffolds for bone tissue engineering. *Materials* **2010**, *3*, 3867–3910. [CrossRef]

43. Punj, S.; Singh, J.; Singh, K. Ceramic biomaterials: Properties, state of the art and future prospectives. *Ceram. Int.* **2021**, *47*, 28059–28074. [CrossRef]

44. Ginebra, M.-P.; Espanol, M.; Maazouz, Y.; Bergez, V.; Pastorino, D. Bioceramics and bone healing. *EFORT Open Rev.* **2018**, *3*, 173–183. [CrossRef]

45. Loh, Q.L.; Choong, C. Three-dimensional scaffolds for tissue engineering applications: Role of porosity and pore size. *Tissue Eng. Part B Rev.* **2013**, *19*, 485–502. [CrossRef]

46. Diez-Escudero, A.; Espanol, M.; Beats, S.; Ginebra, M.-P. In vitro degradation of calcium phosphates: Effect of multiscale porosity, textural properties and composition. *Acta Biomater.* **2017**, *60*, 81–92. [CrossRef]

47. Mastrogiacomo, M.; Scaglione, S.; Martinetti, R.; Dolcini, L.; Beltrame, F.; Cancedda, R.; Quarto, R. Role of scaffold internal structure on in vivo bone formation in macroporous calcium phosphate bioceramics. *Biomaterials* **2006**, *27*, 3230–3237. [CrossRef]

48. Vitale-Brovarone, C.; Baino, F.; Verné, E. High strength bioactive glass-ceramic scaffolds for bone regeneration. *J. Mater. Sci. Mater. Med.* **2009**, *20*, 643–653. [CrossRef]

49. Mouriño, V.; Cattalini, J.P.; Roether, J.A.; Dubey, P.; Roy, I.; Boccaccini, A.R. Composite polymer-bioceramic scaffolds with drug delivery capability for bone tissue engineering. *Expert Opin. Drug Deliv.* **2013**, *10*, 1353–1365. [CrossRef] [PubMed]

50. Sadeghzade, S.; Emadi, R.; Ahmadi, T.; Tavangarian, F. Synthesis, characterization and strengthening mechanism of modified and unmodified porous diopside/baghdadite scaffolds. *Mater. Chem. Phys.* **2019**, *228*, 89–97. [CrossRef]

51. Descamps, M.; Boilet, L.; Moreau, G.; Tricoteaux, A.; Lu, J.; Leriche, A.; Lardot, V.; Cambier, F. Processing and properties of biphasic calcium phosphates bioceramics obtained by pressureless sintering and hot isostatic pressing. *J. Eur. Ceram. Soc.* **2013**, *33*, 1263–1270. [CrossRef]

52. Bellucci, D.; Cannillo, V.; Sola, A. A new potassium-based bioactive glass: Sintering behaviour and possible applications for bioceramic scaffolds. *Ceram. Int.* **2011**, *37*, 145–157. [CrossRef]

53. Perera, F.H.; Martinez-Vazquez, F.J.; Miranda, P.; Ortiz, A.L.; Pajares, A. Clarifying the effect of sintering conditions on the microstructure and mechanical properties of β–tricalcium phosphate. *Ceram. Int.* **2010**, *36*, 1929–1935. [CrossRef]

54. Misra, S.K.; Valappil, S.P.; Roy, I.; Boccaccini, A.R. Polyhydroxyalkanoate (PHA)/inorganic phase composites for tissue engineering applications. *Biomacromolecules* **2006**, *7*, 2249–2258. [CrossRef]

55. Yunos, D.M.; Bretcanu, O.; Boccaccini, A.R. Polymer-bioceramic composites for tissue engineering scaffolds. *J. Mater. Sci.* **2008**, *43*, 4433–4442. [CrossRef]

56. Wöltje, M.; Brünler, R.; Böbel, M.; Ernst, S.; Neuss, S.; Aibibu, D.; Cherif, C. Functionalization of silk fibers by PDGF and bioceramics for bone tissue regeneration. *Coatings* **2020**, *10*, 8. [CrossRef]

57. Gómez-Cerezo, M.N.; Peña, J.; Ivanovski, S.; Arcos, D.; Vallet-Regí, M.; Vaquette, C. Multiscale porosity in mesoporous bioglass 3D-printed scaffolds for bone regeneration. *Mater. Sci. Eng. C* **2021**, *120*, 111706. [CrossRef]

58. De Lacerda Schickert, S.; Jansen, J.A.; Bronkhorst, E.M.; van den Beucken, J.J.; Leeuwenburgh, S.C. Stabilizing dental implants with a fiber–reinforced calcium phosphate cement: An in vitro and in vivo study. *Acta Biomater.* **2020**, *110*, 280–288. [CrossRef]

59. Zhao, X.; Zheng, J.; Zhang, W.; Chen, X.; Gui, Z. Preparation of silicon coated-carbon fiber reinforced HA bio-ceramics for application of load-bearing bone. *Ceram. Int.* **2020**, *46*, 7903–7911. [CrossRef]

60. Zhao, X.; Chen, X.; Gui, Z.; Zheng, J.; Yang, P.; Liu, A.; Wei, S.; Yang, Z. Carbon fiber reinforced hydroxyapatite composites with excellent mechanical properties and biological activities prepared by spark plasma sintering. *Ceram. Int.* **2020**, *46*, 27446–27456. [CrossRef]

61. O'Masta, M.R.; Stonkevitch, E.; Porter, K.A.; Bui, P.P.; Eckel, Z.C.; Schaedler, T.A. Additive manufacturing of polymer-derived ceramic matrix composites. *J. Am. Ceram. Soc.* **2020**, *103*, 6712–6723. [CrossRef]

62. Cui, K.; Zhang, Y.; Fu, T.; Wang, J.; Zhang, X. Toughening mechanism of mullite matrix composites: A Review. *Coatings* **2020**, *10*, 672. [CrossRef]

63. Xing, H.; Zou, B.; Wang, X.; Hu, Y.; Huang, C.; Xue, K. Fabrication and characterization of SiC whiskers toughened Al2O3 paste for stereolithography 3D printing applications. *J. Alloy Compd.* **2020**, *828*, 154347. [CrossRef]

64. Lamon, J. Reinforcement of ceramic matrix composites by ceramic continuous fibers. In *Composite Reinforcements for Optimum Performance*; Elsevier: Amsterdam, The Netherlands, 2021; pp. 55–93.

65. Bahl, S. Fiber reinforced metal matrix composites—A review. *Mater. Today Proc.* **2021**, *39*, 317–323. [CrossRef]

66. Petre, D.G.; Leeuwenburgh, S.C. The use of fibers in bone tissue engineering. *Tissue Eng. Part B Rev.* **2021**. [CrossRef] [PubMed]

67. Nurulhuda, A.; Sudin, I.; Ngadiman, N.H.A. Fabrication a novel 3D tissue engineering scaffold of Poly (ethylene glycol) diacrylate filled with Aramid Nanofibers via Digital Light Processing (DLP) technique. *J. Mech. Eng.* **2020**, *9*, 1–12.

68. Chen, B.; Wu, S.; Ye, Q. Fabrication and characterization of biodegradable KH560 crosslinked chitin hydrogels with high toughness and good biocompatibility. *Carbohydr. Polym.* **2021**, *259*, 117707. [CrossRef]

69. Biswal, T.; BadJena, S.K.; Pradhan, D. Synthesis of polymer composite materials and their biomedical applications. *Mater. Today Proc.* **2020**, *30*, 305–315. [CrossRef]

70. Shesan, O.J.; Stephen, A.C.; Chioma, A.G.; Neerish, R.; Rotimi, S.E. Fiber-matrix relationship for composites preparation. In *Renewable and Sustainable Composites*; IntechOpen: London, UK, 2019; pp. 1–30.

71. Rahman, R.; Syed Putra, S.Z.F.; Abd Rahim, S.Z.; Nainggolan, I.; Jeż, B.; Nabiałek, M.; Musa, L.; Sandu, A.V.; Vizureanu, P.; Al Bakri Abdullah, M.M. The Influence of MMA Esterification on Interfacial Adhesion and Mechanical Properties of Hybrid Kenaf Bast/Glass Fiber Reinforced Unsaturated Polyester Composites. *Materials* **2021**, *14*, 2276. [CrossRef]

72. Shuai, C.; Yu, L.; Feng, P.; Gao, C.; Peng, S. Interfacial reinforcement in bioceramic/biopolymer composite bone scaffold: The role of coupling agent. *Colloids Surf. B Biointerfaces* **2020**, *193*, 111083. [CrossRef] [PubMed]

73. Moussa, H.; El Hadad, A.; Sarrigiannidis, S.; Saad, A.; Wang, M.; Taqi, D.; Al-Hamed, F.S.; Salmerón-Sánchez, M.; Cerruti, M.; Tamimi, F. High toughness resorbable brushite-gypsum fiber-reinforced cements. *Mater. Sci. Eng. C* **2021**, *127*, 112205. [CrossRef] [PubMed]

74. Raut, H.K.; Das, R.; Liu, Z.; Liu, X.; Ramakrishna, S. Biocompatibility of Biomaterials for Tissue Regeneration or Replacement. *Biotechnol. J.* **2020**, *15*, 2000160. [CrossRef] [PubMed]

75. Kumar, P.; Saini, M.; Dehiya, B.S.; Umar, A.; Sindhu, A.; Mohammed, H.; Al-Hadeethi, Y.; Guo, Z. Fabrication and in-vitro biocompatibility of freeze-dried CTS–nHA and CTS–nBG scaffolds for bone regeneration applications. *Int. J. Biol. Macromol.* **2020**, *149*, 1–10. [CrossRef]

76. Foroutan, F.; Nikolaou, A.; Kyffin, B.A.; Elliott, R.M.; Felipe-Sotelo, M.; Gutierrez-Merino, J.; Carta, D. Multifunctional phosphate–based glass fibres prepared via electrospinning of coacervate precursors: Controlled delivery, biocompatibility and antibacterial activity. *Materialia* **2020**, *14*, 100939. [CrossRef]

77. Lopez de Armentia, S.; Del Real, J.C.; Paz, E.; Dunne, N. Advances in Biodegradable 3D Printed Scaffolds with Carbon–Based Nanomaterials for Bone Regeneration. *Materials* **2020**, *13*, 5083. [CrossRef] [PubMed]

78. Li, C.; Wang, H.; Zhao, X.; Fu, Y.; He, X.; Song, Y. Investigation of Mechanical Properties for Basalt Fiber/Epoxy Resin Composites Modified with La. *Coatings* **2021**, *11*, 666. [CrossRef]

79. Yusufu, M.; Ariahu, C.; Igbabul, B. Production and characterization of activated carbon from selected local raw materials. *Afr. J. Pure Appl. Chem.* **2012**, *6*, 123–131. [CrossRef]

80. Bansal, S.A.; Karimi, J.; Singh, A.P.; Kumar, S. Carbon Fibers: Surface Modification Strategies and Biomedical Applications. In *Advanced Manufacturing and Processing Technology*; CRC Press: Boca Raton, FL, USA, 2020; pp. 207–224.

81. Maurer, E.I. Surface Modification of Carbon Structures for Biological Applications. Master's Thesis, Wright State University, Dayton, OH, USA, 2010.

82. Prokopowicz, M.; Żeglinski, J.; Szewczyk, A.; Skwira, A.; Walker, G. Surface-activated fibre-like SBA–15 as drug carriers for bone diseases. *AAPS PharmSciTech* **2019**, *20*, 1–13. [CrossRef] [PubMed]

83. El-Fiqi, A.; Seo, S.-J.; Kim, H.-W. Mineralization of fibers for bone regeneration. In *Biomineralization and Biomaterials*; Elsevier: Amsterdam, The Netherlands, 2016; pp. 443–476.

84. Boda, S.K.; Almoshari, Y.; Wang, H.; Wang, X.; Reinhardt, R.A.; Duan, B.; Wang, D.; Xie, J. Mineralized nanofiber segments coupled with calcium–binding BMP-2 peptides for alveolar bone regeneration. *Acta Biomater.* **2019**, *85*, 282–293. [CrossRef] [PubMed]

85. Metwally, S.; Ferraris, S.; Spriano, S.; Krysiak, Z.J.; Kaniuk, Ł.; Marzec, M.M.; Kim, S.K.; Szewczyk, P.K.; Gruszczyński, A.; Wytrwal–Sarna, M. Surface potential and roughness controlled cell adhesion and collagen formation in electrospun PCL fibers for bone regeneration. *Mater. Des.* **2020**, *194*, 108915. [CrossRef]

86. Rahimizadeh, A.; Sarvestani, H.Y.; Li, L.; Robles, J.B.; Backman, D.; Lessard, L.; Ashrafi, B. Engineering toughening mechanisms in architectured ceramic–based bioinspired materials. *Mater. Des.* **2021**, *198*, 109375. [CrossRef]

87. Caon, M. Skeletal System. In *Examination Questions and Answers in Basic Anatomy and Physiology*; Springer: New York, NY, USA, 2020; pp. 185–212.

88. Office of the Surgeon General. The basics of bone in health and disease. In *Bone Health and Osteoporosis: A Report of the Surgeon General*; Office of the Surgeon General (US): Washington, DC, USA, 2004.

89. Langdahl, B.; Ferrari, S.; Dempster, D.W. Bone modeling and remodeling: Potential as therapeutic targets for the treatment of osteoporosis. *Ther. Adv. Musculoskelet. Dis.* **2016**, *8*, 225–235. [CrossRef]

90. Ronald Watson, V.P. (Ed.) *Bioactive Food as Dietary Interventions for Arthritis and Related Inflammatory Diseases*; Elsevier: Amsterdam, The Netherlands, 2019.

91. Taichman, R.S. Blood and bone: Two tissues whose fates are intertwined to create the hematopoietic stem–cell niche. *Blood* **2005**, *105*, 2631–2639. [CrossRef] [PubMed]

92. Pérez-Amodio, S.; Engel, E. Bone biology and regeneration. *Bio-Ceram. Clin. Appl.* **2014**, 315–342. [CrossRef]

93. Hapidin, H.; Romli, N.A.A.; Abdullah, H. Proliferation study and microscopy evaluation on the effects of tannic acid in human fetal osteoblast cell line (hFOB 1.19). *Microsc. Res. Tech.* **2019**, *82*, 1928–1940. [CrossRef]

94. Maes, C.; Kronenberg, H.M. Postnatal bone growth: Growth plate biology, bone formation, and remodeling. In *Pediatric Bone*; Elsevier: Amsterdam, The Netherlands, 2012; pp. 55–82.

95. Alisafaei, F.; Chen, X.; Leahy, T.; Janmey, P.A.; Shenoy, V.B. Long-range mechanical signaling in biological systems. *Soft Matter* **2021**, *17*, 241–253. [CrossRef] [PubMed]

96. Ribeiro, F.O.; Gómez-Benito, M.J.; Folgado, J.; Fernandes, P.R.; García-Aznar, J.M. In silico mechano-chemical model of bone healing for the regeneration of critical defects: The effect of BMP–2. *PLoS ONE* **2015**, *10*, e0127722. [CrossRef]

97. Wong, J.Y.; Leach, J.B.; Brown, X.Q. Balance of chemistry, topography, and mechanics at the cell-biomaterial interface: Issues and challenges for assessing the role of substrate mechanics on cell response. *Surf. Sci.* **2004**, *570*, 119–133. [CrossRef]

98. Glimcher, M.J. Bone: Nature of the calcium phosphate crystals and cellular, structural, and physical chemical mechanisms in their formation. *Rev. Mineral. Geochem.* **2006**, *64*, 223–282. [CrossRef]

99. Takayanagi, H. Osteoimmunology: Shared mechanisms and crosstalk between the immune and bone systems. *Nat. Rev. Immunol.* **2007**, *7*, 292–304. [CrossRef] [PubMed]

100. Park, H.; Park, K. Biocompatibility issues of implantable drug delivery systems. *Pharm. Res.* **1996**, *13*, 1770–1776. [CrossRef] [PubMed]

101. Civinini, R.; Capone, A.; Carulli, C.; Matassi, F.; Nistri, L.; Innocenti, M. The kinetics of remodeling of a calcium sulfate/calcium phosphate bioceramic. *J. Mater. Sci. Mater. Med.* **2017**, *28*, 1–5. [CrossRef] [PubMed]

102. Hodges, R.M.; MacDonald, N.S.; Nusbaum, R.; Stearns, R.; Ezmirlian, F.; Spain, P.; McArthur, C. The strontium content of human bones. *J. Biol. Chem.* **1950**, *185*, 519–524. [CrossRef]

103. Ducheyne, P.; Qiu, Q.J.B. Bioactive ceramics: The effect of surface reactivity on bone formation and bone cell function. *Biomaterials* **1999**, *20*, 2287–2303. [CrossRef]

104. El-Hajj Fuleihan, G. Strontium ranelate—A novel therapy for osteoporosis or a permutation of the same? *N. Engl. J. Med.* **2004**, *350*, 504–506. [CrossRef]

105. Sips, A.; van der Vijgh, W.; Barto, R.; Netelenbos, J.C. Intestinal strontium absorption: From bioavailability to validation of a simple test representative for intestinal calcium absorption. *Clin. Chem.* **1995**, *41*, 1446–1450. [CrossRef]

106. Steffi, C.; Shi, Z.; Kong, C.H.; Wang, W. Modulation of Osteoclast Interactions with Orthopaedic Biomaterials. *J. Funct. Biomater.* **2018**, *9*, 18. [CrossRef] [PubMed]

107. Osaka, A.; Narayan, R. *Bioceramics: From Macro to Nanoscale*; Elsevier: Amsterdam, The Netherlands, 2020.

108. Heness, G.; Ben-Nissan, B. Innovative bioceramics. *Mater. Forum* **2004**, *27*, 107–114.

109. Pobloth, A.-M.; Mersiowsky, M.J.; Kliemt, L.; Schell, H.; Dienelt, A.; Pfitzner, B.M.; Burgkart, R.; Detsch, R.; Wulsten, D.; Boccaccini, A.R. Bioactive coating of zirconia toughened alumina ceramic implants improves cancellous osseointegration. *Sci. Rep.* **2019**, *9*, 1–16. [CrossRef]

110. Singh, V.K.; Reddy, B.R. Synthesis and characterization of bioactive zirconia toughened alumina doped with HAp and fluoride compounds. *Ceram. Int.* **2012**, *38*, 5333–5340. [CrossRef]

111. Afzal, A. Implantable zirconia bioceramics for bone repair and replacement: A chronological review. *Mater. Express* **2014**, *4*, 1–12. [CrossRef]

112. Moxon, K.A.; Kalkhoran, N.M.; Markert, M.; Sambito, M.A.; McKenzie, J.; Webster, J.T. Nanostructured surface modification of ceramic–based microelectrodes to enhance biocompatibility for a direct brain-machine interface. *IEEE Trans. Biomed. Eng.* **2004**, *51*, 881–889. [CrossRef]

113. Baino, F.; Hamzehlou, S.; Kargozar, S. Bioactive glasses: Where are we and where are we going? *J. Funct. Biomater.* **2018**, *9*, 25. [CrossRef]

114. Yamamuro, T. Bioceramics. In *Biomechanics and Biomaterials in Orthopedics*; Poitout, D.G., Ed.; Springer: London, UK, 2004; pp. 22–33.

115. Hench, L.L. Bioceramics: From concept to clinic. *J. Am. Ceram. Soc.* **1991**, *74*, 1487–1510. [CrossRef]

116. Turnbull, G.; Clarke, J.; Picard, F.; Riches, P.; Jia, L.; Han, F.; Li, B.; Shu, W. 3D bioactive composite scaffolds for bone tissue engineering. *Bioact. Mater.* **2018**, *3*, 278–314. [CrossRef] [PubMed]

117. Gao, C.; Deng, Y.; Feng, P.; Mao, Z.; Li, P.; Yang, B.; Deng, J.; Cao, Y.; Shuai, C.; Peng, S. Current progress in bioactive ceramic scaffolds for bone repair and regeneration. *Int. J. Mol. Sci.* **2014**, *15*, 4714–4732. [CrossRef]

118. Juhasz, J.A.; Best, S.M. Bioactive ceramics: Processing, structures and properties. *J. Mater. Sci.* **2012**, *47*, 610–624. [CrossRef]

119. Kitsugi, T.; Yamamuro, T.; Nakamura, T.; Kokubo, T.; Takagi, M.; Shibuya, T.; Takeuchi, H.; Ono, M. Bonding behavior between two bioactive ceramics in vivo. *J. Biomed. Mater. Res.* **1987**, *21*, 1109–1123. [CrossRef] [PubMed]

120. Baino, F.; Novajra, G.; Vitale-Brovarone, C. Bioceramics and scaffolds: A winning combination for tissue engineering. *Front. Bioeng. Biotechnol.* **2015**, *3*, 202. [CrossRef]

121. Kaya, C.; Singh, I.; Boccaccini, A.R. Multi-walled carbon nanotube-reinforced hydroxyapatite layers on Ti6Al4V medical implants by Electrophoretic Deposition (EPD). *Adv. Eng. Mater.* **2008**, *10*, 131–138. [CrossRef]

122. Furko, M.; Balázsi, C. Calcium phosphate based bioactive ceramic layers on implant materials preparation, properties, and biological performance. *Coatings* **2020**, *10*, 823. [CrossRef]

123. Paital, S.R.; Dahotre, N.B. Calcium phosphate coatings for bio-implant applications: Materials, performance factors, and methodologies. *Mater. Sci. Eng. R Rep.* **2009**, *66*, 1–70. [CrossRef]

124. Gupta, V.; Yuan, J.; Pronin, A. Recent developments in the laser spallation technique to measure the interface strength and its relationship with interface toughness with applications to metal/ceramic, ceramic/ceramic and ceramic/polymer interfaces. *J. Adhes. Sci. Technol.* **1994**, *8*, 713–747. [CrossRef]

125. Adanur, S. *Textile Structural Composites*; Technomic Publishing Company, Inc.: Lancaster, PA, USA, 1995; pp. 231–268.

126. Iannazzo, D.; Pistone, A.; Salamò, M.; Galvagno, S. Hybrid ceramic/polymer composites for bone tissue regeneration. In *Hybrid Polymer Composite Materials*; Elsevier: Amsterdam, The Netherlands, 2017; pp. 125–155.

127. Naslain, R.; Langlais, F. CVD-processing of ceramic-ceramic composite materials. In *Tailoring Multiphase and Composite Ceramics*; Springer: New York, NY, USA, 1986; pp. 145–164.

128. Steiner, P.J.; Kelly, J.R.; Giuseppetti, A.A. Compatibility of ceramic-ceramic systems for fixed prosthodontics. *Int. J. Prosthodont.* **1997**, *10*, 375–380.

129. Deng, Y.; Miranda, P.; Pajares, A.; Guiberteau, F.; Lawn, B.R. Fracture of ceramic/ceramic/polymer trilayers for biomechanical applications. *J. Biomed. Mater. Res. Part A* **2003**, *67*, 828–833. [CrossRef]

130. Abarrategi, A.; Moreno-Vicente, C.; Martínez-Vázquez, F.J.; Civantos, A.; Ramos, V.; Sanz-Casado, J.V.; Martínez-Corriá, R.; Perera, F.H.; Mulero, F.; Miranda, P. Biological properties of solid free form designed ceramic scaffolds with BMP–2: In vitro and in vivo evaluation. *PLoS ONE* **2012**, *7*, e34117. [CrossRef] [PubMed]

131. Elshahawy, W. Biocompatibility. In *Advances in Ceramics-Electric and Magnetic Ceramics, Bioceramics, Ceramics and Environment*; IntechOpen: London, UK, 2011.

132. Chellappa, M.; Anjaneyulu, U.; Manivasagam, G.; Vijayalakshmi, U. Preparation and evaluation of the cytotoxic nature of TiO2 nanoparticles by direct contact method. *Int. J. Nanomed.* **2015**, *10*, 31.

133. Kalaskar, D.; Seifalinan, A.; Salmasi, S.; Prinsloo, N. *Inorganic Biomaterials Characterization*; Smithers Rapra: Shrewsbury, UK, 2014.

134. Durdu, S. Characterization, bioactivity and antibacterial properties of copper-based TiO_2 bioceramic coatings fabricated on titanium. *Coatings* **2019**, *9*, 1. [CrossRef]

135. Yusoff, M.F.M.; Kasim, N.H.A.; Himratul-Aznita, W.H.; Saidin, S.; Genasan, K.; Kamarul, T.; Radzi, Z. Physicochemical, antibacterial and biocompatibility assessments of silver incorporated nano–hydroxyapatite synthesized using a novel microwave-assisted wet precipitation technique. *Mater. Charact.* **2021**, *178*, 111169. [CrossRef]

136. Kumar, P.; Dehiya, B.S.; Sindhu, A. Bioceramics for hard tissue engineering applications: A review. *Int. J. Appl. Eng. Res.* **2018**, *13*, 2744–2752.

137. El–Ghannam, A. Bone reconstruction: From bioceramics to tissue engineering. *Expert Rev. Med. Devices* **2005**, *2*, 87–101. [CrossRef]

138. Sailaja, G.; Ramesh, P.; Vellappally, S.; Anil, S.; Varma, H. Biomimetic approaches with smart interfaces for bone regeneration. *J. Biomed. Sci.* **2016**, *23*, 1–13. [CrossRef]

139. Lam, R.H.; Chen, W. *Biomedical Devices: Materials, Design, and Manufacturing*; Springer: New York, NY, USA, 2019.

140. Wang, K.; Zhou, C.; Hong, Y.; Zhang, X. A review of protein adsorption on bioceramics. *Interface Focus* **2012**, *2*, 259–277. [CrossRef] [PubMed]

141. Furkó, M.; Balázsi, K.; Balázsi, C. Comparative study on preparation and characterization of bioactive coatings for biomedical applications—A review on recent patents and literature. *Rev. Adv. Mater. Sci.* **2017**, *48*, 25–51.

142. Pezzotti, G. Surface chemistry of bioceramics: The missing key. In *Bioceramics*; Elsevier: Amsterdam, The Netherlands, 2021; pp. 297–324.

143. Schmitz, T. Functional Coatings by Physical Vapor Deposition (PVD) for Biomedical Applications. Ph.D. Thesis, Universität Würzburg, Würzburg, Germany, 2016.

144. Awasthi, S.; Pandey, S.K.; Arunan, E.; Srivastava, C. A review on hydroxyapatite coatings for the biomedical applications: Experimental and theoretical perspectives. *J. Mater. Chem. B* **2021**, *9*, 228–249. [CrossRef] [PubMed]

145. Bose, S.; Robertson, S.F.; Bandyopadhyay, A. Surface modification of biomaterials and biomedical devices using additive manufacturing. *Acta Biomater.* **2018**, *66*, 6–22. [CrossRef]

146. Kargozar, S.; Kermani, F.; Mollazadeh Beidokhti, S.; Hamzehlou, S.; Verné, E.; Ferraris, S.; Baino, F. Functionalization and surface modifications of bioactive glasses (BGs): Tailoring of the biological response working on the outermost surface layer. *Materials* **2019**, *12*, 3696. [CrossRef] [PubMed]

147. Fabris, D.; Lasagni, A.F.; Fredel, M.C.; Henriques, B. Direct laser interference patterning of bioceramics: A short review. *Ceramics* **2019**, *2*, 578–586. [CrossRef]

148. Morgan, E.F.; Unnikrisnan, G.U.; Hussein, A.I. Bone mechanical properties in healthy and diseased states. *Annu. Rev. Biomed. Eng.* **2018**, *20*, 119–143. [CrossRef]

149. Dee, P.; You, H.Y.; Teoh, S.-H.; Le Ferrand, H. Bioinspired approaches to toughen calcium phosphate-based ceramics for bone repair. *J. Mech. Behav. Biomed. Mater.* **2020**, *112*, 104078. [CrossRef]

150. Ritchie, R.O. The conflicts between strength and toughness. *Nat. Mater.* **2011**, *10*, 817–822. [CrossRef]

151. Evans, A.G. Perspective on the Development of High-Toughness Ceramics. *J. Am. Ceram. Soc.* **1990**, *73*, 187–206. [CrossRef]

152. Elices, M. *Structural Biological Materials: Design and Structure-Property Relationships*; Elsevier: Amsterdam, The Netherlands, 2000.

153. Ginebra, M.P.; Montufar, E.B. Cements as bone repair materials. In *Bone Repair Biomaterials*; Planell, J.A., Best, S.M., Lacroix, D., Merolli, A., Eds.; Woodhead Publishing: Cambridge, UK, 2009; pp. 271–308.

154. Barralet, J.E.; Hofmann, M.; Grover, L.M.; Gbureck, U.J.A.M. High-strength apatitic cement by modification with α-hydroxy acid salts. *Adv. Mater.* **2003**, *15*, 2091–2094. [CrossRef]

155. Vaidya, A.; Pathak, K. Mechanical stability of dental materials. In *Applications of Nanocomposite Materials in Dentistry*; Elsevier: Amsterdam, The Netherlands, 2019; pp. 285–305.

156. Galeano, M.; Altavilla, D.; Bitto, A.; Minutoli, L.; Calò, M.; Cascio, P.L.; Polito, F.; Giugliano, G.; Squadrito, G.; Mioni, C. Recombinant human erythropoietin improves angiogenesis and wound healing in experimental burn wounds. *Crit. Care Med.* **2006**, *34*, 1139–1146. [CrossRef]

157. Luo, J.-D.; Wang, Y.-Y.; Fu, W.-L.; Wu, J.; Chen, A.F. Gene therapy of endothelial nitric oxide synthase and manganese superoxide dismutase restores delayed wound healing in type 1 diabetic mice. *Circulation* **2004**, *110*, 2484–2493. [CrossRef]

158. Raucci, M.; Giugliano, D.; Ambrosio, L. Fundamental properties of bioceramics and biocomposites. *Handb. Bioceram. Biocomposites* **2016**, 35–58. [CrossRef]

159. Prakasam, M.; Locs, J.; Salma-Ancane, K.; Loca, D.; Largeteau, A.; Berzina-Cimdina, L. Fabrication, properties and applications of dense hydroxyapatite: A review. *J. Funct. Biomater.* **2015**, *6*, 1099–1140. [CrossRef]

160. Salvini, V.R.; Pandolfelli, V.C.; Spinelli, D. Mechanical properties of porous ceramics. *Recent Adv. Porous Ceram.* **2018**, *34*, 171–199. [CrossRef]

161. Kang, T.Y.; Seo, J.Y.; Ryu, J.H.; Kim, K.M.; Kwon, J.S. Improvement of the mechanical and biological properties of bioactive glasses by the addition of zirconium oxide (ZrO_2) as a synthetic bone graft substitute. *J. Biomed. Mater. Res. Part A* **2021**, *109*, 1196–1208. [CrossRef] [PubMed]

162. Kaur, G. *Biomedical, Therapeutic and Clinical Applications of Bioactive Glasses*; Woodhead Publishing: Cambridge, UK, 2018.

163. Shang-Xian, W. Compliance and stress-intensity factor of chevron-notched three-point bend specimen. In *Chevron-Notched Specimens: Testing and Stress Analysis*; ASTM International: West Conshohocken, PA, USA, 1984.

164. Quinn, G.; Salem, J.; Baron, I.; Cho, K.; Foley, M.; Fang, H. Fracture toughness of advanced ceramics at room temperature. *J. Res. Natl. Inst. Stand. Technol.* **1992**, *97*, 579. [CrossRef]

165. Quinn, G.D.; Salem, J.A. Effect of lateral cracks on fracture toughness determined by the surface-crack-in-flexure method. *J. Am. Ceram. Soc.* **2002**, *85*, 873–880. [CrossRef]

166. Tancret, F.; Bouler, J.-M.; Chamousset, J.; Minois, L.-M. Modelling the mechanical properties of microporous and macroporous biphasic calcium phosphate bioceramics. *J. Eur. Ceram. Soc.* **2006**, *26*, 3647–3656. [CrossRef]

167. Bouler, J.M.; Trécant, M.; Delécrin, J.; Royer, J.; Passuti, N.; Daculsi, G. Macroporous biphasic calcium phosphate ceramics: Influence of five synthesis parameters on compressive strength. *J. Biomed. Mater. Res.* **1996**, *32*, 603–609. [CrossRef]

168. Groot, K.; Klein, C.; Wolke, J. *Chemistry of Calcium Phosphate Bioceramics*; CRC Press: Boca Raton, FL, USA, 1990.

169. Dapporto, M.; Tampieri, A.; Sprio, S. Composite Calcium Phosphate/Titania Scaffolds in Bone Tissue Engineering. In *Application of Titanium Dioxide*; IntechOpen: London, UK, 2017; pp. 43–59.

170. Sprio, S.; Guicciardi, S.; Dapporto, M.; Melandri, C.; Tampieri, A. Synthesis and mechanical behavior of β–tricalcium phosphate/titania composites addressed to regeneration of long bone segments. *J. Mech. Behav. Biomed. Mater.* **2013**, *17*, 1–10. [CrossRef]

171. Danilenko, I.; Lasko, G.; Brykhanova, I.; Burkhovetski, V.; Ahkhozov, L. The Peculiarities of Structure Formation and Properties of Zirconia–Based Nanocomposites with Addition of Al_2O_3 and NiO. *Nanoscale Res. Lett.* **2017**, *12*, 1–10. [CrossRef] [PubMed]

172. Benzaid, R.; Chevalier, J.; Saâdaoui, M.; Fantozzi, G.; Nawa, M.; Diaz, L.A.; Torrecillas, R. Fracture toughness, strength and slow crack growth in a ceria stabilized zirconia–alumina nanocomposite for medical applications. *Biomaterials* **2008**, *29*, 3636–3641. [CrossRef]

173. Senatov, F.; Niaza, K.; Stepashkin, A.; Kaloshkin, S. Low-cycle fatigue behavior of 3d-printed PLA-based porous scaffolds. *Compos. Part B Eng.* **2016**, *97*, 193–200. [CrossRef]

174. Zhang, J.; Iwasa, M.; Kotobuki, N.; Tanaka, T.; Hirose, M.; Ohgushi, H.; Jiang, D. Fabrication of hydroxyapatite–zirconia composites for orthopedic applications. *J. Am. Ceram. Soc.* **2006**, *89*, 3348–3355. [CrossRef]

175. Zhao, X.; Yang, J.; Xin, H.; Wang, X.; Zhang, L.; He, F.; Liu, Q.; Zhang, W. Improved dispersion of SiC whisker in nano hydroxyapatite and effect of atmospheres on sintering of the SiC whisker reinforced nano hydroxyapatite composites. *Mater. Sci. Eng. C* **2018**, *91*, 135–145. [CrossRef] [PubMed]

176. Wu, M.; Wang, Q.; Liu, X.; Liu, H. Biomimetic synthesis and characterization of carbon nanofiber/hydroxyapatite composite scaffolds. *Carbon* **2013**, *51*, 335–345. [CrossRef]

177. Wang, X.; Zhao, X.; Zhang, L.; Wang, W.; Zhang, J.; He, F.; Yang, J. Design and fabrication of carbon fibers with needle-like nano–HA coating to reinforce granular nano–HA composites. *Mater. Sci. Eng. C* **2017**, *77*, 765–771. [CrossRef] [PubMed]

178. Guo, G.; Fan, Y.; Zhang, J.-F.; Hagan, J.L.; Xu, X. Novel dental composites reinforced with zirconia–silica ceramic nanofibers. *Dent. Mater.* **2012**, *28*, 360–368. [CrossRef] [PubMed]

179. Hodgson, S.; Cawley, J. The effect of titanium oxide additions on the properties and behaviour of Y–TZP. *J. Mater. Process. Technol.* **2001**, *119*, 112–116. [CrossRef]

180. Kong, Y.-M.; Kim, S.; Kim, H.-E.; Lee, I.-S. Reinforcement of Hydroxyapatite Bioceramic by Addition of ZrO_2 Coated with Al_2O_3. *J. Am. Ceram. Soc.* **1999**, *82*, 2963–2968. [CrossRef]

181. Riu, D.-H.; Kong, Y.-M.; Kim, H.-E. Effect of Cr_2O_3 addition on microstructural evolution and mechanical properties of Al_2O_3. *J. Eur. Ceram. Soc.* **2000**, *20*, 1475–1481. [CrossRef]

182. Launey, M.E.; Hofmann, D.C.; Suh, J.-Y.; Kozachkov, H.; Johnson, W.L.; Ritchie, R.O. Fracture toughness and crack–resistance curve behavior in metallic glass–matrix composites. *Appl. Phys. Lett.* **2009**, *94*, 241910. [CrossRef]

183. Demetriou, M.D.; Launey, M.E.; Garrett, G.; Schramm, J.P.; Hofmann, D.C.; Johnson, W.L.; Ritchie, R.O. A damage–tolerant glass. *Nat. Mater.* **2011**, *10*, 123–128. [CrossRef]

184. Juhasz, J.; Best, S.; Masakazu, K.; Miyata, N.; Kokubo, T.; Nakamura, T.; Bonfield, W. Mechanical properties of glass–ceramic AW–polyethylene composites: Effect of filler content. In *Key Engineering Materials*; Ben-Nissan, B., Sher, D., Walsh, W., Eds.; Trans Tech Publications Ltd.: Bäch, Switzerland, 2003; pp. 947–950.

185. Xu, H.H.; Martin, T.; Antonucci, J.M.; Eichmiller, E. Ceramic whisker reinforcement of dental resin composites. *J. Dent. Res.* **1999**, *78*, 706–712. [CrossRef]

186. De Aza, A.; Chevalier, J.; Fantozzi, G.; Schehl, M.; Torrecillas, R. Crack growth resistance of alumina, zirconia and zirconia toughened alumina ceramics for joint prostheses. *Biomaterials* **2002**, *23*, 937–945. [CrossRef]

187. Ruhle, M.; Krause, B.; Strecker, A.; Waidelich, D. In situ observations of stress–induced phase transformations in ZrO_2–containing ceramics. In *Science and Technology of Zirconia II*; Max Planck Institut fur Metallforschung, Institut fur Werkstoffwissenschaften: Stuttgart, Germany, 1983.

188. Karihaloo, B.L. Contribution of t $\rightarrow$ m Phase Transformation to the Toughening of ZTA. *J. Am. Ceram. Soc.* **1991**, *74*, 1703–1706. [CrossRef]

189. Becher, P.F.; Alexander, K.B.; Bleier, A.; Waters, S.B.; Warwick, W.H. Influence of ZrO_2 grain size and content on the transformation response in the Al_2O_3—ZrO_2 (12 mol % CeO_2) System. *J. Am. Ceram. Soc.* **1993**, *76*, 657–663. [CrossRef]

190. Gregori, G.; Burger, W.; Sergo, V. Piezo–spectroscopic analysis of the residual stresses in zirconia–toughened alumina ceramics: The influence of the tetragonal–to–monoclinic transformation. *Mater. Sci. Eng. A* **1999**, *271*, 401–406. [CrossRef]

191. Kosmac, T.; Swain, M.V.; Claussen, N. Role of tetragonal and monoclinic ZrO/sub 2/particles in the fracture toughness of Al/sub 2/O/sub 3/–ZrO/sub 2/composites. *Mater. Sci. Eng.* **1985**, *71*, 57–64. [CrossRef]

192. Zhigachev, A.O.; Rodaev, V.V.; Zhigacheva, D.V. The effect of titania doping on structure and mechanical properties of calcia–stabilized zirconia ceramic. *J. Mater. Res. Technol.* **2019**, *8*, 6086–6093. [CrossRef]

193. Ślósarczyk, A.; Klisch, M.; Błażewicz, M.; Piekarczyk, J.; Stobierski, L.; Rapacz–Kmita, A. Hot pressed hydroxyapatite–carbon fibre composites. *J. Eur. Ceram. Soc.* **2000**, *20*, 1397–1402. [CrossRef]

194. Dorner–Reisel, A.; Berroth, K.; Neubauer, R.; Nestler, K.; Marx, G.; Scislo, M.; Müller, E.; Slosarcyk, A. Unreinforced and carbon fibre reinforced hydroxyapatite: Resistance against microabrasion. *J. Eur. Ceram. Soc.* **2004**, *24*, 2131–2139. [CrossRef]

195. Zhang, L.; Pei, L.; Li, H.; Zhu, F. Design and fabrication of pyrolytic carbon–SiC–fluoridated hydroxyapatite–hydroxyapatite multilayered coating on carbon fibers. *Appl. Surf. Sci.* **2019**, *473*, 571–577. [CrossRef]

196. Zhang, Y.; Tan, S.; Yin, Y. C–fibre reinforced hydroxyapatite bioceramics. *Ceram. Int.* **2003**, *29*, 113–116. [CrossRef]

197. Chan, K.S. Relationships of fracture toughness and dislocation mobility in intermetallics. *Metall. Mater. Trans. A* **2003**, *34*, 2315–2328. [CrossRef]

198. Po, G.; Cui, Y.; Rivera, D.; Cereceda, D.; Swinburne, T.D.; Marian, J.; Ghoniem, N. A phenomenological dislocation mobility law for bcc metals. *Acta Mater.* **2016**, *119*, 123–135. [CrossRef]

199. Danzer, R.; Lube, T.; Supancic, P.; Damani, R. Fracture of ceramics. *Adv. Eng. Mater.* **2008**, *10*, 275–298. [CrossRef]

200. Philippart, A.; Boccaccini, A.R.; Fleck, C.; Schubert, D.W.; Roether, J.A. Toughening and functionalization of bioactive ceramic and glass bone scaffolds by biopolymer coatings and infiltration: A review of the last 5 years. *Expert Rev. Med. Devices* **2015**, *12*, 93–111. [CrossRef]

201. Chevalier, J.; Gremillard, L.; Virkar, A.V.; Clarke, D.R. The tetragonal-monoclinic transformation in zirconia: Lessons learned and future trends. *J. Am. Ceram. Soc.* **2009**, *92*, 1901–1920. [CrossRef]

202. Budiansky, B.; Truskinovsky, L. On the mechanics of stress–induced phase transformation in zirconia. *J. Mech. Phys. Solids* **1993**, *41*, 1445–1459. [CrossRef]

203. Raut, V.; Glen, T.; Rauch, H.; Yu, H.; Boles, S. Stress–induced phase transformation in shape memory ceramic nanoparticles. *J. Appl. Phys.* **2019**, *126*, 215109. [CrossRef]

204. Jin, X.-J. Martensitic transformation in zirconia containing ceramics and its applications. *Curr. Opin. Solid State Mater. Sci.* **2005**, *9*, 313–318. [CrossRef]

205. Bhattacharyya, A.; Weng, G. An energy criterion for the stress–induced martensitic transformation in a ductile system. *J. Mech. Phys. Solids* **1994**, *42*, 1699–1724. [CrossRef]

206. Marshall, D.; Evans, A.; Drory, M. Transformation toughening in ceramics. *Fract. Mech. Ceram.* **1983**, *6*, 289–307.

207. Wang, C.; Zhou, X.; Wang, M. Influence of sintering temperatures on hardness and Young's modulus of tricalcium phosphate bioceramic by nanoindentation technique. *Mater. Charact.* **2004**, *52*, 301–307. [CrossRef]

208. Li, W.; Rungsiyakull, C.; Zhang, Z.P.; Zhou, S.W.; Swain, M.V.; Ichim, I.; Li, Q. Computational fracture modelling in bioceramic structures. In *Advanced Materials Research*; Trans Tech Publications Ltd.: Bäch, Switzerland, 2011; pp. 853–856.

209. Lin, K.; Sheikh, R.; Romanazzo, S.; Roohani, I. 3D printing of bioceramic scaffolds—Barriers to the clinical translation: From promise to reality, and future perspectives. *Materials* **2019**, *12*, 2660. [CrossRef] [PubMed]

210. Piggott, M. *Load Bearing Fibre Composites*; Springer Science & Business Media: New York, NY, USA, 2002.

211. Rösler, J.; Harders, H.; Bäker, M. *Mechanical Behaviour of Engineering Materials: Metals, Ceramics, Polymers, and Composites*; Springer Science & Business Media: New York, NY, USA, 2007.

212. Shirazi, S.F.S.; Gharehkhani, S.; Metselaar, H.S.C.; Nasiri–Tabrizi, B.; Yarmand, H.; Ahmadi, M.; Abu Osman, N.A. Ion size, loading, and charge determine the mechanical properties, surface apatite, and cell growth of silver and tantalum doped calcium silicate. *RSC Adv.* **2016**, *6*, 190–200. [CrossRef]

213. Guo, D.; Mao, M.; Qi, W.; Li, H.; Ni, P.; Gao, G.; Xu, K. The influence of Sr and H 3 PO 4 concentration on the hydration of SrCaHA bone cement. *J. Mater. Sci. Mater. Med.* **2011**, *22*, 2631–2640. [CrossRef]
214. Guo, D.; Xu, K.; Zhao, X.; Han, Y. Development of a strontium–containing hydroxyapatite bone cement. *Biomaterials* **2005**, *26*, 4073–4083. [CrossRef]
215. Just, A.; Yang, M. Attrition dry milling in continuous and batch modes. In Proceedings of the Powder and Bulk Solids Conference/Exhibition, Chicago, IL, USA, 6–8 May 1997.
216. Wang, K.; Wang, D.; Han, F. Effect of crystalline grain structures on the mechanical properties of twinning–induced plasticity steel. *Acta Mech. Sin.* **2016**, *32*, 181–187. [CrossRef]
217. Wachtman, J.B.; Cannon, W.R.; Matthewson, M.J. *Mechanical Properties of Ceramics*; John Wiley & Sons: Hoboken, NJ, USA, 2009.
218. Kambale, K.; Mahajan, A.; Butee, S. Effect of grain size on the properties of ceramics. *Met. Powder Rep.* **2019**, *74*, 130–136. [CrossRef]
219. Casellas, D.; Nagl, M.; Llanes, L.; Anglada, M. Microstructural Coarsening of Zirconia-Toughened Alumina Composites. *J. Am. Ceram. Soc.* **2005**, *88*, 1958–1963. [CrossRef]
220. Elias, C.N.; Duailibi Filho, J.; Oliveira, L.G.D. Mechanical properties of alumina–zirconia composites for ceramic abutments. *Mater. Res.* **2004**, *7*, 643–649.
221. Shah, M.B. Mechanistic Aspects of Fracture and Fatigue in Resin Based Dental Restorative Composites. Ph.D. Thesis, Oregon State University, Corvallis, OR, USA, 2009.
222. Wiederhorn, S. Brittle fracture and toughening mechanisms in ceramics. *Annu. Rev. Mater. Sci.* **1984**, *14*, 373–403. [CrossRef]
223. Basu, B.; Balani, K. *Advanced Structural Ceramics*; John Wiley & Sons: Hoboken, NJ, USA, 2011.
224. Bhat, M.; Kaur, B.; Kumar, R.; Bamzai, K.; Kotru, P.; Wanklyn, B. Effect of ion irradiation on dielectric and mechanical characteristics of ErFeO3 single crystals. *Nucl. Instrum. Methods Phys. Res. Sect. B Beam Interact. Mater. At.* **2005**, *234*, 494–508. [CrossRef]
225. Parthasarathy, S.; Mertens, J.; Moore, C.; Weisner, C. Utilization and cost impact of integrating substance abuse treatment and primary care. *Med. Care* **2003**, *41*, 357–367. [CrossRef] [PubMed]
226. Rice, R.W. *Mechanical Properties of Ceramics and Composites: Grain and Particle Effects*; CRC Press: Boca Raton, FL, USA, 2000.
227. Souza, R.C.; Barboza, M.J.R.; Baptista, C.A.R.P.; Elias, C.N.; Dos Santos, C.; Strecker, K. Performance of 3Y–TZP bioceramics under cyclic fatigue loading. *Mater. Res.* **2008**, *11*, 89–92. [CrossRef]
228. Haddad, Y.M. *Mechanical Behaviour of Engineering Materials: Volume 2: Dynamic Loading and Intelligent Material Systems*; Springer Science & Business Media: New York, NY, USA, 2013.
229. Pezzotti, G.; Enomoto, Y.; Zhu, W.; Boffelli, M.; Marin, E.; McEntire, B.J. Surface toughness of silicon nitride bioceramics: I, Raman spectroscopy–assisted micromechanics. *J. Mech. Behav. Biomed. Mater.* **2016**, *54*, 328–345. [CrossRef]
230. Krüger, R.; Groll, J. Fiber reinforced calcium phosphate cements—On the way to degradable load bearing bone substitutes? *Biomaterials* **2012**, *33*, 5887–5900. [CrossRef] [PubMed]
231. Fu, S.-Y.; Feng, X.-Q.; Lauke, B.; Mai, Y.-W. Effects of particle size, particle/matrix interface adhesion and particle loading on mechanical properties of particulate–polymer composites. *Compos. Part B Eng.* **2008**, *39*, 933–961. [CrossRef]
232. Elghazel, A.; Taktak, R.; Bouaziz, J. Investigation of Mechanical Behaviour of a Bioceramic. In *Fracture Mechanics: Properties, Patterns and Behaviours*; Alves, L., Ed.; Intech: London, UK, 2016; p. 277.
233. Shokrieh, M.; Ghoreishi, S.; Esmkhani, M. Toughening mechanisms of nanoparticle–reinforced polymers. In *Toughening Mechanisms in Composite Materials*; Elsevier: Amsterdam, The Netherlands, 2015; pp. 295–320.
234. Da Costa Neto, C.A. *Creep and Mechanical Behavior of Silicon Nitride Whisker–Reinforced Silicon Nitride Composite Ceramics*; Illinois Institute of Technology: Chicago, IL, USA, 1996.
235. Rödel, J. Interaction between crack deflection and crack bridging. *J. Eur. Ceram. Soc.* **1992**, *10*, 143–150. [CrossRef]
236. Gu, W.-H.; Faber, K.; Steinbrech, R. Microcracking and R–curve behavior in SiC–TiB2 composites. *Acta Metall. Mater.* **1992**, *40*, 3121–3128. [CrossRef]
237. Evans, A.; Williams, S.; Beaumont, P. On the toughness of particulate filled polymers. *J. Mater. Sci.* **1985**, *20*, 3668–3674. [CrossRef]
238. Evans, A.; Faber, K. Crack growth resistance curve of non–phase transforming ceramics. *J. Am. Ceram. Soc.* **1984**, *67*, 255–260. [CrossRef]
239. Evans, A.G.; Faber, K.T. Toughening of ceramics by circumferential microcracking. *J. Am. Ceram. Soc.* **1981**, *64*, 394–398. [CrossRef]
240. Dolgopolsky, A.; Karbhari, V.; Kwak, S. Microcrack induced toughening—An interaction model. *Acta Metall.* **1989**, *37*, 1349–1354. [CrossRef]
241. Kachanov, M.; Montagut, E. Interaction of a crack with certain microcrack arrays. *Eng. Fract. Mech.* **1986**, *25*, 625–636. [CrossRef]
242. Bengisu, M.; Inal, O.T. Whisker toughening of ceramics: Toughening mechanisms, fabrication, and composite properties. *Annu. Rev. Mater. Sci.* **1994**, *24*, 83–124. [CrossRef]
243. Becher, P.F. Crack bridging processes in toughened ceramics. In *Toughening Mechanisms in Quasi–Brittle Materials*; Springer: New York, NY, USA, 1991; pp. 19–33.
244. Becher, P.F. *Reinforced Ceramics Employing Discontinuous Phases*; Oak Ridge National Lab.: Oak Ridge, TN, USA, 1990.
245. Logsdon, W.; Liaw, P.K. Tensile, fracture toughness and fatigue crack growth rate properties of silicon carbide whisker and particulate reinforced aluminum metal matrix composites. *Eng. Fract. Mech.* **1986**, *24*, 737–751. [CrossRef]
246. Ohji, T.; Jeong, Y.K.; Choa, Y.H.; Niihara, K. Strengthening and toughening mechanisms of ceramic nanocomposites. *J. Am. Ceram. Soc.* **1998**, *81*, 1453–1460. [CrossRef]

247. Phillips, D. Interfacial bonding and the toughness of carbon fibre reinforced glass and glass–ceramics. *J. Mater. Sci.* **1974**, *9*, 1847–1854. [CrossRef]

248. Basista, M.; Węglewski, W. Modelling of damage and fracture in ceramic matrix composites—An overview. *J. Theor. Appl. Mech.* **2006**, *44*, 455–484.

249. Clegg, R.E.; Paterson, G. Ductile particle toughening of hydroxyapatite ceramics using platinum particles. In Proceedings of the SIF2004 Structural Integrity and Fracture, Brisbane, Australia, 26–29 September 2004.

250. Barenblatt, G.I. The mathematical theory of equilibrium cracks in brittle fracture. In *Advances in Applied Mechanics*; Elsevier: Amsterdam, The Netherlands, 1962; Volume 7, pp. 55–129.

251. Schumacher, M.; Gelinsky, M. Strontium modified calcium phosphate cements–approaches towards targeted stimulation of bone turnover. *J. Mater. Chem. B* **2015**, *3*, 4626–4640. [CrossRef]

252. Shi, H.; Zhou, Z.; Li, W.; Fan, Y.; Li, Z.; Wei, J. Hydroxyapatite based materials for bone tissue engineering: A brief and comprehensive introduction. *Crystals* **2021**, *11*, 149. [CrossRef]

253. Kulkarni, M.; Mazare, A.; Gongadze, E.; Perutkova, Š.; Kralj–Iglič, V.; Milošev, I.; Schmuki, P.; Iglič, A.; Mozetič, M.J.N. Titanium nanostructures for biomedical applications. *Nanotechnology* **2015**, *26*, 062002. [CrossRef]

254. Milewski, J.V. Whiskers and short fiber technology. *Polym. Compos.* **1992**, *13*, 223–236. [CrossRef]

255. Jones, R.M. *Mechanics of Composite Materials*; CRC Press: Boca Raton, FL, USA, 2018.

256. Belitskus, D. *Fiber and Whisker Reinforced Ceramics for Structural Applications*; CRC Press: Boca Raton, FL, USA, 1993.

257. Hillig, W. Strength and toughness of ceramic matrix composites. *Annu. Rev. Mater. Sci.* **1987**, *17*, 341–383. [CrossRef]

258. Wereszczak, A.A. Toughening Mechanisms in Whisker/Short Fiber–Reinforced Ceramic Matrix Composites at Elevated Temperatures. Ph.D. Thesis, University of Delaware, Newark, DE, USA, 1992.

259. Miyamoto, A.; Lee, S.; Cooray, N.F.; Lee, S.; Mori, M.; Matsuhisa, N.; Jin, H.; Yoda, L.; Yokota, T.; Itoh, A.; et al. Inflammation–free, gas–permeable, lightweight, stretchable on–skin electronics with nanomeshes. *Nat. Nanotechnol.* **2017**, *12*, 907–913. [CrossRef]

260. Phakatkar, A.H.; Shirdar, M.R.; Qi, M.-L.; Taheri, M.M.; Narayanan, S.; Foroozan, T.; Sharifi-Asl, S.; Huang, Z.; Agrawal, M.; Lu, Y.-P.; et al. Novel PMMA bone cement nanocomposites containing magnesium phosphate nanosheets and hydroxyapatite nanofibers. *Mater. Sci. Eng. C* **2020**, *109*, 110497. [CrossRef]

261. Wu, J.; Cao, L.; Liu, Y.; Zheng, A.; Jiao, D.; Zeng, D.; Wang, X.; Kaplan, D.L.; Jiang, X. Functionalization of silk fibroin electrospun scaffolds via BMSC affinity peptide grafting through oxidative self–polymerization of dopamine for bone regeneration. *ACS Appl. Mater. Interfaces* **2019**, *11*, 8878–8895. [CrossRef] [PubMed]

262. Li, X.; Zhao, Y.; Li, D.; Zhang, G.; Long, S.; Wang, H. Hybrid dual crosslinked polyacrylic acid hydrogels with ultrahigh mechanical strength, toughness and self–healing properties via soaking salt solution. *Polymer* **2017**, *121*, 55–63. [CrossRef]

263. Fujie, K.; Yamada, T.; Ikeda, R.; Kitagawa, H. Introduction of an ionic liquid into the micropores of a metal–organic framework and its anomalous phase behavior. *Angew. Chem.* **2014**, *126*, 11484–11487. [CrossRef]

264. Peng, S.; Cao, L.; He, S.; Zhong, Y.; Ma, H.; Zhang, Y.; Shuai, C. An overview of long noncoding RNAs involved in bone regeneration from mesenchymal stem cells. *Stem Cells Int.* **2018**, *2018*, 1–11. [CrossRef] [PubMed]

265. Fujie, T.; Ricotti, L.; Desii, A.; Menciassi, A.; Dario, P.; Mattoli, V. Evaluation of substrata effect on cell adhesion properties using freestanding poly (l–lactic acid) nanosheets. *Langmuir* **2011**, *27*, 13173–13182. [CrossRef] [PubMed]

266. Feng, L.-L.; Yu, G.; Wu, Y.; Li, G.-D.; Li, H.; Sun, Y.; Asefa, T.; Chen, W.; Zou, X. High–index faceted Ni3S2 nanosheet arrays as highly active and ultrastable electrocatalysts for water splitting. *J. Am. Chem. Soc.* **2015**, *137*, 14023–14026. [CrossRef] [PubMed]

267. Saranya, N.; Moorthi, A.; Saravanan, S.; Devi, M.P.; Selvamurugan, N. Chitosan and its derivatives for gene delivery. *J. Biol. Macromol.* **2011**, *48*, 234–238. [CrossRef]

268. Feng, P.; Kong, Y.; Yu, L.; Li, Y.; Gao, C.; Peng, S.; Pan, H.; Zhao, Z.; Shuai, C. Molybdenum disulfide nanosheets embedded with nanodiamond particles: Co–dispersion nanostructures as reinforcements for polymer scaffolds. *Appl. Mater. Today* **2019**, *17*, 216–226. [CrossRef]

269. Zhang, X.; Xie, X.; Wang, H.; Zhang, J.; Pan, B.; Xie, Y. Enhanced photoresponsive ultrathin graphitic–phase C3N4 nanosheets for bioimaging. *J. Am. Chem. Soc.* **2013**, *135*, 18–21. [CrossRef]

270. Zhang, C. Understanding the wear and tribological properties of ceramic matrix composites. In *Advances in Ceramic Matrix Composites*; Elsevier: Amsterdam, The Netherlands, 2014; pp. 312–339.

271. Mittal, G.; Rhee, K.Y.; Mišković–Stanković, V.; Hui, D. Reinforcements in multi–scale polymer composites: Processing, properties, and applications. *Compos. Part B Eng.* **2018**, *138*, 122–139. [CrossRef]

272. Prashanth, S.; Subbaya, K.; Nithin, K.; Sachhidananda, S. Fiber reinforced composites—A review. *J. Mater. Sci. Eng.* **2017**, *6*, 2–6.

273. Berguiga, L.; Bellessa, J.; Vocanson, F.; Bernstein, E.; Plenet, J. Carbon nanotube silica glass composites in thin films by the sol–gel technique. *Opt. Mater.* **2006**, *28*, 167–171. [CrossRef]

274. Huang, Q.; Gao, L.; Sun, J. Effect of adding carbon nanotubes on microstructure, phase transformation, and mechanical property of BaTiO3 ceramics. *J. Am. Ceram. Soc.* **2005**, *88*, 3515–3518. [CrossRef]

275. Wang, X.; Padture, N.P.; Tanaka, H. Contact–damage–resistant ceramic/single–wall carbon nanotubes and ceramic/graphite composites. *Nat. Mater.* **2004**, *3*, 539–544. [CrossRef]

276. Zakharov, N.; Safonova, A.; Orlov, M.; Demina, L.; Aliev, A.; Kiselev, M.; Matveev, V.; Shelekhov, E.; Zakharova, T.; Kuznetsov, N. Synthesis and properties of calcium hydroxyapatite/carbon fiber composites. *Russ. J. Inorg. Chem.* **2017**, *62*, 1162–1172. [CrossRef]

277. Tian, W.-B.; Kan, Y.-M.; Zhang, G.-J.; Wang, P.-L. Effect of carbon nanotubes on the properties of ZrB_2–SiC ceramics. *Mater. Sci. Eng. A* **2008**, *487*, 568–573. [CrossRef]

278. Williams, P.T.; Reed, A.R. Development of activated carbon pore structure via physical and chemical activation of biomass fibre waste. *Biomass Bioenergy* **2006**, *30*, 144–152. [CrossRef]

279. Sarkar, C.; Sahu, S.K.; Sinha, A.; Chakraborty, J.; Garai, S. Facile synthesis of carbon fiber reinforced polymer–hydroxyapatite ternary composite: A mechanically strong bioactive bone graft. *Mater. Sci. Eng. C* **2019**, *97*, 388–396. [CrossRef]

280. Shen, L.; Yang, H.; Ying, J.; Qiao, F.; Peng, M. Preparation and mechanical properties of carbon fiber reinforced hydroxyapatite/polylactide biocomposites. *J. Mater. Sci. Mater. Med.* **2009**, *20*, 2259–2265. [CrossRef] [PubMed]

281. Uddin, M.N.; Dhanasekaran, P.S.; Asmatulu, R. Mechanical properties of highly porous PEEK bionanocomposites incorporated with carbon and hydroxyapatite nanoparticles for scaffold applications. *Prog. Biomater.* **2019**, *8*, 211–221. [CrossRef] [PubMed]

282. Nayak, L.; Rahaman, M.; Giri, R. Surface modification/functionalization of carbon materials by different techniques: An overview. In *Carbon–Containing Polymer Composites*; Springer: Singapore, 2019; pp. 65–98. [CrossRef]

283. Rasheed, A.; Howe, J.Y.; Dadmun, M.D.; Britt, P.F. The efficiency of the oxidation of carbon nanofibers with various oxidizing agents. *Carbon* **2007**, *45*, 1072–1080. [CrossRef]

284. Agrawal, D. Microwave sintering of ceramics, composites and metal powders. In *Sintering of Advanced Materials*; Elsevier: Amsterdam, The Netherlands, 2010; pp. 222–248.

285. Ring, T.A. *Fundamentals of Ceramic Powder Processing and Synthesis*; Elsevier: Amsterdam, The Netherlands, 1996.

286. Hench, L.L. *An Introduction to Bioceramics*; World Scientific: Singapore, 1993; Volume 1.

287. Rahaman, M.; Aldalbahi, A.; Bhagabati, P. Preparation/Processing of Polymer–Carbon Composites by Different Techniques. In *Carbon–Containing Polymer Composites*; Springer: Singapore, 2019; pp. 99–124.

288. Zhang, H.; Liu, J.; Yao, Z.; Yang, J.; Pan, L.; Chen, Z. Biomimetic mineralization of electrospun poly(lactic–co–glycolic acid)/multi-walled carbon nanotubes composite scaffolds in vitro. *Mater. Lett.* **2009**, *63*, 2313–2316. [CrossRef]

289. Rahaman, M.N. *Sintering of Ceramics*; CRC Press: Boca Raton, FL, USA, 2007.

290. Angelo, P.; Subramanian, R. *Powder Metallurgy: Science, Technology and Applications*; PHI Learning Pvt. Ltd.: New Delhi, India, 2008.

291. Boda, S.K.; Basu, B.; Sahoo, B. Structural and magnetic phase transformations of hydroxyapatite–magnetite composites under inert and ambient sintering atmospheres. *J. Phys. Chem. C* **2015**, *119*, 6539–6555. [CrossRef]

292. Porz, L.; Klomp, A.J.; Fang, X.; Li, N.; Yildirim, C.; Detlefs, C.; Bruder, E.; Höfling, M.; Rheinheimer, W.; Patterson, E.A.; et al. Dislocation–toughened ceramics. *Mater. Horiz.* **2021**, *8*, 1528–1537. [CrossRef]

Journal of
Composites Science

Article

Mechanically Stable β-TCP Structural Hybrid Scaffolds for Potential Bone Replacement

Matthias Ahlhelm [1,*], Sergio H. Latorre [2], Hermann O. Mayr [2], Christiane Storch [3], Christian Freytag [4], David Werner [4], Eric Schwarzer-Fischer [4] and Michael Seidenstücker [2,*]

[1] Fraunhofer Institute for Ceramic Technologies and Systems, IKTS, Winterbergstraße 28, 01277 Dresden, Germany

[2] G.E.R.N. Center for Tissue Replacement, Regeneration & Neogenesis, Department of Orthopedics and Trauma Surgery, Medical Center-Albert-Ludwigs-University of Freiburg, Faculty of Medicine, Albert-Ludwigs-University of Freiburg, Hugstetter Straße 55, 79106 Freiburg, Germany; sergio.latorre@uniklinik-freiburg.de (S.H.L.); hermann.mayr@uniklinik-freiburg.de (H.O.M.)

[3] Fraunhofer Institute for Cell Therapy and Immunology, IZI, Perlickstraße 1, 04103 Leipzig, Germany; christiane.storch@izi.fraunhofer.de

[4] Fraunhofer Institute for Ceramic Technologies and Systems, IKTS, Maria-Reiche-Str. 2, 01109 Dresden, Germany; Christian.Freytag@ikts.fraunhofer.de (C.F.); david.werner@ikts.fraunhofer.de (D.W.); eric.schwarzer@ikts.fraunhofer.de (E.S.-F.)

* Correspondence: matthias.ahlhelm@ikts.fraunhofer.de (M.A.); michael.seidenstuecker@uniklinik-freiburg.de (M.S.); Tel.: +49-351-888-15778 (M.A.); +49-761-270-26104 (M.S.)

Citation: Ahlhelm, M.; Latorre, S.H.; Mayr, H.O.; Storch, C.; Freytag, C.; Werner, D.; Schwarzer-Fischer, E.; Seidenstücker, M. Mechanically Stable β-TCP Structural Hybrid Scaffolds for Potential Bone Replacement. *J. Compos. Sci.* **2021**, *5*, 281. https://doi.org/10.3390/jcs5100281

Academic Editors: Corrado Piconi and Simone Sprio

Received: 25 August 2021
Accepted: 9 October 2021
Published: 17 October 2021

Publisher's Note: MDPI stays neutral with regard to jurisdictional claims in published maps and institutional affiliations.

Abstract: The authors report on the manufacturing of mechanically stable β-tricalcium phosphate (β-TCP) structural hybrid scaffolds via the combination of additive manufacturing (CerAM VPP) and Freeze Foaming for engineering a potential bone replacement. In the first step, load bearing support structures were designed via FE simulation and 3D printed by CerAM VPP. In the second step, structures were foamed-in with a porous and degradable calcium phosphate (CaP) ceramic that mimics porous *spongiosa*. For this purpose, Fraunhofer IKTS used a process known as Freeze Foaming, which allows the foaming of any powdery material and the foaming-in into near-net-shape structures. Using a joint heat treatment, both structural components fused to form a structural hybrid. This bone construct had a 25-fold increased compressive strength compared to the pure CaP Freeze Foam and excellent biocompatibility with human osteoblastic MG-63 cells when compared to a bone grafting Curasan material for benchmark.

Keywords: Freeze Foam; hybrid bone; biocompatibility; bone replacement

1. Introduction

As reported in the U.S., 7.9 million fractures occur annually, of which 5–10% develop non-unions and/or delayed unions, which are major sources of complications in the treatment of bone fractures [1]. In 2005, 17 billion dollars in medical costs were attributed to the treatment of fractures caused by osteoporosis alone. By 2025, costs are estimated to rise to 25 billion dollars [2,3]. These numbers highlight the importance of achieving early mechanical stability and load-bearing capability in long weight-bearing bones. For this reason, the successful treatment of bone defects is of great importance. Vascularity and mechanical stability need to be taken into account. The "gold standard" of many surgical techniques used to reconstruct bone for critical-sized bone defects is the use of autologous bone tissue [4]. However, the use of autografts has limitations, like donor-side morbidity, additional operations, or limited availability of tissue, as well as geometric mismatch between the harvested bone and the defect site, which can result in voids and poor integration [5,6]. Further alternative substitutes are allografts and xenografts [7]. Xenograft (animal-derived material) approaches often carry risks, like inflammation and rejection of the transplant due to physiological incompatibility of animal organs in human beings [8]. Therefore, research activity concerning bone-grafting approaches has shifted from natural grafts to synthetic bone graft substitutes and the use of biological factors [9]. Among these materials, next to metals (e.g., titanium, titanium alloys [10]) and bioglasses [11], ceramics like calcium phosphates (CaP; e.g., tricalcium phosphates (TCP), hydroxyapatite (HAp)) as well as added active growth factor recombinant human bone morphological proteins (e.g., rhBMPs) are the typical materials of choice, either alone or in combination [9,12]. Calcium phosphate ceramics are among the most commonly used and effective synthetic bone replacement materials. For example, β-tricalcium phosphate (β-TCP) is osteoconductive and is integrated into the bone without a disturbing connective tissue layer [13,14]. This property combined with its cell-mediated resorption enables the complete regeneration of bone defects. Pores and especially micropores (0.1 to 10 μm) promote bone ingrowth and can give β-TCP osteoinductive properties [13,15–18]. From this, it can be deduced that the bone implants should have similar properties, in terms of porosity, strength and stiffness, to the piece of bone to be replaced. If the strength of the implant is too low, there is a risk that the component will fail after implantation. However, if the strength and stiffness of the implant is too high, the surrounding bone will degrade. This process is known as "stress-shielding" and, like component failure, should be avoided at all costs [19,20]. The compressive strength of spongy bone is between 2–20 MPa, depending on the literature reference. With regard to porosity, cancellous bone has values between 50–90%, which explains the low mechanical load capacity. At the same time, however, the bone becomes light and the pores enable the supply of nutrients to the bone and the removal of metabolic products [21,22]. The majority of studies on CaP scaffolds focus on bone growth in the macropores (>100 μm), where bone structures such as osteones and trabeculae can form. However, more and more studies show that micropores (<50 μm) also play an important role. Not only do they improve bone growth in the macropores, but they also provide additional space for bone growth [23]. Bone growth in the micropores offers great mechanical advantages in CaP scaffolds, as it optimizes the properties of otherwise brittle materials by further stabilizing the implant and improving load transfer.

To date, specific material combinations have been examined in order to combine the tissue engineering advantages of organic materials with the mechanical load resistance of inorganic materials. Examples of such composite biomaterials are given in [24] and [25]. The latter reports on inorganic-organic hybrid scaffolds. Polyethylene glycol (PEG) and star poly(dimethylsiloxane) were mixed with bone-like matrices collagen type I, CaP, and osteocalcin, indicating that developed hybrid gels may prove promising for osteochondral regeneration. However, the compressive strength was limited by these polymers. In addition, no specific 3D construct was achievable. Ref. [24] reports on the fabrication of porous SF/β-TCP hybrid scaffolds for bone tissue reconstruction by a freeze-drying process. The manufactured scaffolds demonstrated high porosity (>60%) with good inter-pore connectivity and showed good biocompatibility. However, compressive strength and modulus were relatively low (<1 MPa), and no complex 3D scaffold was achieved.

Before the development of additive manufacturing technologies, ceramic bone replacement structures were usually manufactured using so-called dip coating processes in order to be able to approximately reproduce the filigree and highly porous structures [26]. Foams made of polyurethane, for example, serve as a lost form in this process. The polymer foam is cut to size and dipped into a ceramic slurry, which penetrates the pores. In a subsequent processing step, the foam mold is then burned out, and the scaffold is sintered. Although this approach can be used to produce highly porous structures, the resulting geometry is not greatly influenced but is rather predetermined by the (PU)foam. This disadvantage is overcome by the use of specific direct foaming or additive manufacturing technologies, since the mold geometry can be specifically modeled or is not needed at all. An example of an additively manufactured SiO_2- and zinc-doped β-TCP scaffold is given in [27]. Although the results indicate that addition of dopants to the TCP scaffolds enhanced early stages of bone formation and implant fixation when compared to pure

TCP alone in a rabbit tibia model, the compressive strength of the achieved scaffolds only amounted to around 6 MPa.

The human bone consists of a dense and solid outer shell (*Substantia corticalis*) and an inner porous filling (*Substantia spongiosa*). In order to be able to reproduce such bone architectures with different structures, which could be used as implants in the future, two technologies were recently intelligently combined. The outer shell of the bone was produced using a commercial three-dimensional (3D) printer, and the sponge-like inner bone structure was reproduced by a ceramic foam [28,29]. For the foam production, so-called Freeze Foaming was used. In this approach, in a freeze dryer, the ambient pressure around an aqueous ceramic suspension is lowered, causing the suspension to first foam and then to suddenly freeze. Ongoing pressure reduction lets the frozen water sublimate, i.e., it evaporates without becoming liquid beforehand. A subsequent heat treatment produces a solid ceramic foam. In the next step, the porous bone-like structures are fitted to a customized, complex outer ceramic shell and, thereby, made mechanically more stable. This is where additive manufacturing (AM) comes into play. One of the best-known processes in AM is the conventional stereolithography (SLA) process. This process basically allows photopolymerizable suspensions, which are filled with ceramic particles, to be cured by a UV laser. Today, the commercially available material portfolio using lithography-based ceramic manufacturing (LCM) for high-performance components also works with β-TCP, thus playing a role in this contribution. The LCM technology as a projection-based (PSL) top-down process with a light source in the blue range (452–465 nm) is representative of the so-called ceramic additive manufacturing vat photopolymerization (CerAM VPP) process (Ceramic Additive Manufacturing Vat Photopolymerization). This allows a digital micro-mirror unit, which splits a light beam into individual pixels and then projects a digital image pixel-by-pixel onto the building platform. This makes it possible to image the entire contour of the component cross-section without a mask. Thus, layer by layer, a complex 3D structure is created. In a last hybridization step, the two methods can be combined to produce porous-dense, graded, structural hybrids by a joint sintering process. However, it is possible to not only foam within additively manufactured structures but also to foam them in. This solution makes it possible to provide a porous and sponge-like scaffold as the lead structure for cells to grow into, and at the same time, AM parts serve as load-bearing support structures. In this current study, advanced scaffolds made of β-TCP were manufactured and analyzed in terms of their biocompatibility in vitro and in vivo and tested for their mechanical behavior. The authors postulated that such a complex inorganic hybrid structure, due to the combination of load-bearing support and porous cell-ingrowth-allowing interior, will eventually allow the manufacturing of bone-like mechanically stable implants that are potentially applicable for long-bone defects.

2. Materials and Methods

2.1. Freeze Foaming

Hydroxyapatite (Sigma-Aldrich, now Merck KGaA, Darmstadt, Germany; BET = 70.01 m^2/g, d_{50} = 2.64 µm) was chosen as the raw material. Prior to suspension, it was calcined at 900 °C for 2 h to reduce the BET (now only 5.9 m^2/g). The ceramic suspensions consisted of water, Dolapix CE 64 (Co. Zschimmer & Schwarz Mohsdorf GmbH & Co. KG, Burgstädt, Germany) as a dispersing agent, the ceramic powder, polyvinyl alcohol as the binder and a rheological modifier (Tafigel AP15, Co. Münzing Chemie GmbH, Heilbronn, Germany) in combination with 2-Amino-2-methyl-1-propanol—AMP (Merck KGaA, 64,271 Darmstadt, Germany) for pH adjustment. The following processing route was used: 49 wt.% deionized water, 1.3 wt.% polyvinyl alcoholic binder, hydroxyapatite and 4.6 wt.% dispersing agent, referring to powder content, were mixed in a centrifugal vacuum mixer (ARV310, Thinky Corporation, Fukuoka, Japan). To disperse the particles and reduce agglomeration, the mixture was exposed to a high stirring rate (2000 rpm, mixing time 1 min, with 3 ZrO$_2$ mixing spheres of 10 mm diameter). The spheres were then separated, and 1.9 wt.% rheological modifier together with 1.5% wt. AMP was added. To distribute the modifier, the suspension was mixed for 2 min at 1500 rpm. Afterwards, the suspensions were filled into specific molds (see Section 2.3) and transferred to a freeze dryer (Lyo Alpha 2–4, LSCplus, Co. Martin Christ Gefriertrocknungsanlagen GmbH, Osterode, Germany) for Freeze Foaming.

Freeze Foaming for In Vivo Studies

For in vivo studies, the potential of using the Freeze Foaming process and manufacturing structural hybrids was assessed for the manufacture of artificial rat bones. A femoral bone from an 11-week-old rat was scanned using computer tomography followed by CAD file reverse-engineering. From this, a femoral bone segment was chosen, and single Freeze Foams as well as structural hybrids were manufactured (see Section 2.4). In addition to using HAp as the initial material, ZrO_2 (TZ-3YS-E; Co. TOSOH with a $d_{50} = 0.7$ µm) was used as a bioinert counterpart for manufacturing porous Freeze Foams. In doing so, we hoped to be able to differentiate between material effects and structural effects regarding possible in vitro/vivo results. Ideally, the Freeze Foam's characteristic porous structure/pore morphology would perform independently from the bioceramic materials.

2.2. FE Analysis and Material Failure Model

For providing a mechanical stability of porous bioceramics sufficient to address long-bone defects, a support structure is needed. This is offered either by providing a shell-like structure that mimics a complete artificial *corticalis* and/or by an outside, broad, accessible support structure that provides adequate strength. Both approaches are presented. However, the focus was on an outside accessible support. A simple column geometry was chosen as such support (Figure 1a, left). Its additive manufacturing by the VPP process should be uncomplicated and provide required mechanical support as well as load balance for inside lying foams. A FE (Finite Element) analysis was made (ANSYS v. 2020 R2, ANSYS Inc., Pittsburgh, PA, USA) to approximate mechanical loads appearing in the structure and, additionally, to identify possible locations of material failure. Therefore, a geometrical model was created with respect to the existing symmetry conditions (Figure 1a, right).

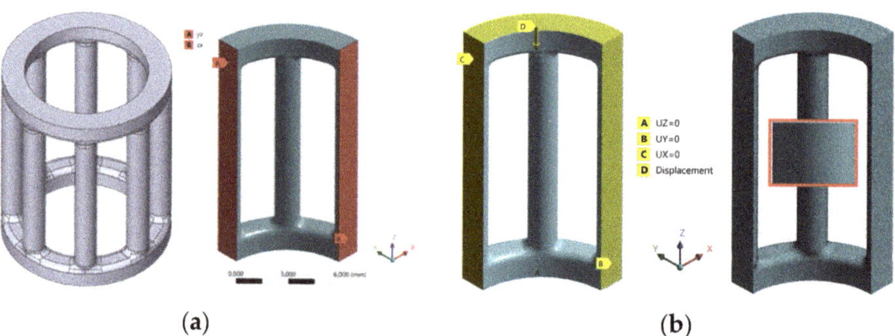

(a)

(b)

Figure 1. Simple column structure and boundary conditions: (**a**) column structure (left) for simulation using geometry regarding symmetry planes (right); (**b**) boundary conditions (left); meshed geometry (right).

Corresponding boundary conditions were applied to the model (Figure 1b, left), which implies fixation of translational DOF (Degree of Freedom) in the z-direction at the bottom face, symmetry conditions in the x- and y-direction and a given displacement at the top face of the structure. A displacement instead of a pressure load was used to match experimental conditions where a press specifies a defined displacement, and the resulting pressure (reaction force) was measured. Displacement values for the model were selected to achieve reaction forces of about 800 N for 1/4 of the geometry and 3200 N for the complete geometry. This load was well estimated because experimental data indeed showed that the full column structure withstood average loads up to 3200 N (see Section 3.2 on mechanical characterization). Table 1 shows the displacements and reaction forces used for the simulation. Three load cases were chosen to simulate the complex deformation of the ceramic VPP structure. Because of the well-known brittleness of ceramic hydroxyapatite/TCP [30], the load was slowly increased, and the deformation was determined starting from a 780 N reaction force. Material failure should occur suddenly at a certain point, as shown by simulation results.

Table 1. The displacement, reaction forces and respective maximum principal stresses used in the structure.

Load Case	Displacement/mm	Reaction Force/N	Maximum Principal Stress/MPa
1st	−0.41	780	61
2nd	−0.45	806	80
3rd	−0.48	809	128

The chosen material for the model geometry was β-TCP. Young's modulus can be assumed to be 5.37 GPa at a porosity of 18% (previously measured [31]). Previously measured porosities of sintered VPP column structures varied between 5 and 20%, because process conditions, not yet thoroughly optimized to the β-TCP suspension, led to micro delamination between layers and/or cavities within the bulk material. For simulations, a porosity of 5% was selected, representing a stable and load-bearing material. Young's modulus, depending on porosity, can be estimated using the following Equation [32]:

$$E(\phi) = E_0(1 - 2\phi) \tag{1}$$

where $E(0.05)$ = 7.552 GPa for porosity of 5% (E is Young's modulus and ϕ is porosity). There was no given data for Poisson's ratio. To be reasonable, 0.22 was chosen. The model geometry was discretized with a tetrahedral mesh of an appropriate density to get reliable results for a mechanical solution (Figure 1b, right). It was solved considering non-linear mechanical behavior (large deflection effects) and a high-resolution load stepping for simulation runtime to capture complex deformation of the model. Whether those material failure studies fit the experimental compressive strength tests is further shown in Section 3.2.

2.3. CerAM VPP

As bioceramic material, the same calcined HAp was used as for the Freeze Foams. HAp (content: 40 vol.-%) was dispersed in a fluid (polyethylene glycol, Sigma-Aldrich, now Merck KGaA, Darmstadt, Germany) with a dispersing agent (BYK-Chemie), and various monomers were used a as binder (a mixture of acrylic resins) and a photoiniator (combination of a camphor derivate and an amine). A planetary centrifugal high-speed vacuum mixer (Thinky ARV310, Thinky Corporation, Tokyo, Japan) was used for the stepwise (three times 5 min at 2000 rpm) preparation of the suspensions. Following CerAM VPP manufacturing, the column geometry as defined in Section 2.2 was chosen.

2.4. Mold Filling, Hybridization and Part Characterization

For achieving the hybrid parts, cylindrical rubber molds were made, in which the CerAM VPP parts fit in very closely. In the first step, the ceramic suspension filled the molds. Then the column structure was pushed into the cavity. All molds were then transferred to the freeze dryer and foamed at once. A crucial step in creating the structural hybrids is the adjustment of the shrinkage of the two different structures. At the beginning, different suspensions were developed having different contents of water, rheological modifier and binder. With those suspensions, Freeze Foams were manufactured, which shrank between 30–46% (determined by thermo dilatometry DIL 402 C/7/G Netzsch-Gerätebau, Selb, Wunsiedel, Germany). The VPP-manufactured column structure shrank around 30% altogether. However, it was found that the VPP part shrank around 5.4% at the beginning of the heating process (the debinding), whereas the foam did not. As a result, the foams would have shrunk onto the VPP part, leading to a part failure. To compensate for the overall shrinkage, the VPP columns were pre-sintered prior to being used for in situ Freeze Foaming. This pre-sintering amounted to a shrinkage of around 5% volume. Thus, carefully adjusted for shrinkage, the hybrid parts were sintered at 1250 K (+50 K overheating effect) for 1.5 h. Afterwards, the column-including foams were dismantled and evaluated regarding porosity and microstructure. It must be noted that, after sintering, the initial hydroxyapatite was changed to β-TCP. Among many other references reporting about the transition of HA to TCP during heat treatment, similar Freeze Foams with the same initial HA powder were analyzed via XRD in a previous work [22], showing the HAp to TCP transformation. For microstructure analysis, the resulting Freeze Foams were characterized by SEM (Ultra 55, Co. Carl Zeiss, Oberkochen, Germany). By measuring the height and diameter of three different foam positions of manufactured Freeze Foams and deriving the average, geometrical porosities were calculated according to (P = porosity, ρ_{th} = theoretical density, ρ_{bulk} = bulk density):

$$P = 1 - (\rho_{th}/\rho_{bulk}) \tag{2}$$

In addition, porosity was determined via a foam structure analysis tool based on computer tomographic images of the manufactured parts. The allocation of that 3D volumetric pore morphology information (foam cell size) was managed using VGStudio Max v3.0 (Volume Graphics GmbH, Heidelberg, Germany). For X-ray computed tomography, a CT-Compact (Procon X-ray, max. 150 kV power) was used. The universal electromechanical testing machine Instron 8562 (Norwood, MA, USA) was used for the compression strength tests (load cell 10 kN, 1-Taster). To determine the surface roughness, the hybrid foam samples and the Cerasorb M (Kleinostheim, Germany) control were examined using a KEYENCE 3D Laser Scanning Microscope VK-X210 (Keyence, Osaka, Japan). The surface roughness (Sa) was determined using KEYENCE VK analysis software version 3.5.0.0. At least 3 different samples of each were analyzed. Five different positions were determined for each sample. The specimens were measured using 400x magnification.

2.5. In Vitro Biocompatibility

For comparison purposes, commercially available β-TCP ceramics, cylindrical Cerasorb M moldings with a diameter of 7 mm and a length of 25 mm, were purchased from Curasan (Kleinostheim, Germany). In all experiments, we used our hybrid foam scaffolds, consisting of (TCP) and, for comparison, β-TCP scaffold (Cerasorb M) cubes with an edge length of 15 mm and regular parallel macropores (1.2–1.4 mm), purchased from Curasan (Kleinostheim, Germany). Biocompatibility experiments were performed using the human osteoblastic MG-63 cells (ATCC CRL 1427). The cells were first thawed from the liquid nitrogen tank (at $-196\,^{\circ}$C) in passage 15 and cultured in a Dulbecco's Modified Eagle Medium (DMEM) with an F12 nutrient content and additives consisting of 1% penicillin/streptomycin (P/S) and 10% fetal bovine serum (FBS). Cells were maintained in a New Brunswick Galaxy 170R incubator (Eppendorf, Hamburg, Germany) at 37 $^{\circ}$C, with a CO_2 saturation of 5%. The cells were passaged twice a week and then split 1:10 and 1:5. For all biocompatibility tests and experiments with SBF, the scaffolds were heat-sterilized at 200 $^{\circ}$C for 4 h in a UF500 drying oven (Memmert, Schwabach, Germany). Each experiment was repeated 3 times.

2.5.1. Live/Dead Assay

The live/dead examinations were performed after 3, 7, and 10 days. Three samples per scaffold (DD, Curasan) per time period were placed in cell culture plates. Subsequently, 50,000 cells each, which were in 200 µL of medium, were placed directly onto the samples and incubated for 2 h at 37 $^{\circ}$C, with a CO_2 saturation of 5% in the incubator so that the cells could adhere to the surface of the samples. After two hours, 2.5 mL of a DMEM-F12 (Art.

No. BE12-719F, Lonza, Basel, Switzerland) complete medium was added to each well and incubated in the incubator for a defined time (3, 7, and 10 days). After this, the samples were prepared for staining. The staining solution was first prepared by adding 2 mL of DPBS (Art. No. 14190-094, Gibco, Grand Island, NE, USA) to a Falcon and 4 μL of Ethidium Homodimer III (Eth D-III) solution. The solution was then mixed. Then, 1 μL of calcein dye was added and mixed again. Finally, the prepared solution was covered with aluminum foil due to the sensitive fluorescent dye. For staining after the first cultivation, the medium was removed, and the cells were washed to eliminate serum esterase activity. Subsequently, the cells were stained according to the protocol [19]. After incubation, the cells were inspected under a fluorescence microscope. For evaluation, images were taken with an Olympus fluorescence microscope (BX51, Olympus, Tokyo, Japan) from five different positions, with $5\times$ and $10\times$ magnification on the scaffolds. Then, the ceramics were cut horizontally and viewed at the same three positions with the known magnifications. Living cells fluoresced green under blue light, and dead cells fluoresced red.

2.5.2. Cell Proliferation Assay

Three samples of each of the differently sized scaffolds were examined after 3, 7 and 10 days using the WST-1 test. A Nunc™ Thermanox™ Coverslip (Thermo Fisher Scientific, Waltham, MA, USA) membrane served as the positive control. All samples and controls were equally covered with 50,000 cells in 200 μL. The cells were incubated for 2 h at 37 °C, with a CO_2 saturation of 5% in the incubator so that they could adhere to the surface of the sample. At the end of this period, 2.5 mL of the DMEM-F12 complete medium was added to each sample and incubated. A medium change with the DMEM-F12 with the 10% FBS and 1% P/S additives was performed for days 7 and 10. The plate from day 3 was prepared for the WST evaluation. The medium was aspirated, and the wells were washed three times with PBS. The samples and the Thermanox coverslips were then transferred to a new well, and then 2.5 mL of the DMEM-F12 phenol red free (Art. No. 11039-021, Gibco, Grand Island, NE, USA) with the 1% P/S and 1% FBS additives were added to the wells with the sample (TCP + R). A total of 400 microliters of the medium was added to the previously used empty sample wells (TCP), positive control (C + R), empty control well (C+) and the blank. The blank contained only the DMEM medium without phenol red and was measured to account for background absorption. A 10% WST reagent (Art. No. 05015944001, Roche, Basel, Switzerland) was added to the corresponding volume of medium. Thus, 250 μL WST was added to the wells with sample (TCP + R), and 40 μL was added to the old wells (TCP and C+), the blank wells and the positive control (C+). This was incubated in an incubator at 37 °C for 2 h. After this time, the liquids were transferred into a 96-well plate. Three times in a row, 100 μL of each solution was added to the wells. The absorption was then measured at 450 nm using a Spectrostar Nano microplate reader (BMG Labtech, Ortenberg, Germany). The experiment was performed at least three times for each time point (3, 7, and 10 days).

2.5.3. Lactate Dehydrogenase (LDH) Assay

The scaffolds for use in the lactate dehydrogenase LDH experiment were seeded in three 12-well plates. Each experiment assessed three scaffolds from each size, three Thermanox coverslips each as controls, a positive control, a negative control, and a blank to account for background absorbance in the ELISA reader. The experiments were repeated at least three times. A 200 μL cell solution containing 50,000 cells was seeded onto each scaffold, and a 100 μL cell solution containing 50,000 cells was seeded onto the Thermanox coverslips and additionally into two empty wells to act as the positive and negative controls, respectively. One well was left empty for use as a blank. The well plate was placed in an incubator at 37 °C with 5% CO_2 for 2 h. Following incubation, 2.5 mL of DMEM-F12 phenol red free with the 1% P/S and 1% FBS additives was added into the samples wells and negative control wells. Since FBS itself contains LDH, a concentration of 10% in the medium might have triggered background absorption. Therefore, only a concentration of 1% FBS was added to the medium. For the positive controls, 1% Triton X 100 (Art. No. X100, Sigma Aldrich, Saint Louis, MO, USA) was added to the DMEM-F12 medium with 1% P/S and 1% FBS to 100% to kill the cells. The LDH experiments were carried out at 24, 48 and 72 h following seeding, and the same procedure was repeated at each interval. Three 100 µL samples were taken from each well into a 96-well plate. An LDH reagent (100 µL) was added to each well in use, and the plate was incubated in darkness at room temperature for 30 min. Following incubation, the plate was placed in a Spectrostar Nano microplate reader, and absorbance was measured at a λ of 490 nm with a reference λ of 600 nm.

2.5.4. GIEMSA Staining

MG-63 cells were seeded onto the samples analogous to Sections 2.5.1–2.5.3 and stained after 3, 7 and 10 days with GIEMSA solution (GIEMSA Azure Eosin Methylene Blue, Merck). For this purpose, the samples were washed with PBS and incubated with 1 mL GIEMSA solution (diluted 1:10 with deionized water) for 10 min at room temperature. Samples were subsequently rinsed with deionized water. Microscopy was performed on an OLYMPUS SZ-61 stereo microscope.

2.6. In Vivo Preparations

A total of 30 female Wistar rats aged between 12 and 15 weeks were used to test the TCP implants. They were divided in six groups of 5 rats each. One group was a negative control (SHAM-surgery without any implant), and the zirconium oxide group served as a positive control along with the four experimental groups. Before we started, the animals were housed in the IVC cages, with two daily feedings and water *ad libitum*, in the Animal Testing Center of the Fraunhofer IZI. Animals were examined for release 5 days before surgery. On surgery day, the rats were anesthetized with a fully antagonizable cocktail containing Medetomidin (0.15 mg/kg), Midazolam (2.0 mg/kg) and Fentanyl (0.005 mg/kg) i.m. The artificial bones were implanted into a prepared subdermal pocket of the rat's flanks and closed with clips. The sham was treated the same way but was put into the subderm of the flank. The antagonizing was done with a cocktail containing Atipamezol (0.75 mg/kg), Flumazenil (0.2 mg/kg) and Naloxon (0.12 mg/kg) i.m. On the surgery day, and up 2 days after this intervention, the rats were treated with meloxicam 0.2–0.5 mg/kg s.c. During the controls on day 2, 7 and 14, the rats were anesthetized in a box with 2.0–3.0% isoflurane (0.8–1.5 L/min oxygen) and kept in this unconscious status with 2.0% isoflurane (0.4–0.8 L/min oxygen). On surgery day (day 0) and every control day, 500 µL of blood was taken to test the liver (ALT, AST, GGT) and kidney values (Urea, Creatinine). The test was performed by the Clinic for Ungulates of the University of Leipzig after centrifugation of the blood at $10,000 \times g$ for 5 min at room temperature. The serum was then stored at -20 °C. On days 7 and 14, we also took a fine needle biopsy of the implant location averted from the incision and fixed it with 4% PFA. On day 14, the animal tests were finalized using deep isoflurane anesthesia. In the following necropsy, tissue was removed for further investigation from the liver, kidney, spleen implant location and local lymph nodes of the implant location, which was preserved in 4% PFA. Implants were transferred into a 15 mL BlueCup with saline, photographed with a Leica camera 2.0 after 1 to 3 h and evaluated semi-objectively according to vascularization/tissue ingrowth, removability of the tissue and loss of substance/tendency for break using a numerical system. In the case of an indifferent vascularization/tissue ingrowth, we evaluated two halves of each of the bones and took the average value.

0 = no
1 = minimal
2 = minimal–moderate
3 = moderate
4 = moderate–high
5 = high

This semi-objective data and the serum parameters were collected for each group in an Excel table and evaluated via box-plot. To assess the anticipated differences of tissue attachment to the porous foam section and the denser, rather smooth CerAM VPP-manufactured surface, whole artificial rat bones were cut in half. Each half was further analyzed.

2.7. Statistics

The collected data were analyzed descriptively using SPSS statistics software (Version 25, IBM, Armonk, NY, USA). Based on the raw data, the mean value and the standard deviation were calculated. The Mann-Whitney U-Test was used to evaluate the differences between experimental and control samples. p-values < 0.05 were considered to indicate statistical significance.

3. Results

The subsequent figure shows one exemplary manufactured near-net shaped hybrid foam (Figure 2a, right side) consisting of the foamed-in additively manufactured support structures (Figure 2a, left side) and the porous Freeze Foam, which together make up the support structure case. Figure 2b illustrates the Curasan control. Figure 3 displays the workflow from the rat bone to the reverse-engineered CAD file to the manufactured single Freeze Foams and bioceramic artificial *corticalis* (i.e., *corticalis* case).

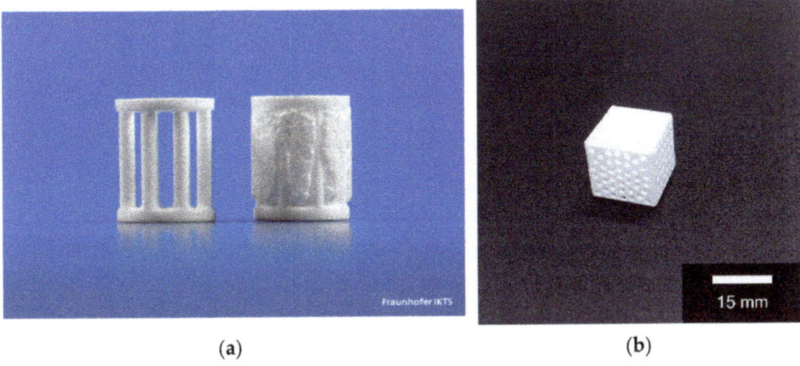

(a)   (b)

Figure 2. Overview samples vs. Curasan control: (**a**) VPP-manufactured column support structure (left) and hybrid with Freeze Foam enclosing the VPP support (same figure, right); (**b**) Curasan control.

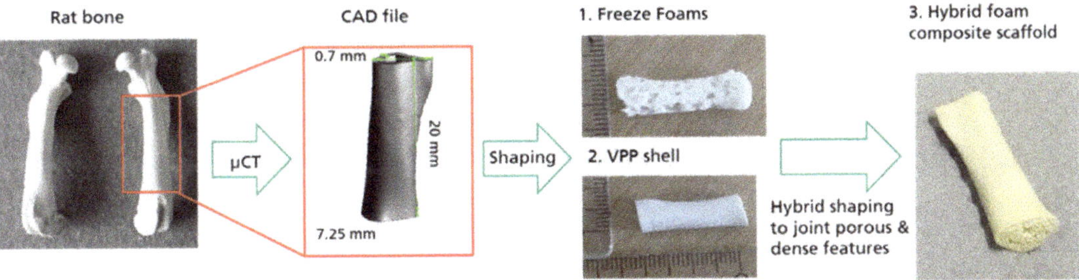

Figure 3. Reverse-engineering workflow from rat bone to a CAD file to the manufactured Freeze Foams and hybrid foams (artificial *corticalis* case).

3.1. Microstructural Characterization

Computer tomographic images of an exemplary hybrid structure confirmed that the form and material fit between the columns and the foam (Figure 4). From left to right, first the VPP column and then the hybrid, both in the green state, are displayed, followed by the sintered hybrid.

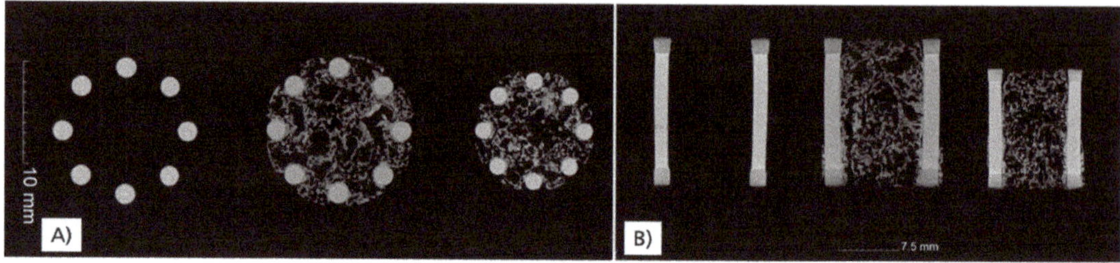

Figure 4. CT images of an exemplary hybrid structure; from left to right: left: VPP support structure (green state); middle: hybrid (green state); right: sintered hybrid (**A**: plan view, **B**: side view).

SEM images clearly showed the denser VPP-manufactured round column and the porous Freeze Foam (Figure 5). At the junction between them, several gaps appear. Foam and the VPP part fused together, but only partially and only in a few spots. However, this SEM analysis only shows one specific location within one hybrid structure. More hybrids need to be manufactured and examined to come to a general conclusion.

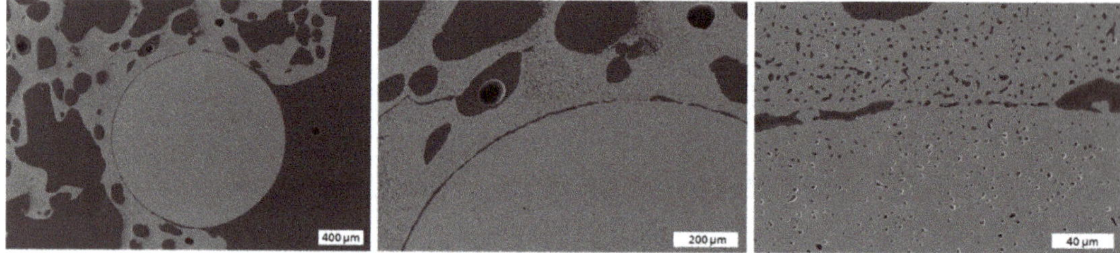

Figure 5. FESEM cross-cutting images displaying interface between VPP-manufactured column and Freeze Foam; magnification/HFW 35×/3334 µm (**left**); 100×/1160 µm (**middle**); 500×/242 µm (**right**); 8 kV acceleration voltage.

The gap between column and foam was measured at one location (Figure 6), which varied between 1 and 13 µm. In general, the column was much denser than the foam, with macropores of around 100 to <600 µm. A closer look at a higher magnification showed mesopores of around 1–2 µm in the foam and in the struts (Figures 5 and 6).

One hybrid foam was analyzed in the fractured view (Figure 7). The gap at the interface between the foam and column is obvious as well as the interconnected pores in the Freeze Foam. At this location, a higher magnification showed that the material fit between the foam and the column and formed the TCP microstructure, with mesopores of around 1–2 µm.

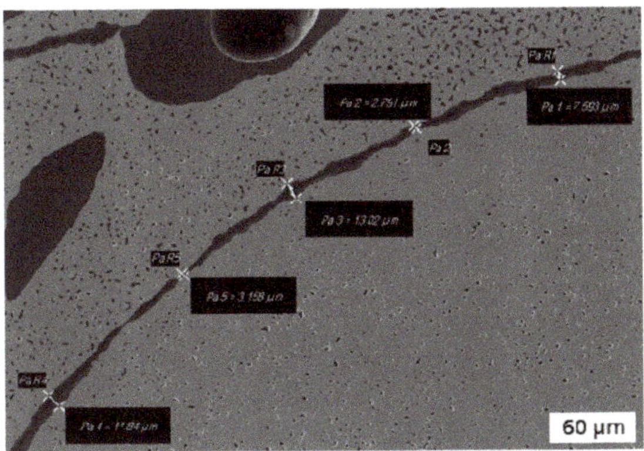

Figure 6. FESEM image: measured distance in the gap interface of VPP column and Freeze Foam.

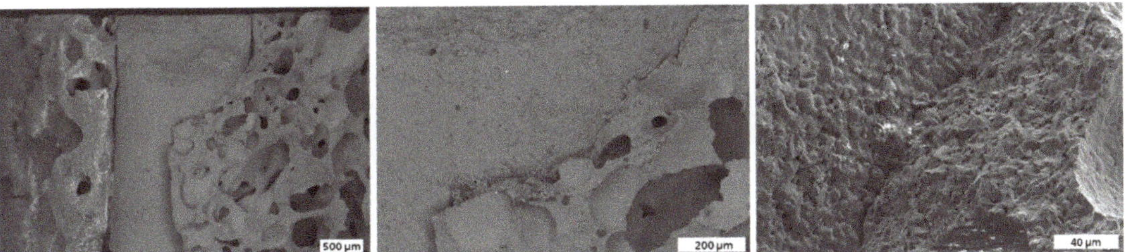

Figure 7. SEM fractured view of gap interface of VPP column and Freeze Foam: magnification/HFW 26×/4446 µm (**left**); 100×/1109 µm (**middle**); 500×/227 µm (**right**); 6 kV acceleration voltage.

Fractured and cross-sectioned images (Figure 8) once again show a good material fit between the column and the foam. As stated before, the interface connection was not thoroughly complete, and its state/appearance depended on the location in the hybrid (referring to each column and each column length).

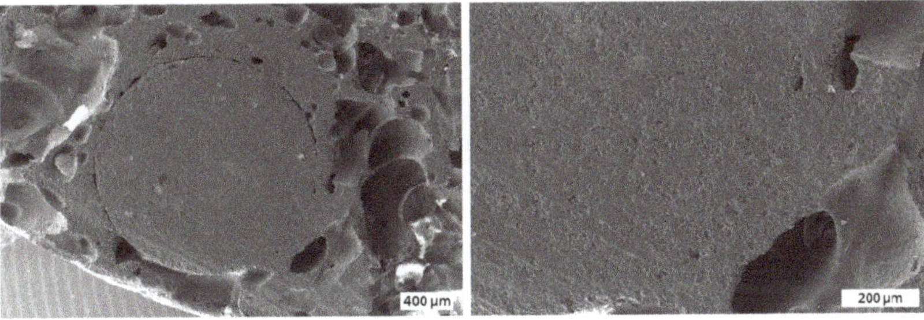

Figure 8. SEM fractured and cross-section view of gap interface of VPP column and Freeze Foam: magnification/HFW 35×/3550 µm (**left**); 100×/1109 µm (**right**); 8 kV acceleration voltage.

The surface roughness Sa was determined to be 5.99 ± 1.43 µm for the Curasan control, 3.73 ± 1.94 µm for the outer ceramics (column's) ring of the hybrid foam and 6.54 ± 2.93 µm for the inner foam structure of the hybrid foams (Figure 9).

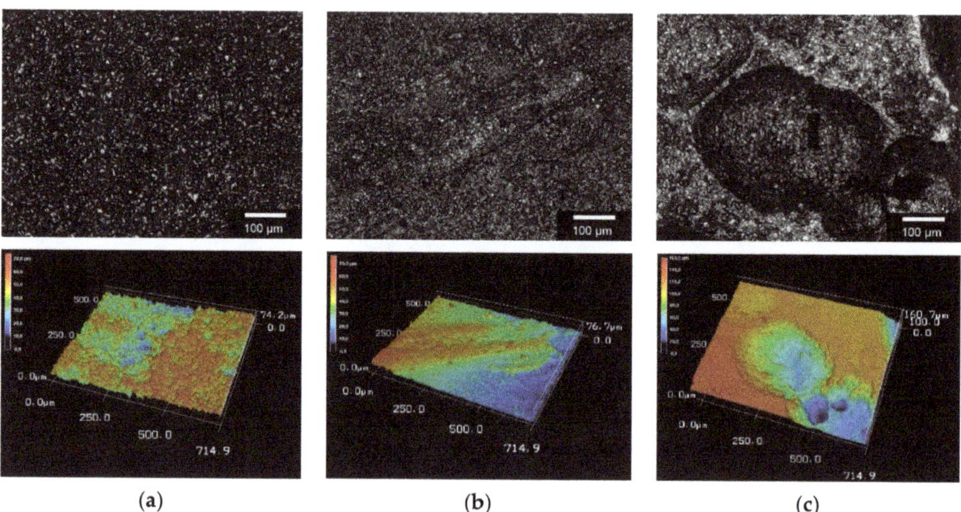

(a)   (b)   (c)

Figure 9. 3D Laser scanning image (top) and 3D reconstruction (bottom) of: (**a**) Curasan control; (**b**) Hybrid Foam, outer ceramic ring; (**c**) Hybrid Foam, inner foam. Images were taken with the KEYENCE VK-X210 3D Laser scanning microscope, 400× magnification.

3.2. Mechanical Characterization and Comparison to FE Simulation

One of the aims of this study was to enhance the mechanical stability of the TCP. Therefore, Freeze Foams and hybrid foams were tested for compressive strength. The following table summarizes the compressive strength (relating to the cylindrical cross-section) and porosity (geometrical and foam cells determined via foam structure analysis based on CT images) of tested samples (five each, mean values shown in Table 2).

Table 2. Compressive strength of manufactured components (foams, hybrids, and columns).

Sample	Geometrical Porosity (%)	Porosity of the Foam Cells (%)	Fmax (N)	Compressive Strength (MPa)
Freeze Foam	80 ± 0.5	76.1 ± 1.4	101 ± 53	0.9 ± 0.5
Hybrid Foam	74.4 ± 0.5	69.9 ± 0.9	2641 ± 452	23 ± 4
VPP Column	$16.5 \pm 0.7 *$		3199 ± 831	31 ± 8
Curasan	$55 \pm 2 *$		693 ± 89	3 ± 0.4

* Archimedes method.

Freeze Foams and hybrid foams exhibited similar porosity. However, the hybrid's compressive strength was 25 times higher (23 MPa) than the Freeze Foam alone (0.9 MPa). Surprisingly, the VPP columns alone showed an even higher compressive strength. Those values lie, however, within the standard deviation. It must be noted that the standard deviation was quite large. There were microdefects leading to failures in the macrostructure and/or the loaded surface was not plane, leading to varying forces upon contact with the compression stamp. The Curasan component provided the lowest porosity of all CaP scaffolds and showed much lower compressive strength than the hybrid foam (roughly one-seventh).

For interpretation of the simulation results, the maximum principal stress was considered because of the known brittleness of the support structure's ceramic material. Tensile load cases are critical for ceramics. Results of the first load case (Figure 10a) showed the largest maximum principal stress at the ringed segments. Maximum tensile stress appeared at the bottom surface of the top ring at around 61 MPa. Its origin can be assumed by the expansion of the rings by given external loads. This leads to an increase of a tangential component of normal stress in the ring. For the first load case, a reaction force of 780 N was considered. In experiments, a structural failure occurred at an average load of 3200 N for the support structure. This corresponds to 800 N for one-quarter of the structure. Therefore, reaction force was increased for the second load case, up to 806 N (results displayed in Figure 10b). The largest maximum principal stress reached 80 MPa. Maximum tensile stress appeared at the same location, similar to the first load case. However, in the middle region of the columns (front), a tensile stress suddenly evolved, which was likely due to the buckling sensitivity of the support structure. At a certain uniaxial load, the structure will collapse because of buckling if the tensile strength of the bulk material is larger compared to this. In the third load case, the reaction force was increased to 809 N. Results are shown in Figure 10c. In this load case, the maximum tensile stress appeared in the middle of the columns at 128 MPa. The maximum principal stress was also high in the ring segments but not at this level. This led to the conclusion that, in third load case, the structure started to collapse by buckling. In Figure 11, a comparison of the three load cases for the maximum principal stress is shown. This figure clearly demonstrates the buckling sensitivity of the support structure. Between the first and the second load cases, the axial force was increased by 26 N.

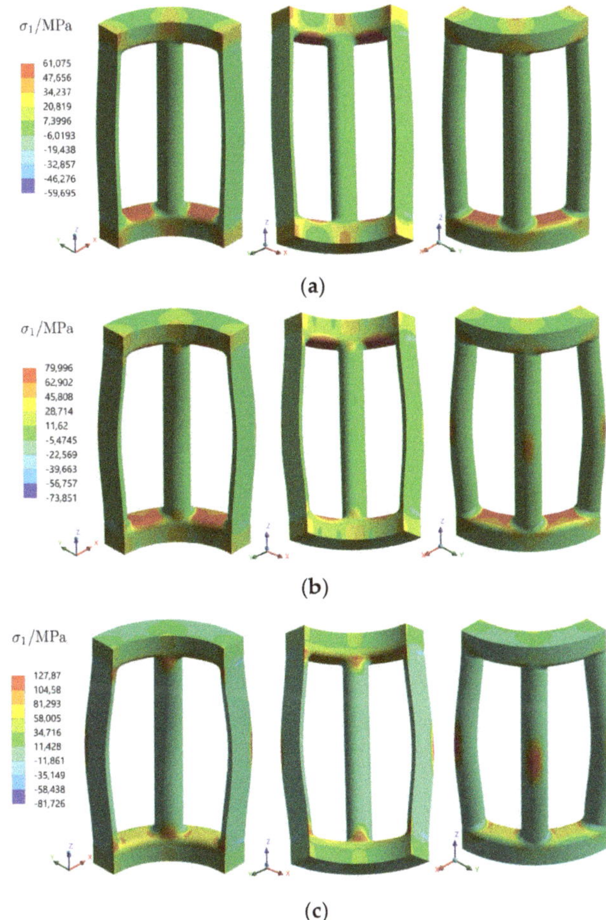

Figure 10. Results of analysis for the maximum principal stress: (**a**) @780 N load; (**b**) @806 N load; (**c**) @809 N load.

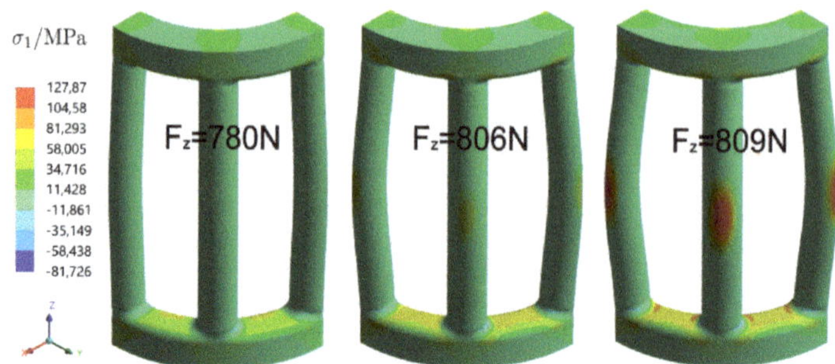

Figure 11. Comparison of uniaxial load sets (deformation shown in true scale).

This contributed to an expected increase of the maximum principal stress in the ringed segments but also led to an increase of tensile stress in the column of the structure. From the second to the third load case, a further small increase of axial force of 3 N significantly changed the load conditions. In the center region of the columns, maximum principal stress grew proportionally. The location of the maximum tensile stress changed from the ringed segments to the center region of the columns.

In Figure 12, experimental observed defects of the broken columns are shown. The red marked defects (primary defects) show those defects that initially occurred during experimental compression testing when the structure collapsed; the blue marked defects show the secondary defects that followed.

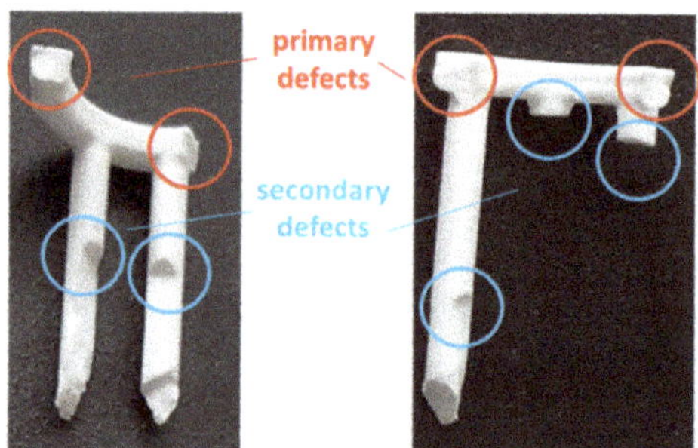

Figure 12. Experimental observed defects at the column structure.

The experimental failure pattern shown above leads to the assumption that the support structure collapsed by exceeding the tensile strength of the VPP-manufactured material, which corresponds to the FE analysis (first load case), and not by buckling, because cracks appeared at locations of maximum tensile stress in the ringed segments. However, some aspects must be considered. First, the simulation showed that the critical buckling loads appeared in the middle of the columns (third load case), which is very close to the experimental testing conditions (i.e., secondary defects shown in Figure 12). Therefore, complete structural failure by buckling should not be excluded. Second, the distribution of micropores in the VPP-manufactured column structure was considered homogenous for the simulated model. However, manufactured structures probably also include microdefects leading to the observed failure pattern.

Third, technological tolerances (by 3D printing, sintering, etc.) can lead to geometric imperfections (e.g., flatness of the ringed cross-section), which induce critical tensile stresses. Such effects were not included in the described FE model. At this stage of the presented research, there is still opportunity for adjustment if indeed microdefects appear, e.g., adjusting CerAM VPP exposure parameter, suspension parameters, debinding/sintering regime or the VPP design. By neglecting these aspects, a tensile strength of our manufactured material with 5% porosity could be obtained by further optimizing the VPP process. The tensile strength of a β-TCP 5% porous material should lie between 61 MPa and 80 MPa and take the mentioned restrictions into account.

3.3. In Vitro Biocompatibility

3.3.1. Live/Dead Assay

Human osteoblastic MG-63 cells were counted using Image-J (Fiji, Version 1.52 h), through which the cell number/mm^2 of living and dead cells was determined. Figure 13 shows representative samples with live/dead staining of the inner surface of the 500 µm scaffold as compared to the Curasan control after 3, 7 and 10 days. Long-term studies (>4 weeks) were not assessed.

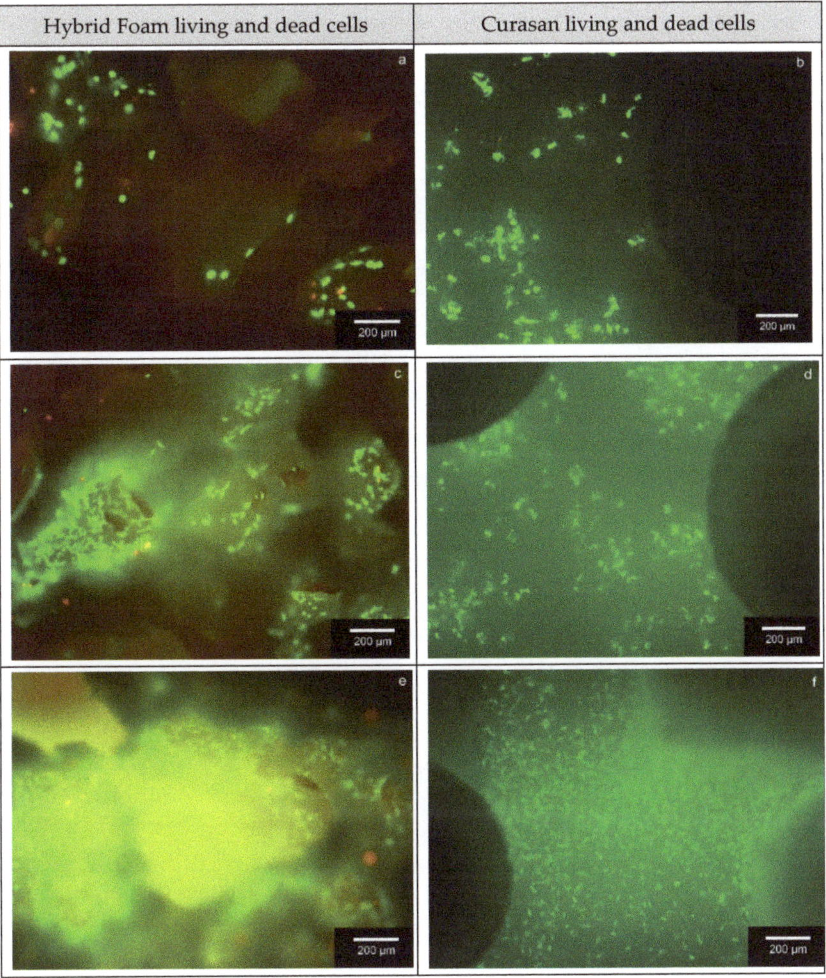

Figure 13. *Cont.*

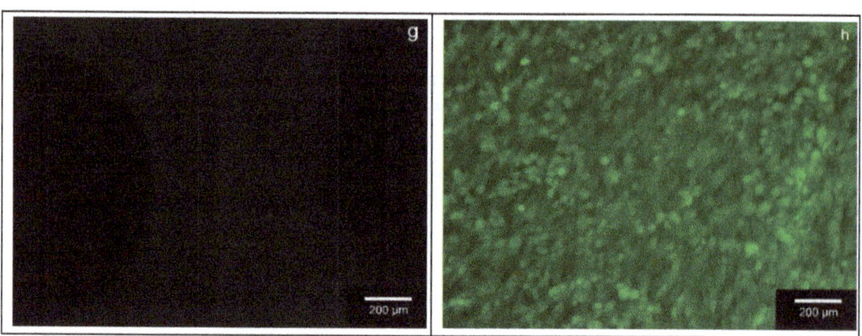

Figure 13. Living/dead cells on the inner surface of the ceramic; 500 µm scaffold after three days ((**a**): Hybrid foam; (**b**): Curasan), seven days ((**c**): Hybrid foam; (**d**): Curasan) and 10 days ((**e**): Hybrid foam; (**f**): Curasan); (**g**): Auto-fluorescence of the ceramics; (**h**): Thermanox membrane (pos. control, 10 days); white bar = 200 µm. Green indicates living cells; red indicates dead cells.

Quantitative results of the number of living and dead cells per mm^2 is shown in Figure 14a,b, respectively. The number of living cells increased over the course of the experiment in both the scaffold and the Curasan control, with no significant differences between our scaffold and the Curasan control.

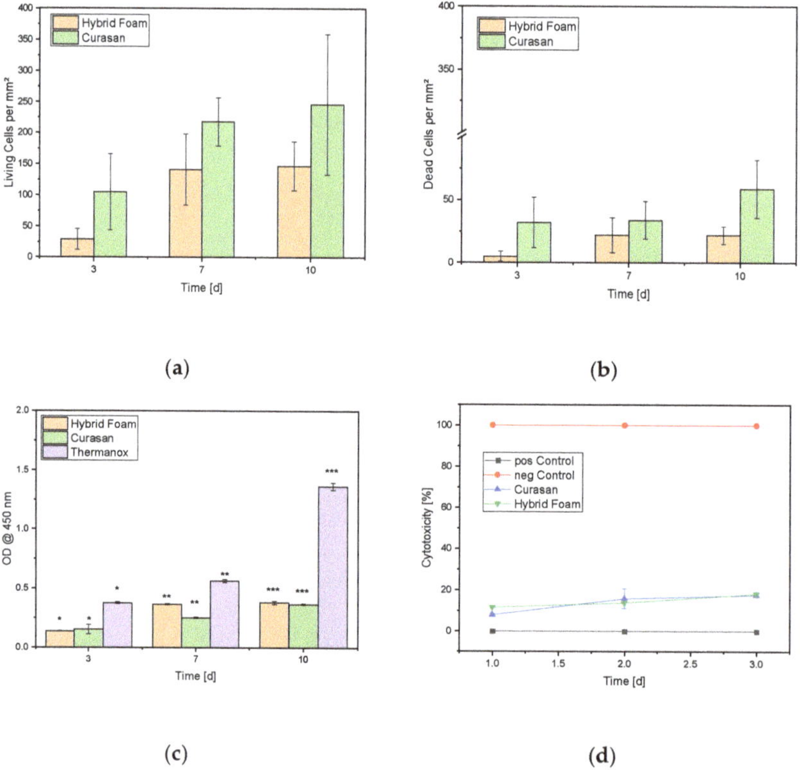

Figure 14. Overview of the biocompatibility tests: cell counts of (**a**) living cell numbers and (**b**) dead cell numbers per mm^2 on the materials after 3, 7 and 10 days; (**c**) WST assay to demonstrate proliferation of MG 63 on the samples. Means and controls were statistically compared to assess the material effect. Significances set at $p < 0.05$ are assigned the same symbol. (**d**) Cytotoxicity of hybrid foams compared to the Curasan control; pos. Control = cells, neg. Control = TritonX.

3.3.2. Cell Proliferation Assay

Figure 14c shows that the growth rate of the cells on the scaffold and Curasan control compared to the growth of cells on a Thermanox cover slip as a positive control. The growth rate of the cells on the scaffolds in the cell culture plates increased only up to seven days and stagnated thereafter, while the cells on the Thermanox cover slip continuously proliferated. No significant difference in cell proliferation was observed between the hybrid foam and the Curasan control.

3.3.3. LDH Assay

The cytotoxicity for both the hybrid foam and the Curasan control was slightly above the positive control (cells on the Thermanox cover slip) and very clearly below the negative control (Triton X), with no significant differences between our scaffold, the Curasan control and the positive control noted. The graphs of cytotoxicity over time were nearly congruent for the Curasan control and the Hybrid Foam (see Figure 14d).

3.3.4. GIEMSA Staining

In the GIEMSA staining, it was evident that the MG-63 cells only colonized the inner sponge area of the hybrid foams after 3 and 10 days, but only sporadically on the surface of the CerAM VPP shell (see Figure 15). Once again we saw mainly complete material and a form fit but also a gap between the ring structures and the Freeze Foam (Figure 15, upper right-hand side.)

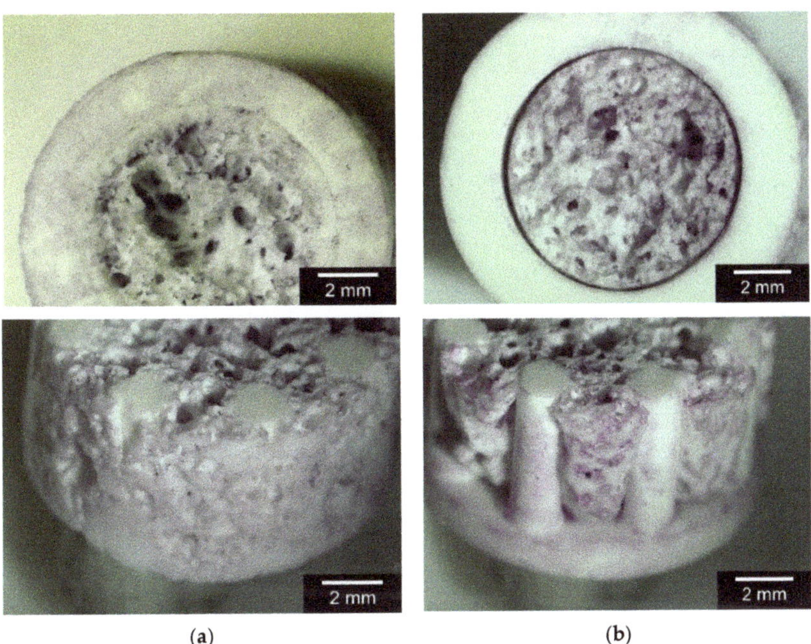

Figure 15. GIEMSA staining after (**a**) 3 and (**b**) 10 days; top and side view.

3.4. *In Vivo Studies*

Figure 16 displays manufactured TCP Freeze Foams (porosities between 83–85%, with an average of 84%), TCP hybrid scaffold halves (porosities around 80%) and single ZrO_2 Freeze Foams (porosities between 70–72%, with an average of 71%).

a.   Clinical Examination

After implantation, all rats had a score of "0" overall. This score means, clinically, we had no signs that the surgery or the implants had negative influences on the rats. However, the evaluation of histological samples (fine needle aspiration and explanted implantation area) had not been commenced. However, for all of the presented in vivo results, the authors think that the gathered serum parameters are more important regarding our following in vivo assessments than histological analyses would be at this stage of the research. We base our argument on studies of Trevisani et al. [33]. The main findings of this study were that the agreement between chronic histological kidney damage (CKD) and CKD staging was poor. In fact, about 30–40% of patients with CKD stage 3 had mild or no lesions in the histological evaluation (Chronicity Score = 0–1), whereas 7 to 10% of cases with CKD stage 1 (eGFR > 90 mL/min/1.73 m^2) had moderate or even severe histological lesions (Chronicity Score $\geq$ 3). Moreover, different patients with the same eGFR values may have had either severe (Chronicity Score $\geq$ 3) or no histological damage (Chronicity Score = 0) (eGFR = estimated glomerular filtration rate).

b.   Serum Parameters

The serum parameters of day 0, 2, 7, and 14 were measured by the accredited laboratory of the Clinic for Ungulates of the Veterinary Faculty of the University of Leipzig. These results were evaluated in a box-plot diagram to adjust them according to the physiological parameters as described in Charles River 2008 [34] and Boehm et al. [35] (Figures 21 and 22). Regarding these analyses, a photometric measurement (extinction determination) was executed. The photometric method is applicable for multi-species analyses [36]. Since no references were sent by the clinic, we had to compare the determined values, especially the creatinine and urea values, with the literature references of Charles River [34] and Boehm et al. [35].

Figure 16. Comparison of the different manufactured scaffolds: left—TCP freeze foams, middle—TCP hybrid scaffold halves and right—ZrO$_2$ Freeze Foams.

3.4.1. ALT

It is clearly visible that all alanine transaminase (ALT) values of all rats on the surgery and the control days are greater than the references in the literature. Furthermore, there is one higher aspartate aminotransferase (AST) value in the negative control group, while all the other AST values of the other four rats and the median of this group stay in the reference area over the experimental time. Although all ALT and AST values in one rat of the negative control were this high over the entire time period, there is no sign of liver damage as proven by Gamma-Glutamyl Transferase (GGT), as the liver-specific value is located well within the reference in all rats all over the experiment. However, the chosen physiological ALT reference may not be specific enough for rats, or it might be increased by food containing a higher amount of proteins. Furthermore, the indicated references in different literature sources varied between 25 U/L and 163 U/L [35] in the mean for ALT and from 26 U/L to 155 U/L in mean for AST. They also varied based on age and sex. That is why we took a critical view of these parameters according to their values. In addition, it is well known in veterinary medicine that ALT and AST are also produced in other organs, e.g., in muscles and kidneys both are produced quickly in response to medical/toxic agents, and values three times the upper limit attract attention in practice. As the rats had such a medical supply during anesthesia, analgesic treatment, and anesthesia during surgery, and analgesics were injected, the increase of ALT and AST might be a result of this treatment (see Figure 17). However, it is noteworthy that, with the exception of ALT levels on day 14, AST and ALT concentrations were within the physiological range in all of the experimentally treated groups.

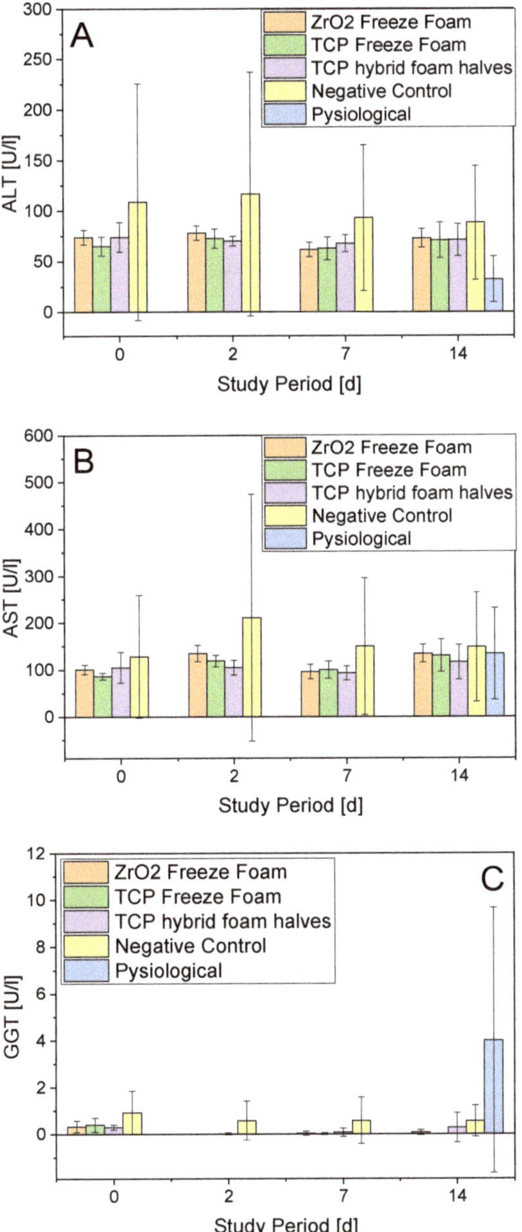

Figure 17. Course of (**A**) ALT, (**B**) AST and (**C**) GGT over the study period of 14 d; for clarity, the physiological control values were only entered for day 14. Physiological values taken from Giknis et al. [34] (ALT, AST) and Boehm et al. [35] (GGT).

3.4.2. Creatinine

Kidney creatinine (crea, long-term) and urea (short-term) values stayed under the references in all rats of all groups at all time points (Figure 18). As values depend heavily on the method used for their determination, the engaged laboratories have their own references. Since no references were sent by the clinic as stated above, we compared the determined values of creatinine and urea with references [34,35]. This might be an explanation for the lower values. However, as the kidney values did not show any increases over the time, this shows that the implants did not have a negative influence on the kidneys. Nevertheless, regarding proof of a non-toxic effect, long-term additional histological examinations of the removed kidneys, in line with the 3R principles, must be executed.

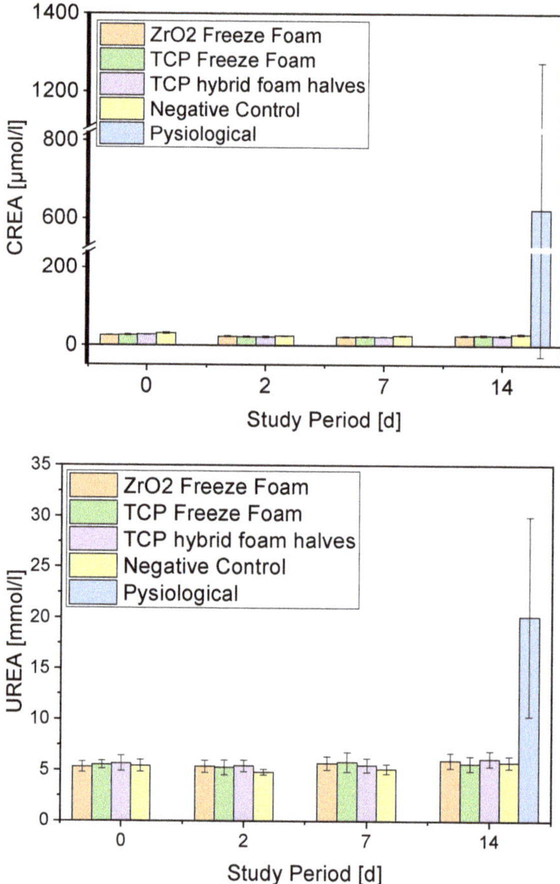

Figure 18. Concentrations of creatinine and urea from days 0 to 14. For clarity, the physiological control values were only entered for CREA and UREA at day 14. The physiological values for creatinine (CREA) and UREA were taken from Boehm et al. [35].

3.4.3. Necropsy

All rats in all treatment groups showed no macroscopic alterations of the spleen, kidney or liver during the necropsy. One rat treated with the ZrO_2 foams showed a minimally increased spleen, which was probably caused by post-mortal blood congestion. All rats in all treatment groups showed a slight to moderate increase of the regional lymph nodes, which is likely due to the resorption and healing processes after surgery.

3.4.4. Implant Parameters

As written above we judged the implants according to different semi-objective parameters. One of them was their ability to be vascularized or rather the adherence and ingrowth of tissue to the implants after 14 days. Vascularization is important for the supply of oxygen, nutrients, the transport of metabolic products and immune cells. As the goal is to help bone to grow into this pattern, a high vascularization results in a high metabolic rate and growth factors in the area. As shown in the following figure, the more porous the artificial bones the more tissue grows into them (84% TCP Freeze Foam > 81% TCP hybrid foam halves > 71% ZrO_2 Freeze Foams). The zirconium oxide group showed less tissue adherence than the TCP specimen (Figure 19A). Additionally, we noted a non-adherence of host tissue to the part of the hybrid foam halves that were made by CerAM VPP for the artificial *corticalis*. As we do not only want vessels to grow into the implants but ultimately, host tissue to replace the implant and later giving stability, we evaluated the removability of tissue from the implants after 14 days. It is obvious that not only the porosity seems to be an important factor for the surrounding tissue but also the material. As expected, TCP was more integrative than zirconium oxide (Figure 19B). We also observed that parts of broken implants were held together by the immigrated tissue giving them additional stability. On the other hand, this adherent growing can be seen critical in case of a removal of an implant for instance due to incompatibility or failure. Though this growth was very invasive, we detected no signs of macroscopic fibrosis, capsular formation, inflammation, or calcification in the implantation area. An additional statement could be given after the evaluation of histological samples (fine needle aspiration and explanted implantation area). As the biodegradable implants are developed to replace bone in the short to midterm, they have to provide enough stability until the hosts own bone material is calcified. Therefore, we also looked at the scaffolds loss of stability and the tendency to break after 14 days (Figure 19C). In necropsy we did not find evidence of a broken implant in the ZrO_2 group, likely due to a slightly decreased porosity as well as its general material properties), one broken implant in the TCP Freeze Foams and one nearly broken implant in the hybrid bone, whereby the fracture was located at the connection between porous and additively manufactured shell part. These tendencies likely reflect the material and porosity properties.

3.4.5. In Vivo Conclusion

From the macroscopic and clinical point of view, and according to incompatibility and toxicity, we had no sign that any of the implants, independent of the material, the manufacturing or the handling before implantation, negatively influenced the results of this oriented and leveled study. This well-founded statement is based on proven literature references as discussed above and as demonstrated in [36,37]. In accordance, our in vivo results are unobtrusive. In that regard, we can make a recommendation that the scaffolds be developed further as a result of their vascularization/tissue ingrowth tendency, which is an important factor for an implant in the muscoskeletal system. The TCP Freeze Foams are the most promising scaffolds for a use in artificial trabecular bones according to the determined parameters in the study. The TCP hybrid foam halves (artificial *corticalis* case) showed insufficient connection of additively manufactured parts to the tissue. In accordance with the in vitro analyses, where the cells only sporadically colonized/attached to the CerAM VPP-manufactured columns, the tissue did not adhere to the CerAM VPP shell structure but only to the porous artificial *spongiosa* acting Freeze Foams (see Figure 15). Roughness measurements indicated that there were clear differences between the manufactured components, with the additively manufactured one likely being too smooth for cell attachment (see Figure 9). However, they still may be good candidates for further development, considering the fact that the shell part is very stable, likely for a long time, and thus this implant could potentially allow bridging of very big/long bone defects. However, a solution must be found to enhance cell attachment capability (e.g., chemical and/or physical surface modification and/or adding porosity). Alternatively, the support structure case might be chosen.

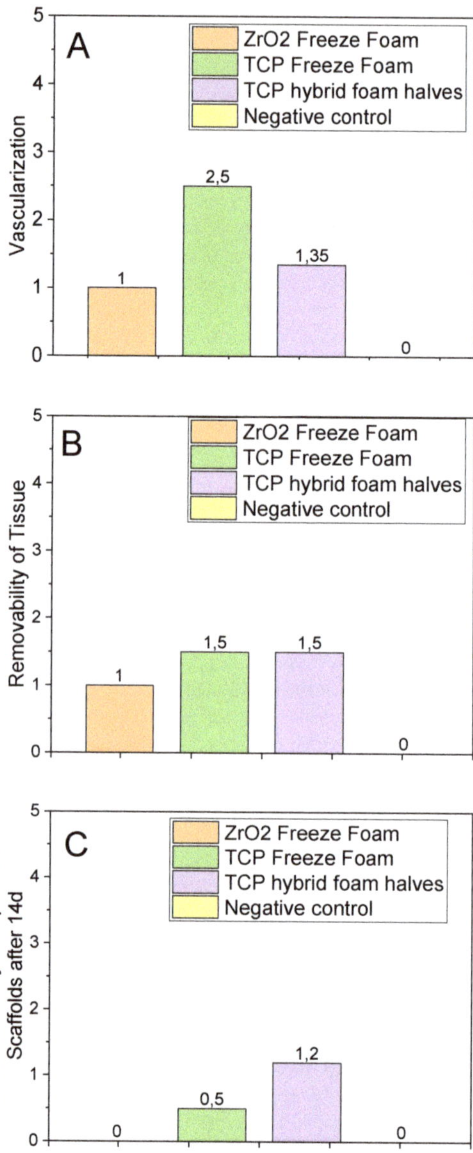

Figure 19. Implant parameters after 14 days: (**A**) tissue adherence/ingrowth into the artificial bones; (**B**) removability of tissue from the implants; (**C**) break-tendency of implanted scaffolds.

4. Discussion

An ideal engineered tissue scaffold for the regeneration of load-bearing bones should possess appropriate mechanical functions to provide structural support, share the biomechanical load, and distribute stress that stimulates bone growth and remodeling. Due to excellent biocompatibility, calcium phosphate scaffolds have been successfully used for non-load-bearing bone restoration in recent decades. Current bioceramic scaffolds cannot re-establish massive load-bearing bones. The mechanical properties of materials decrease with increasing porosity and pore size. For optimal new bone building, scaffolds normally require an interconnected macroporous structure, with a high porosity of over 90% and a pore size ranging from 100 to 1000 μm. Such porous constructs typically have low mechanical properties. Therefore, in this study, porous bone-like foams were fitted around/in a customized additively manufactured support structure to manufacture bioceramic hybrid foams that are mechanically much more stable than single porous components. The scaffolds were made of β-TCP and were analyzed in terms of their biocompatibility and mechanical behavior. As a result, the authors postulate that these complex structural hybrids, due to the combination of load-bearing support and porous cell-ingrowth allowing interior growth, will eventually allow the manufacturing of bone-mimicking, mechanically stable implants.

4.1. Microstructural and Mechanical Characterization

All manufactured Freeze Foams and hybrid foams showed the microstructural characteristics necessary for use as potential bone replacement materials and implants, including macro and micro/mesoporosity of the right size as well as interconnectivity. Manufactured hybrid foams showed connected/joined porous and dense sections similar to a real bone. However, more parts must be analyzed in order to obtain a general overview of the success rate of materials as well as form fit. On the other hand, it is yet not clear to what extent apparent gaps between additively manufactured and porous components have influenced the mechanical as well as biocompatibility behaviors. More research needs to be conducted. Regarding the already enhanced compressive strength, for possible further improvements and increased failure tolerance, the column structure should be reengineered. For example, this can be done by adding a further ring in the structure's middle or making the columns meet each other in the center of the structure (reduction buckling length). Optimization of the VPP process would further result in less porosity and fewer microdefects, assuming that the same material and thermal treatment is applied. However, through improving loads, the biodegradability will most likely decline. A balance between good biocompatibility and sufficient mechanical strength must be found. The maximum failure load was 29 ± 9.0 N for the commercially available Curasan β-TCP ceramic [38], 693 ± 85 N for the Curasan cube, which served as a control, and 2641 ± 452 N for the hybrid foam. The hybrid's failure load was 91 times higher than the commercially available Curasan ceramic and four times higher than the control. There were comparable values for compressive strength: 23 ± 4 MPa for the hybrid foam and 24 ± 6 MPa for the Curasan β-TCP ceramic. Freeze Foams and hybrid foams exhibited similar porosity. However, the hybrids compressive strength was 25 times higher (23 MPa) than the Freeze Foam alone (0.9 MPa). The additively manufactured bioceramic support structures made the porous spongious structures mechanically more stable. Bone is structurally complex and hierarchically designed. Cortical bone is stronger and stiffer in comparison to trabecular bone. The material behavior of cortical bone is anisotropic. The compressive moduli of cortical bone along the longitudinal direction (193 MPa) are greater than those along the transverse direction (133 MPa) [39]. The compressive moduli of trabecular bone is 50 MPa. Trabecular bone is a highly porous material with anisotropic mechanical properties. Due to its high porosity versus that of cortical bone, the mechanical properties of trabecular bone are determined primarily by its porosity. The mechanical properties of the bone are thus still higher than the measured hybrid foams, which still only have a strength support structure and no surrounding *corticalis*. With the help of FE analysis, we were able to approximate mechanical loads appearing in the structure and to analyze and predict failure mechanisms that then also occurred in the mechanical tests.

4.2. Biocompatibility

Regarding the biocompatibility experiments, the most noticeable aspect was that the cells did not attach to the VPP-manufactured parts. There was a clear gap present. We now need to work out whether that gap was correlated with a possible mismatch/non-material fit between the VPP part and the Freeze Foam, as shown before. Alternatively, the cells might behave like this because the VPP part is too smooth to "hold" onto. Despite this, the cells clearly attached to the foam surface and even grew into the Freeze Foam, as proven in in vitro and in vivo results. In the WST-1 experiment, the ceramics and hybrid foams and the control group showed comparable vitality values and a constant cell growth over the examined period of time. These results are congruent with the ones seen in the live-dead assay. The hybrid foam sample showed a similar high biocompatibility comparable to the Curasan ceramic. This is not surprising, since both samples consist of β-TCP. We were able demonstrate the high biocompatibility of β-TCP in various studies in the past [15,16,40]. However, the proliferation values for the Curasan sample were slightly lower than those of the hybrid foam, which may be due to the fact that cells generally prefer a structured surface [41,42]. In addition, the Curasan sample had a lower porosity. In terms of cytotoxicity, both samples, the hybrid foam and the Curasan sample, are on an equal footing, with partly even congruent curves. This was also not surprising, as both studies involved β-TCP. In line with our previous studies, β-TCP is non-cytotoxic [16,40,43].

Regarding the manufactured zirconia Freeze Foams, we showed that the specific porous structure/pore morphology resulting from the Freeze Foaming process allows tissue ingrowth independent of the bioceramic materials used.

5. Conclusions

As mentioned in the introduction and in the discussion, previous studies show that β-TCP is a performing bone replacement material, not only as pure material and shaped by conventional methods (e.g., freeze drying) but also as composite material (e.g., with polymers) and shaped by additive manufacturing. The resulting compressive strength is always relatively low, ruling out load-bearing clinical indications/bone defects. In contrast, our results show that we can address load-bearing bone defects using the same material as previously reported, by advancing known composites to become complexly shaped structural composites that not only unite the structural features of a real bone (dense and porous sections) but also reach similar and improved compressive strengths (of trabecular bone [44,45]), while at the same time providing degradability as given by the material. By fine-tuning the support structure design and working on composite materials to develop new structural and material composites for potential bone replacements, we might be able to further develop mechanical properties, aligning our approach with a variety of bone defects, especially long-bone and load-bearing ones. Offering the same biocompatibility, the bioceramic hybrid foams have significant mechanical advantages over the Curasan benchmark. The hybrid's failure load is 91 times higher in comparison to the commercially available β-TCP ceramic. To summarize, the compressive strength of the bone-mimicking hybrid bones was significantly enhanced, while high biocompatibility was maintained as proven on the Curasan material. At present, the BMBF-funded project "Hybrid-Bone" (03VP07633) is in progress, which builds on these results and strives for the evaluation and validation of materials, processes and hybrid scaffolds for use as compressive-strength-enhanced biodegradable jaw-bone replacements. Within the framework of this project, the comprehensive bone-forming performance tests of similar scaffolds, with a focus on hybrid foams, are carried out in animal models. The authors hope to report on these results soon.

Author Contributions: Conceptualization, M.A.; Data curation, M.A., S.H.L., C.S., C.F., D.W., E.S.-F. and M.S.; Investigation, C.F., D.W. and E.S.-F.; Methodology, M.A. and H.O.M.; Project administration, M.A.; Software, C.F.; Supervision, M.A.; Validation, S.H.L., C.S., C.F., D.W. and E.S.-F.; Visualization, S.H.L. and M.S.; Writing—Original draft, M.A.; Writing—Review and editing, M.A. and M.S. All authors have read and agreed to the published version of the manuscript.

Funding: This research received no external funding.

Institutional Review Board Statement: The animal study (file number 47/18) was reviewed and approved by the Landesdirektion Leipzig, Germany. The approval date was January 15th in 2019 and the study was conducted according to Directive 2010/63/EU.

Informed Consent Statement: Not applicable.

Acknowledgments: This contribution originally was intended to be published together with Anke Bernstein, for Musculoskeletal Biomaterials of the Research Center for Tissue Replacement, Regeneration and Neogenesis (G.E.R.N.) and the Deputy Director of the G.E.R.N in the Department of Orthopaedics and Trauma Surgery at the University Hospital Freiburg in Breisgau, Germany. With our deepest regrets, Anke died in a mountaineering accident at the end of June 2021. Bernstein's research revolved around biomaterials-based therapies for repair and regeneration of musculoskeletal tissues. Bernstein was a dedicated member of the German Society for Biomaterials and held the role of the President since 2019. Anke, you will be remembered as a friendly person with a sharp mind, a great sense of humor and lots of laughter. Thank you, indeed, for walking some of the way together. We offer Anke Bernstein's family, friends and colleagues our sincere condolences on her passing. Many thanks as well to Melanie Lynn Hart for spell-checking this contribution.

Conflicts of Interest: The authors declare no conflict of interest.

References

1. Mills, L.A.; Aitken, S.A.; Simpson, A.H.R.W. The risk of non-union per fracture: Current myths and revised figures from a population of over 4 million adults. *Acta Orthop.* **2017**, *88*, 434–439. [CrossRef] [PubMed]
2. Solomon, D.H.; Patrick, A.R.; Schousboe, J.; Losina, E. The potential economic benefits of improved postfracture care: A cost-effectiveness analysis of a fracture liaison service in the us health-care system. *J. Bone Miner. Res.* **2014**, *29*, 1667–1674. [CrossRef]
3. Burge, R.; Dawson-Hughes, B.; Solomon, D.H.; Wong, J.B.; King, A.; Tosteson, A. Incidence and economic burden of osteoporosis-related fractures in the united states, 2005–2025. *J. Bone Miner. Res. Off. J. Am. Soc. Bone Miner. Res.* **2007**, *22*, 465–475. [CrossRef] [PubMed]
4. Gómez-Barrena, E.; Rosset, P.; Lozano, D.; Stanovici, J.; Ermthaller, C.; Gerbhard, F. Bone fracture healing: Cell therapy in delayed unions and nonunions. *Bone* **2015**, *70*, 93–101. [CrossRef]
5. Dimitriou, R.; Jones, E.; McGonagle, D.; Giannoudis, P.V. Bone regeneration: Current concepts and future directions. *BMC Med.* **2011**, *9*, 66. [CrossRef] [PubMed]
6. Reichert, J.C.; Wullschleger, M.E.; Cipitria, A.; Lienau, J.; Cheng, T.K.; Schütz, M.A.; Duda, G.N.; Nöth, U.; Eulert, J.; Hutmacher, D.W. Custom-made composite scaffolds for segmental defect repair in long bones. *Int. Orthop.* **2011**, *35*, 1229–1236. [CrossRef]
7. de Grado, G.F.; Keller, L.; Idoux-Gillet, Y.; Wagner, Q.; Musset, A.-M.; Benkirane-Jessel, N.; Bornert, F.; Offner, D. Bone substitutes: A review of their characteristics, clinical use, and perspectives for large bone defects management. *J. Tissue Eng.* **2018**, *9*, 2041731418776819. [CrossRef]
8. Fishman, J.A.; Scobie, L.; Takeuchi, Y. Xenotransplantation-associated infectious risk: A who consultation. *Xenotransplantation* **2012**, *19*, 72–81. [CrossRef] [PubMed]
9. Zwingenberger, S.; Nich, C.; Valladares, R.D.; Yao, Z.; Stiehler, M.; Goodman, S.B. Recommendations and considerations for the use of biologics in orthopedic surgery. *BioDrugs* **2012**, *26*, 245–256. [CrossRef]
10. Takemoto, M.; Fujibayashi, S.; Neo, M.; Suzuki, J.; Kokubo, T.; Nakamura, T. Mechanical properties and osteoconductivity of porous bioactive titanium. *Biomaterials* **2005**, *26*, 6014–6023. [CrossRef]
11. Rainer, A.; Giannitelli, S.M.; Abbruzzese, F.; Traversa, E.; Licoccia, S.; Trombetta, M. Fabrication of bioactive glass–ceramic foams mimicking human bone portions for regenerative medicine. *Acta Biomater.* **2008**, *4*, 362–369. [CrossRef]
12. Voss, P.; Sauerbier, S.; Wiedmann-Al-Ahmad, M.; Zizelmann, C.; Stricker, A.; Schmelzeisen, R.; Gutwald, R. Bone regeneration in sinus lifts: Comparing tissue-engineered bone and iliac bone. *Br. J. Oral Maxillofac. Surg.* **2010**, *48*, 121–126. [CrossRef] [PubMed]
13. LeGeros, R.Z. Properties of osteoconductive biomaterials: Calcium phosphates. *Clin. Orthop. Relat. Res.* **2002**, *395*, 81–98. [CrossRef] [PubMed]
14. Miranda, P.; Saiz, E.; Gryn, K.; Tomsia, A.P. Sintering and robocasting of β-tricalcium phosphate scaffolds for orthopaedic applications. *Acta Biomater.* **2006**, *2*, 457–466. [CrossRef] [PubMed]
15. Bernstein, A.; Nobel, D.; Mayr, H.O.; Berger, G.; Gildenhaar, R.; Brandt, J. Histological and histomorphometric investigations on bone integration of rapidly resorbable calcium phosphate ceramics. *J. Biomed. Mater. Res. B Appl. Biomater.* **2008**, *84*, 452–462. [CrossRef]
16. Mayr, H.O.; Dietrich, M.; Fraedrich, F.; Hube, R.; Nerlich, A.; von Eisenhart-Rothe, R.; Hein, W.; Bernstein, A. Microporous pure beta-tricalcium phosphate implants for press-fit fixation of anterior cruciate ligament grafts: Strength and healing in a sheep model. *Arthrosc. J. Arthrosc. Relat. Surg.* **2009**, *25*, 996–1005. [CrossRef]
17. Mayr, H.O.; Klehm, J.; Schwan, S.; Hube, R.; Südkamp, N.P.; Niemeyer, P.; Salzmann, G.; von Eisenhardt-Rothe, R.; Heilmann, A.; Bohner, M.; et al. Microporous calcium phosphate ceramics as tissue engineering scaffolds for the repair of osteochondral defects: Biomechanical results. *Acta Biomater.* **2013**, *9*, 4845–4855. [CrossRef]
18. Mayr, H.O.; Suedkamp, N.P.; Hammer, T.; Hein, W.; Hube, R.; Roth, P.V.; Bernstein, A. Beta-tricalcium phosphate for bone replacement: Stability and integration in sheep. *J. Biomech.* **2015**, *48*, 1023–1031. [CrossRef]
19. Kuiper, J.H. *Numerical Optimization of Artificial Hip Joint Designs*; Radboud University Nijmegen: Nijmegen, The Netherlands, 1993.

20. Paul, J.P. Strength requirements for internal and external prostheses. *J. Biomech.* **1999**, *32*, 381–393. [CrossRef]
21. Clarke, B. Normal bone anatomy and physiology. *Clin. J. Am. Soc. Nephrol.* **2008**, *3*, S131. [CrossRef]
22. Alaribe, F.N.; Manoto, S.L.; Motaung, S.C.K.M. Scaffolds from biomaterials: Advantages and limitations in bone and tissue engineering. *Biologia* **2016**, *71*, 353–366. [CrossRef]
23. Rustom, L.E.; Poellmann, M.J.; Johnson, A.J.W. Mineralization in micropores of calcium phosphate scaffolds. *Acta Biomater.* **2019**, *83*, 435–455. [CrossRef]
24. Park, S.A.; Lee, S.H.; Kim, W.D. Fabrication of porous polycaprolactone/hydroxyapatite (pcl/ha) blend scaffolds using a 3d plotting system for bone tissue engineering. *Bioprocess Biosyst. Eng.* **2011**, *34*, 505–513. [CrossRef]
25. Munoz-Pinto, D.J.; McMahon, R.E.; Kanzelberger, M.A.; Jimenez-Vergara, A.C.; Grunlan, M.A.; Hahn, M.S. Inorganic–organic hybrid scaffolds for osteochondral regeneration. *J. Biomed. Mater. Res. Part A* **2010**, *94*, 112–121. [CrossRef]
26. Karl, S.; Somers, A.V. Method of Making Porous Ceramic Articles 1963. Google Patents.
27. Nandi, S.K.; Fielding, G.; Banerjee, D.; Bandyopadhyay, A.; Bose, S. 3d printed β-tcp bone tissue engineering scaffolds: Effects of chemistry on in vivo biological properties in a rabbit tibia model. *J. Mater. Res.* **2018**, *33*, 1939–1947. [CrossRef] [PubMed]
28. Ahlhelm, M.; Schwarzer, E.; Scheithauer, U.; Moritz, T.; Michaelis, A. Novel ceramic composites for personalized 3d structures. *J. Ceram. Sci. Technol.* **2017**, *8*, 91–100. [CrossRef]
29. Ahlhelm, M.; Günther, P.; Scheithauer, U.; Schwarzer, E.; Günther, A.; Slawik, T.; Moritz, T.; Michaelis, A. Innovative and novel manufacturing methods of ceramics and metal-ceramic composites for biomedical applications. *J. Eur. Ceram. Soc.* **2016**, *36*, 2883–2888. [CrossRef]
30. Willmann, G. Materialeigenschaften von hydroxylapatit-keramik. *Mater. Und Werkst.* **1992**, *23*, 107–110. [CrossRef]
31. Ahlhelm, M. *Entwicklung Zellularer Strukturen für Vielfältige Anwendungen*; Fraunhofer Verlag: Dresden, Germany, 2016; ISBN 978-3-8396-0977-4.
32. Živcová, Z.; Černý, M.; Pabst, W.; Gregorová, E. Elastic properties of porous oxide ceramics prepared using starch as a pore-forming agent. *J. Eur. Ceram. Soc.* **2009**, *29*, 2765–2771. [CrossRef]
33. Trevisani, F.; Marco, F.D.; Capitanio, U.; Dell'Antonio, G.; Cinque, A.; Larcher, A.; Lucianò, R.; Bettiga, A.; Vago, R.; Briganti, A.; et al. Renal histology across the stages of chronic kidney disease. *J. Nephrol.* **2021**, *34*, 699–707. [CrossRef]
34. Giknis, C. *Clinical Laboratory Parameters for Crl:Wi (han)*; Charles Rivers Laboratories: Wilmington, MA, USA, 2008.
35. Boehm, O.; Zur, B.; Koch, A.; Tran, N.; Freyenhagen, R.; Hartmann, M.; Zacharowski, K. Clinical chemistry reference database for wistar rats and C57/BL6 mice. *Biol. Chem.* **2007**, *388*, 547–554. [CrossRef]
36. Ozer, J.; Ratner, M.; Shaw, M.; Bailey, W.; Schomaker, S. The current state of serum biomarkers of hepatotoxicity. *Toxicology* **2008**, *245*, 194–205. [CrossRef]
37. Ramaiah, S.K. A toxicologist guide to the diagnostic interpretation of hepatic biochemical parameters. *Food Chem. Toxicol.* **2007**, *45*, 1551–1557. [CrossRef]
38. Seidenstuecker, M.; Schmeichel, T.; Ritschl, L.; Vinke, J.; Schilling, P.; Schmal, H.; Bernstein, A. Mechanical properties of the composite material consisting of β-tcp and alginate-di-aldehyde-gelatin hydrogel and its degradation behavior. *Materials* **2021**, *14*, 1303. [CrossRef] [PubMed]
39. Hart, N.H.; Nimphius, S.; Rantalainen, T.; Ireland, A.; Siafarikas, A.; Newton, R.U. Mechanical basis of bone strength: Influence of bone material, bone structure and muscle action. *J. Musculoskelet. Neuron. Interact.* **2017**, *17*, 114–139.
40. Seidenstuecker, M.; Ruehe, J.; Suedkamp, N.P.; Serr, A.; Wittmer, A.; Bohner, M.; Bernstein, A.; Mayr, H.O. Composite material consisting of microporous β-tcp ceramic and alginate for delayed release of antibiotics. *Acta Biomater.* **2017**, *51*, 433–446. [CrossRef] [PubMed]
41. Cai, S.; Wu, C.; Yang, W.; Liang, W.; Yu, H.; Liu, L. Recent advance in surface modification for regulating cell adhesion and behaviors. *Nanotechnol. Rev.* **2020**, *9*, 971–989. [CrossRef]
42. Zhao, G.; Zinger, O.; Schwartz, Z.; Wieland, M.; Landolt, D.; Boyan, B.D. Osteoblast-like cells are sensitive to submicron-scale surface structure. *Clin. Oral Implant. Res.* **2006**, *17*, 258–264. [CrossRef]
43. Bernstein, A.; Niemeyer, P.; Salzmann, G.; Südkamp, N.P.; Hube, R.; Klehm, J.; Menzel, M.; von Eisenhart-Rothe, R.; Bohner, M.; Görz, L.; et al. Microporous calcium phosphate ceramics as tissue engineering scaffolds for the repair of osteochondral defects: Histological results. *Acta Biomater.* **2013**, *9*, 7490–7505. [CrossRef]
44. Misch, C.E.; Qu, Z.; Bidez, M.W. Mechanical properties of trabecular bone in the human mandible: Implications for dental implant treatment planning and surgical placement. *J. Oral Maxillofac. Surg.* **1999**, *57*, 700–706. [CrossRef]
45. Morgan, E.F.; Unnikrisnan, G.U.; Hussein, A.I. Bone mechanical properties in healthy and diseased states. *Annu. Rev. Biomed. Eng.* **2018**, *20*, 119–143. [CrossRef] [PubMed]

Journal of
Composites Science

MDPI

Review

Calcium-Based Biomineralization: A Smart Approach for the Design of Novel Multifunctional Hybrid Materials

Elisabetta Campodoni *, Margherita Montanari, Chiara Artusi, Giada Bassi, Franco Furlani, Monica Montesi, Silvia Panseri, Monica Sandri and Anna Tampieri *

Institute of Science and Technology for Ceramics-National Research Council (CNR), 48018 Faenza, Italy; margherita.montanari@istec.cnr.it (M.M.); chiara.artusi@istec.cnr.it (C.A.); giada.bassi@istec.cnr.it (G.B.); franco.furlani@istec.cnr.it (F.F.); monica.montesi@istec.cnr.it (M.M.); silvia.panseri@istec.cnr.it (S.P.); monica.sandri@istec.cnr.it (M.S.)
* Correspondence: elisabetta.campodoni@istec.cnr.it (E.C.); anna.tampieri@istec.cnr.it (A.T.);
  Tel.: +39-0546-699761 (E.C.); +39-0546-699753 (A.T.)

Abstract: Biomineralization consists of a complex cascade of phenomena generating hybrid nanostructured materials based on organic (e.g., polymer) and inorganic (e.g., hydroxyapatite) components. Biomineralization is a biomimetic process useful to produce highly biomimetic and biocompatible materials resembling natural hard tissues such as bones and teeth. In detail, biomimetic materials, composed of hydroxyapatite nanoparticles (HA) nucleated on an organic matrix, show extremely versatile chemical compositions and physical properties, which can be controlled to address specific challenges. Indeed, different parameters, including (i) the partial substitution of mimetic doping ions within the HA lattice, (ii) the use of different organic matrices, and (iii) the choice of cross-linking processes, can be finely tuned. In the present review, we mainly focused on calcium biomineralization. Besides regenerative medicine, these multifunctional materials have been largely exploited for other applications including 3D printable materials and in vitro three-dimensional (3D) models for cancer studies and for drug testing. Additionally, biomineralized multifunctional nano-particles can be involved in applications ranging from nanomedicine as fully bioresorbable drug delivery systems to the development of innovative and eco-sustainable UV physical filters for skin protection from solar radiations.

Keywords: calcium-based biomineralization; hydroxyapatite nanoparticles; biomimicry; multifunctional materials

Citation: Campodoni, E.; Montanari, M.; Artusi, C.; Bassi, G.; Furlani, F.; Montesi, M.; Panseri, S.; Sandri, M.; Tampieri, A. Calcium-Based Biomineralization: A Smart Approach for the Design of Novel Multifunctional Hybrid Materials. *J. Compos. Sci.* **2021**, *5*, 278. https://doi.org/10.3390/jcs5100278

Academic Editor: Francesco Tornabene

Received: 23 August 2021
Accepted: 8 October 2021
Published: 15 October 2021

1. Introduction

Biomineralization is a naturally occurring process in which organisms form minerals and consist in a complex cascade of phenomena generating hybrid nanostructured materials based on organic and inorganic matter [1–3]. These components are hierarchically organized from the nanoscale to the macroscopic scale to create a protective and/or load-bearing structure [4–9]. Resulting structures combine the hardness and pressure resistance, due to the inorganic phase, and elasticity and tensile strength, due to the organic one. Indeed, the inorganic phase helps to protect the living organisms (e.g., mollusk shells or crustacean exoskeleton) and to support organisms (e.g., bones, teeth, and coral skeleton) [10–12]. Due to the strict interaction between biomineralized crystals and organic matter, natural structures are usually very different to the synthetic ones. In detail, the high level of control over the composition, structure, size, and morphology of natural structures allows to create very fascinating properties that often overtake those of the synthetic analogues [13–15]. Organisms use macromolecules (e.g., collagen and chitin) to control the nucleation and growth of biominerals as well as crystalline form and shape of inorganic crystals in a process called molecular recognition [2,16,17].

Biomineralization can be subdivided in two main categories, namely biological induction and biological control. These processes differ for the fine regulation of size, shape

173

and arrangement of resulting biominerals [2,8,18]. It is no surprise, then, that scientists are strongly intrigued by these processes that have become a source of inspiration for the development of highly organized materials with customized properties [19–21].

Mimicking Biomineralization in the Lab

Biominerals compared to natural or synthetic minerals often display excellent mechanical and other properties due to their multi-level order, hierarchically organized from the nanoscale to the microscale. For this reason, in the last decades, researchers have been trying to reproduce the calcium-based biomineralization processes in laboratory, inducing the heterogeneous nucleation of the inorganic phase into organic matrix through fine mechanisms driven by the organic matrix itself. The chemical and physical interaction between phases confers unique features to the resulting hybrid materials, in a similar way compared to what happens in the natural biomineralization process (Figure 1). These peculiar properties cannot be obtained through a simple mixing of the phases [22–24].

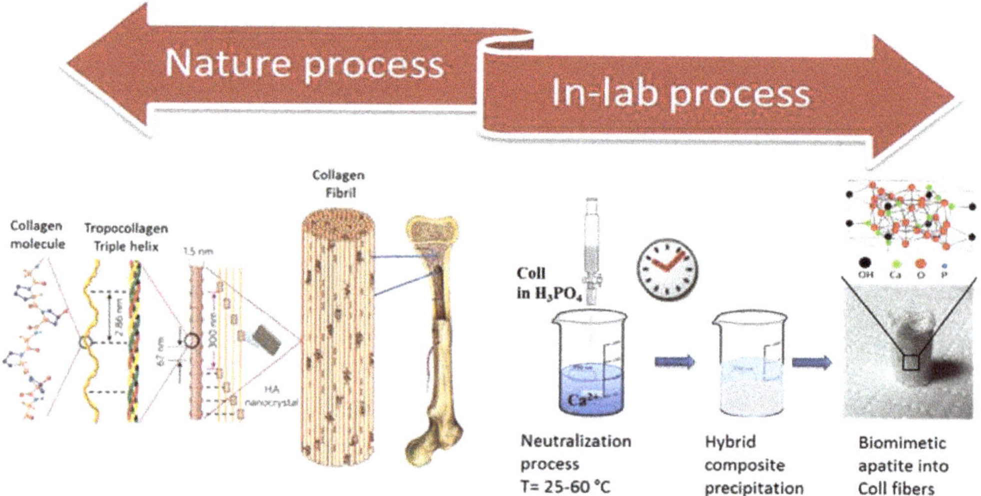

Figure 1. Schematic illustration of the naturally occurring structure of bone microstructure and the synthetic approach used to reproduce it. More in detail, in the natural occurring structure of bone, collagen fibers are organized in a triple helix, forming tropocollagen fibrils; these fibrils are tightly tied and reinforced by hydroxyapatite crystals; the organization and association of these fibrils confers peculiar structural and mechanical properties to the bone. Within the lab process, a collagen acidic solution containing phosphate ions (e.g., phosphoric acid) is dropwise added and mixed to a basic solution containing calcium ions (e.g., calcium hydroxide), promoting the formation of nano-hydroxyapatite crystals within the collagen fibers. Reproduced from "Biomineralization process generating hybrid nano-and micro-carriers" by E. Campodoni et al., 2018, Core-Shell Nanostructures for Drug Delivery and Theranostics: Challenges, Strategies, and Prospects for Novel Carrier Systems, 19–24, (doi:10.1016/C2016-0-03458-7) (Under a Creative Commons Attribution 4.0 International License).

As a consequence, the biomineralization process study, together with other emerging technologies to synthesize new nanomaterials, has spread into many fields in our life such as mechanical, electrical [25,26], and environmental [27,28], as well as biomedical engineering [29,30].

In this review focused on the biomedical field, we aim to provide an overview of different materials mimicking the natural calcium-based biomineralization process to prove that, finely tuning some process variables, it is possible to design multifunctional materials. These materials can be exploited in several applications in order to obtain customized and precise medical tools [31–34]. We will first provide a brief overview of the biomineralization process: how it happens in nature, and how scientists have translated this natural process to an in-lab process. Taking into account the wide chosen topic, we decided to focus on calcium-based biomineralization, more specifically on different applications aside from bone regeneration, that were poorly or not considered in other reviews. Specifically, we will discuss biomimetic hybrid material features that can be obtained by modulating different process parameters, focusing on the materials chemical–physical and biological features which are essential to make them suitable for biomedical field. Finally, we will discuss the several applications of these materials besides tissue regeneration, such as their use for the creation of 3D cancer predictive models or drug testing, as well as on their use as innovative physical filters against solar radiations or as nano and micro drug delivery systems.

2. Features of Biomimetic and Hybrid Biomaterials

From its first understanding, calcium-based biomineralization process uniqueness has attracted high attention due to its applicability in many different fields, especially for bone tissue engineering and regenerative medicine as well as the mild conditions, i.e., physiological temperature, pressure, and pH, in which this process occurs. Thus, the first step has been trying to translate the natural process into an in-lab process, mimicking the formation of the natural hierarchically structured organic–inorganic composites. Natural mineralization is commonly divided into two different groups recognized as "biologically induced" and "biologically controlled".

As reported by Weiner and Dove [35], the precipitation of minerals that occurs as result of interactions between biological components and the environment is termed "biologically induced" mineralization. In this situation, the chemical conditions of the environment, such as pH and CO_2, indirectly favor the formation of a specific mineral type. The mineral composition varies consistently with the variation of the environments in which they form, resulting in different morphologies, water content, element composition, structure, and particle size.

In "biologically controlled" mineralization, the organism uses cells that actively take part in the nucleation, growth, morphology, and final location of the mineral that is deposited. While the degree of control varies across species, almost all controlled mineralization processes occur in an isolated environment. The result can be remarkably sophisticated, species-specific products that give the organism specialized biological functions. Calcium-based biomineralization can be described with subsequent stages, the first of which consist of inorganic molecules pre-assembling into ordered structures. Then, the molecular recognition among the organic and inorganic interfaces controls crystal nucleation and growth, allowing the formation of subunits. Finally, cells will take part in the process, forming biominerals with multilevel structure by assembling subunit minerals. All these steps are controlled by synergistic action of various environmental factors such as pH, temperature, and organic matrix chemistry [2]. Regarding the matrix, the interaction between inorganic and organic matrices leads the entire process, resulting in widely different results depending on the stereo-chemical and physical interaction occurring among the two components. This process is affected by different parameters; nevertheless, taking into account the influencing factors, the process can be transferred into an "in-lab" process and can even be controlled and directed towards the development of novel materials endowed with intriguing features. Speaking of organic matrices, it has been proven that the structural organization complexity highly affects the final biomineralization product, which might result in a three-dimensional scaffold in the case of highly organized matrices, or well as flakes or powders in the case of poorly organized matrices [36]. A clear example of this is given by the mineralization of collagen compared to gelatin. Collagen, with its multi-level organization, can undergo the biomineralization process without losing its structure, resulting in a three-dimensional scaffold with different properties depending on the biomineralization grade [10]. Gelatin, on the other hand, being a lower assembled polymer, can be exploited to form low structured biomineralized materials, composed of hybrid micro-flakes made of clusters of nano-particles [37].

In addition to collagen and gelatin, it is possible to find examples in the literature of mineralization with different polymer matrices as cellulose, chitosan, alginate, and silk (Table 1). Ahmed Salama has shown, for instance, how cellulose and its derivatives are promising candidates for developing and constructing smart organic/inorganic hybrid biomaterials through a calcium-based biomineralization process [38]. Cellulose/calcium phosphate hybrid materials, for example, combine the properties of both components, the functionality and flexibility of the cellulose with the heat resistance and stability of the inorganic material, to generate compounds that are being exploited for different applications such as bone regeneration, drug delivery vehicles, dental repair, and adsorption. Chitosan can be efficiently exploited to produce hybrid composites based on hydroxyapatite [39,40]. These hybrid composites are mainly devised for bone regeneration. Indeed, chitosan shares some peculiar features with collagen (e.g., a comparable role in the exo- vs. endo-skeleton and flexibility) [41]. Regarding alginate, a recent work [42]. reported the effect of alginate and its well-defined oligomers with defined structure on brushite nucleation and growth for the synthesis of hybrid materials, useful in bioactive agent delivery, wound healing, and tissue engineering. Growth experiments showed that molecular weight and additives of alginate affect the crystal growth rates and the growth mechanisms. Finally, Yang and collaborators [22] have described as silk fibroin as able to facilitate nucleation of the hydroxyapatite crystals through its molecular self-assembly, and to create efficient biomaterial with mechanical and functional properties for biomedical applications.

The ideal scaffold useful to obtain a robust in vivo affinity must be designed with the final purpose to resemble the 3D architecture, nanostructure, chemical composition and mechanical properties of the extracellular matrix (ECM) of the native tissue [21]. The scaffold must be composed of biocompatible materials able to guarantee cells viability, support of cellular functions and induction of molecular and mechanical signals without eliciting adverse effects on cells and, consequently, local or systemic undesired responses in the host [43,44]. Biomineralized substrates influence cells biochemistry by the exchange in proteins and ions that, subsequently, conditions the 3D microenvironment. This implies that proteins and ions composition (e.g., calcium and phosphate ratio, collagen, proteoglycans, etc.) of native tissue is a crucial parameter to be consider for scaffold design and synthesis [45]. If materials quickly degrade in vivo, the scaffold fails in its mechanical support role; conversely, an inflammatory response could be provoked by the foreign material if the scaffold has an excessively long biodegradation time. Therefore, a controlled biodegradability is crucial in scaffolds fabrication [46]. Overall, the microarchitecture of the support should be highly porous and interconnected to provide inwards diffusion of oxygen and nutrients and elimination of waste products in order to meet cellular requirements for adhesion, growth, differentiation and migration [47].

Porosity generally supports cell migration into the scaffold by promoting interaction between cells and the available surface area of the scaffold. Pores density and size influence cellular behavior in a inversely proportional way [48]; as pores size decreases, pores density increases as well as the surface area available on the scaffold for cells interactions. However, if pores are too small, cells are not able to penetrate and migrate inside the structure. Indeed, the pores dimension is able to affect vascularization of scaffolds. A different vascularization can consequently tune cells differentiation [49–51]. Consequently, scaffolds composition, but also scaffold porosity, need to be optimized on dependence of the tissue type to interact with; as example, for bone tissue engineering scaffolds need to contain a mixture of macro- and micro-pores that allow cells to grow in vivo, facilitating cell-material interaction and complete scaffold colonization, respectively [52].

Additionally, in vivo mechanical signals (Young's modulus, compressive strength and fatigue strength mechanical forces) for cells must be replicated in the scaffold to induce the correct cellular differentiation pathway. Mechanical stiffness and porosity are often conflicting physical properties, as the first is inversely related to the other. Consequently, finding a good compromise between the scaffold properties for to promote the correct cellular activity and mechanical integrity is often a hard challenge [53]. The scaffold surface can be bio-decorated with specific proteins/biomolecules or biomotives to improve its outcomes on cell adhesion, proliferation, and differentiation. Tissue formation and ECM deposition are regulated and performed by a wide variety of biomolecules. For this reason, scaffolds can be enriched by growth factors during their production, assuming a significant role in tissue engineering applications [54–56], such as improving tissue formation or the reward of specific biomolecules in pathological conditions [21].

The set of all above-mentioned properties makes the scaffold bioactive, osteoconductive and osteoinductive in tissue engineering, regeneration and modelling of mineralized tissue (bones, tooth, tendons, and cartilage). Osteoconductive scaffold can support the ingrowth of cells into pores, channel or pipes, while osteoinduction refers to the activity of a contact or soluble material to induce the differentiation of pluripotent stem cells towards specific cells lineage. Achieving these demands, calcium-based biomineralization process represents an excellent biomimetic manufacturing strategy to obtain hierarchically designed scaffolds with appropriate mechanical stability, flexibility and a highly porous and interconnected structure [12,15,57].

Biomineralized nanomaterials represent also a promising tool for different clinical applications. Nanoparticles as drug carriers for drug delivery involves the linking of drug molecules to the nanomaterial to guarantee a controlled distribution and release rate of the drug [58,59]. In this case a good biological activity and biodegradability together with good stability under close physiological conditions are required. Most of the components of biomineralized nanomaterials (e.g., calcium phosphate, carbonate and iron oxide) are naturally present in the body and easily and normally metabolized and absorbed by it; (3) good biological activity. Biomineralized nanomaterials facilitate loading of active molecules into the mineral phase by the interaction between biominerals and biomolecules, making possible to use them for drugs and molecules loading to be used inseveral biomedical practices [1].

Table 1. Influence of physical chemical properties of hybrid materials on in vitro/in vivo behavior.

Physic-Chemical Parameter	Scaffold	Nanosystems
Biocompatibility	Absence of cytotoxicity [44]; Support and stimulation of cellular activity [43]	Absence of cytotoxicity Support and stimulation of cellular activity [1]
Biodegradability	Controlled biodegradability [46]	High bioabsorption and biodegradability Absence of bioaccumulation of ions [1]
Architecture	Stability under physiological condition Highly porous and interconnected [47] Hierarchical design structure [9,57]	Stability under physiological conditions [1]
Porosity and pore size Mechanical properties	Mixture of macro- and micro-porosity [48–52] Mechanical integrity [53]	/ Stability under physiological conditions [1]
Surface properties	Support and stimulation of cellular activity Tissue-specific functionalization [54–56]	Tissue-specific functionalization Target-specific functionalization [1]
Bioactivity	Osteoinductive Osteoconductive [6,9,57]	Controlled drug release and distribution [1,58,59]

3. Applications in Biomedical Field: Tissue Regeneration and Many More

Bone is the second most transplanted tissue with four million operations per year, using different bone alternatives worldwide [21,60]. Calcium-based biomineralization process is responsible for the formation of natural hard tissues such as bone and teeth [2,9,36,61] that are composed of extracellular matrix (ECM), several cells types, and water, which through a fascinating process are able to create a highly organized and hierarchically structured nanocomposite [13,36,62]. In detail, biomineralized ECM is mainly composed of plate-like nanocrystalline hydroxyapatite (the inorganic phase) and several organic components among which collagen is prevailing [63–65].

ECM highly influences adhesion, proliferation and differentiation of several cells types such as osteoblasts, bone lining cells, osteocytes as well as osteoclasts [65–67]. Several factors determine the fascinating and unique characteristic of bone, starting from composition arriving to structure and porosity, features essential for bone remodeling process. More precisely, bone tissue is characterized by a dynamic nature which allows to preserve its healthy condition, thanks to bone remodeling which consist in a dynamic and fine equilibrium between bone resorption by osteoclasts and bone formation by osteoblasts in response to biomechanical stimuli [68,69].

However, although this process preserve bone, its self-healing capacity is enough only to repair small bone damages. Subsequently, in case of extensive bone damage, bone substitutes are often required. For several decades, the gold standards were autografts or allografts [36,44]. In the last decade, instead, tissue engineering and regenerative medicine attention shifted towards development of biological replacements able to mimic and regenerate the tissues itself. In this perspective, scaffolds are becoming a fundamental tool for tissue engineering and regeneration, they indeed act as substrates able to structurally guide cells and provide anchorage sites, making possible to develop engineered structure made of a combination of materials and living cells. More precisely, the ideal bone scaffold is a transient implant that must create an adequate environment for cells to stay viable, attach, proliferate, differentiate and deposit ECM to replace the damaged or impaired one [70–72]. To do what describe above, scaffolds should resemble the natural bone tissue under different points of view, such as composition, structure, mechanical requirements and biological features. As consequence, to mimic the chemistry and structure of natural bone, biomineralization process is absolutely the most suitable in-lab process to produce biomimetic biomaterials, in which low-crystalline hydroxyapatite is mineralized on collagen fibrils, and highly porous scaffolds suitable to promote bone tissue regeneration [13,16,73].

Although this process has touched several fields of our life, biomedical engineering has surely received more specific attention, ending to be the most investigated. Nonetheless, recently, the attention has moved towards new advanced applications of mineralized materials. These applications include: (i) development of 3D predictive models for cancer study or drug testing, (ii) study of physical filters against solar radiation, (iii) creation of nano and micro controlled drug delivery systems. Additionally, the introduction of new fabrication technologies, e.g., 3D (bio) printing, are further extending the applications in other fields. These features will be discussed in the next sections.

3.1. Biomineralization, 3D Printing and 3D Bio-Printing

Biomineralization has been proven to be an incredible versatile process, exploitable for multiple purposes in different fields. The same could be stated for 3D printing, that with its simple basic principle paved the way for realization of incredibly ambitious and elaborated projects. The combination of these two elements have been giving rise, in the last decades, to incredible results.

The importance of scaffolds geometry at multiple level, from nano- to macro- scale, for cells attachment, spreading and viability has been so far well assessed and recognized. At the macro-scale, it is desirable for the scaffold to assume the defective part shape of the tissue or organ meant to be replaced or repaired, in order to help the neo tissue to organize into the required three-dimensional structure [74].

From the macro- point of view, especially for pursuing highly challenging goals, as it can be the regeneration of long bone defects, additive manufacturing has been attracting great interest. This, as it allows both a precise and controlled spatial-deposition of materials as well as their easy combination in complex multi-material structure, in addition to the possibility of creating shape-customized scaffold based on the anatomical site targeted [75].

If the scaffold shape and geometry play a crucial role at the macroscale, the same can be said for the composition of the material used as ink for the realization of the three-dimensional structure at the micro- and nanoscale

Speaking of bone regeneration, the key element lies in the mineral phase, in nature commonly produced by cells through the calcium-based biomineralization process.

Many approaches have been used through the years to mimic, reproduce, or induce this process in order to exploit its unique ability within the creation of new materials suitable for bone tissue regeneration. 3D printing tries to combine the ultimate technology with the ancient principle of biomimesis. One of the most used approaches has been the use of different mineral phases, such as hydroxyapatite (HA), within the composition of the ink or as post-process modification. This, with the intention to not only to emulate, but also to stimulate the biomineralization, triggers the process by making the cells perceive a suitably functionalized environment and transforming the material from just osteoconductive to osteoinductive (Figure 2) [20,76]. In order to stimulate bone regeneration, other biomineralized polymers, including silk, a well-known high strength polymer, were exploited as biocompatible filler for 3D printing [77].

An alternative approach was devised by Romanazzo and collaborators. According to this strategy, it is possible to promote the formation of mineralized constructs in a support bath containing live cells and microgels, mimicking the complex and hierarchical structure of native bone. In these conditions, cells were able to differentiate at the interface of the constructs, while remaining multipotent in the intervening spaces, opening the potential for fabricating gradient tissue structures and for future in situ fabrication of bone-like tissues [78].

In this context, it is easy to understand how coatings to improve prosthetic implants biomimetic features have been the first to arise, but in recent years this trend has reached upgraded levels. For instance, Park and co-workers reported an osteoinductive coating of polydopamine, biomineralized HA, and bone morphogenetic protein-2 of a 3D printed polycaprolactone (PCL) structure elegantly exploits the advantages of three different strategies to reach noteworthy results. A different approach, which departs from post-process modifications and moves to the inclusion of biomineralization boosting elements into the ink formulation, is proposed by Hernandez and collaborators. A complex 3D printed multimaterial system made of PCL and HA-loaded hydrogel for long bone defect regeneration was developed. In this work, the PCL provided the appropriate mechanical support, while the hydrogel composition supports cells, promoting biomineralization [75]. Recently, more unconventional mineral phases, such as graphene oxide (GO) and black phosphorus (BP) have been used with the same purpose, with promising results [79].

Indeed, Yang and co-workers used an innovative combination of bio-glass and black phosphorus to devise three-dimensional therapeutic structures for localized photothermal osteosarcoma treatment and subsequent bone regeneration (Figure 3), profit from black phosphorus degradation to induce a phosphorus-driven in situ biomineralization [80].

A similar approach, but with different applications was exploited by Lin et al. They used a biomineralization inspired process to create a synthetic graphene composite with reshaping and self-healing features which can be used in a large variety of applications, including energy storage to actuators [81].

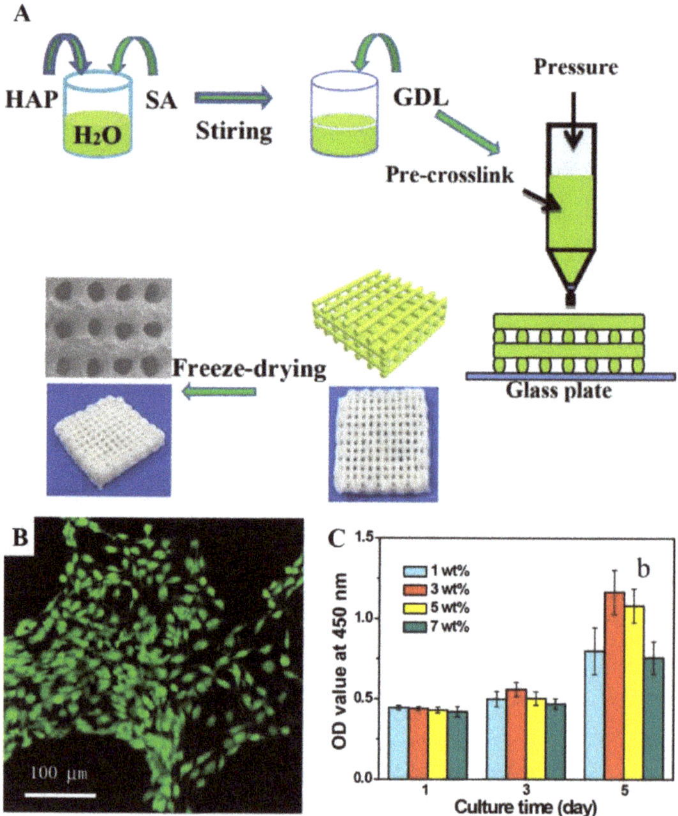

Figure 2. (**A**) Schematic illustration of the 3D printing of hydrogels based on sodium alginate and hydroxyapatite (SA/HAP). In this case, alginate and hydroxyapatite were solubilized in water and extensively mixed together. Glucono-delta-lactone (GDL), a controlled acidifying agent able to promote partial calcium release from hydroxyapatite, was then added to the alginate mixture (pre-crosslinking). The resulting mixture was then extruded. The progressive calcium release was able to promote the gelation of alginate, and thus the formation of hybrid hydrogels based on alginate and hydroxyapatite. These hydrogels were freeze-dried obtaining scaffolds. (**B**) Confocal Laser Scanning Microscopy (CLSM) micrographs of fluorophore labelled Bone Marrow Stem Cells (BMSCs) after 5 days of culture on the same scaffolds; (**C**) BMSCs proliferation at different timeframes (1, 3, and 5 days of culture) on the porous scaffolds with different HAP amount. Reproduced with permission from *Bioactive and Biocompatible Macroporous Scaffolds with Tunable Performances Prepared Based on 3D Printing of the Pre-Crosslinked Sodium Alginate/Hydroxyapatite Hydrogel Ink* by S. Liu et al., 2019, Macromolecular Materials and Engineering, 304 (4), 11 (doi:10.1002/mame.201800698). Copyright 2019 by John Wiley and Sons.

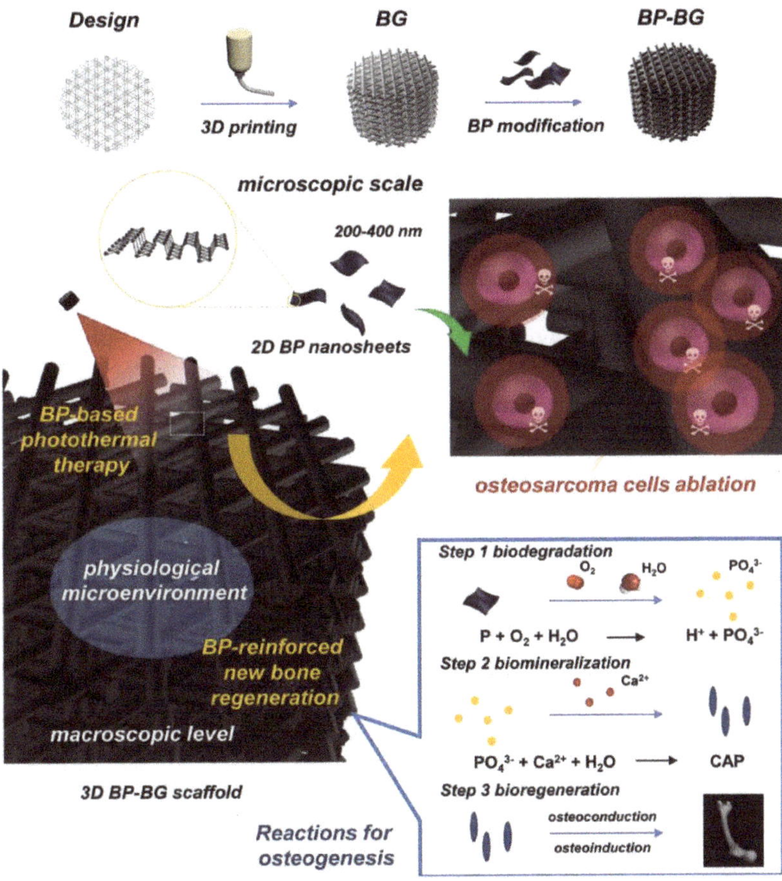

Figure 3. Schematic illustration of scaffolds fabrication based on black phosphorous/bio-glass (BP-BG) and devised for the elimination of osteosarcoma and the subsequent osteogenesis. These scaffolds were produced by 3D printing a black phosphorus and bio-glass mixture (BP-BG). These scaffolds can be exploited to promote osteosarcoma cells ablation upon light exposure. Subsequently, the scaffolds, due to their osteoconductive and osteoinductive properties, can be degraded into the main components of bone, promoting biomineralization and formation of new bone. Reproduced with permission from *2D-Black-Phosphorus-Reinforced 3D-Printed Scaffolds: A Stepwise Countermeasure for Osteosarcoma* by Shi et al., 2018, Advanced Materials, 30 (10), 12 [80]. Copyright 2018 by John Wiley and Sons.

3.2. 3D Predictive Models: From Cancer Study to Drug Testing

The scientific community has recently focused on the design and bioengineering of innovative 3D culture systems able to overcome the well-recognized inadequacy of conventional bi-dimensional (2D) in vitro models in recapitulating the complexity of in vivo microenvironment [82–84]. 3D tools and technologies reproduce cellular heterogeneity, tissue-specific ECM, and biological interactions in a more biomimetic way, providing in vitro platforms which closely resemble native microenvironment. These biomimetic materials can be exploited for basic biological studies, drug screening, and reproduction of viable biological niches for in vivo transplantation. The biomineralization process is an excellent strategy for the development of advanced biomimetic models, including 3D biomaterials, cellular coatings, and nanoplatforms with flexibility, diversity, and utility of frameworks for a wide variety of applications [85,86].

Ye and co-workers [87] established a rapid biomimetic mineralization approach to obtain a 3D porous and mineralized hydroxyapatite/collagen composite scaffold for bone regeneration. By a custom synthesis process based on self-assembled collagen fibrils as fixed template, the authors created an in vitro 3D bone-like niche seeded with human Umbilical Cord Mesenchymal Stem Cells (hUCMSCs) with high cell viability, adhesion, proliferation, and differentiation into osteoblasts due to the mineralized scaffold. A rabbit femoral condyle defect model was tested to confirm the ability of the viable niche to facilitate bone regeneration and repair over a period of 6–12 weeks. The mineralized collagen scaffold seeded with hUCMSCs successfully promoted the healing of bone defect in vivo; as new bone tissue formed, the scaffold gradually degraded and was absorbed, confirming the promising use of the hUCMSCs-loaded bone-like niche for in vivo transplantation for bone tissue regeneration.

The same concept was exploited by Menale et al. [71] in 2019 for a cell-therapy based strategy. In this case, a biomineralized bone-like scaffold was used as a rationally designed device conceived to be seeded with cells and subsequently transplanted in vivo to restore or replace a missing function that cannot be completely renewed by only cells [88]. The authors used the scaffold as productive factory of bioactive soluble osteogenic Receptor Activator of Nuclear Factor k B Ligand (RANKL) directly secreted by seeded Mesenchymal Stem Cells (MSCs) on the 3D support [89,90]. The scaffold, obtained through direct nucleation of magnesium-doped hydroxyapatite (HA) nano-crystals on self-assembling collagen fibrils (MgHA/Coll) by a pH-driven biomineralization process, showed structural, compositional, and morphological similarities to the native bone ECM. The biomineralized scaffold guarantees the development of an in vitro, viable, bone-like niche able to compensate the RANKL factor deficit in Autosomal Recessive Osteopetrosis (RANKL-ARO) once transplanted in vivo due to the continuous secretion by MSCs; the MgHA/Coll scaffold promoted the differentiation of MSCs towards osteoclasts [91,92], helping to restore the physiological functions of bone cells in a RANKL-/- mice.

Recently, the same MgHA/Coll scaffold was used as bone-like ECM to be seeded with tumor spheroids, called sarcospheres, and parental cells of MG63 and SAOS-2 osteosarcoma cell lines as enriched Cancer Stem Cells (CSCs) models with the final purpose of obtaining a 3D in vitro CSC-niche of osteosarcoma (Figure 4) [73]. The material provided specific physical-chemical and biomechanical stimuli to the critical pluripotent stem cell population, giving birth to a 3D predictive in vitro model of CSC-niche of osteosarcoma with enhanced stemness and niche-related properties compared to those seeded with parental cells. Through an in-depth cellular and molecular characterization of sarcospheres, and an optimization of the scaffold resembling tumor ECM, the authors were able to provide a closely mimetic in vitro platform for tumor studies and CSC-specific drug screening [93].

A novel biomineralization-inspired cancer therapy has recently been developed as proof-of-concept of advanced nanotechnological therapy. Natural mineral accumulation is a significant biological process that, in abnormal cases, causes the excessive deposition of calcium ions in damaged or defective tissues, leading to common pathologies, such as kidney stones and vascular calcification [94]. The anomalous mineralization can be exploited as "biomimetic pathological mineralization" onto some tumor cells, such as human cervical cancer cell line (HeLa), which can selectively assimilate, folate, and concentrate calcium ions by the overexpression of folate receptor in cancer cells, creating a Cancer Cell-Targeting Calcification-based therapy (CCTC) as reported by Zhao and co-workers [95]. On this trajectory, a biomineralization-inspired drug free strategy can be used to promote cell death by creating a calcium phosphate (CaP) mineral cell coating that leads to the agglutination of tumor cell nuclei without inducing normal cell death [95]. This approach also showed promising results on metastasis, where the survival rate of pathological mice improved significantly (up to 80%) due to the suppression of the metastasis by selective calcification-based substitution of the tumor with curable sclerosis. However, the required concentration of calcium ions exceeds physiological levels, thus identifying an innovative biomineralization-inspired material able to specifically accumulate ions in the target tissue to facilitate calcium mineral nucleation is still a challenge that may be addressed, for example, by the exploitation of specific ligand/antigens interactions on cancer cell membranes [95]. Without a doubt, this concept can be used to eventually create in vitro 3D biomineralized-based scaffolds able to specifically induce tumor cell death.

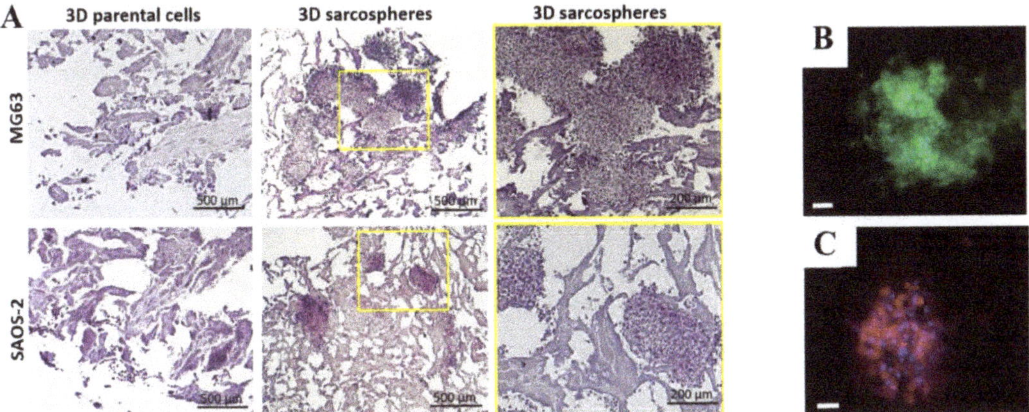

Figure 4. Panel of figures of in vitro 3D model of Cancer Stem Cells (CSCs)-niche of osteosarcoma from [73]. (Under a Creative Commons Attribution 4.0 International License) using biomineralized scaffolds based on collagen and magnesium-doped hydroxyapatite (MgHA/Coll scaffold) as bone-like ExtraCellular Matrix (ECM). (**A**) Histological analysis after Haematoxylin–Eosin (H&E) staining of the MgHA/Coll scaffold seeded with both cellular phenotypes, parental and spheroidal, of MG63 and SAOS-2 osteosarcoma cell lines. The morphological features and the interaction behavior of the sarcospheres and parental cells with the scaffold is shown, with image enlargements of 200 μm on the right of the figure. (**B**,**C**) Immunofluorescence analysis of the 3D MgHA/Coll models with sarcospheres. Representative image of OCT-4 immunolocalization in SAOS-2 sarcospheres in image (**B**); scale bar 50 μm. SOX-2 immunolocalization in MG63 sarcospheres in image (**C**); scale bar 25 μm. Blue DAPI: cell nuclei; green: OCT-4; and red: SOX-2.

Microcalcifications (MCs) also serve as diagnostic markers for breast cancer; breast cancer screenings (e.g., mammography) frequently rely on MCs, and their chemical composition (e.g., calcium phosphate, apatite, calcium oxalate, etc.) is associated with tumor malignancy [96,97]. However, due to the absence of sufficiently predictive 3D tumor models, little is known about how they form in the body, their effective role in cancer progression, or how cancer cells are involved in the mineralization process. Therefore, Vidavsky and co-workers [98] exploited the role of biological induction of biomineralization [99] for developing in vitro 3D model of breast tumor MCs to study the cellular pathways involved in MCs formation as a function of malignancy potential. Mammary multicellular spheroids were obtained by parent MCF10A benign human breast epithelial cell line, MCF10DCIS.com [100] and MCF10CA1a [101] which derived from MCF10A and possessed ductal carcinoma in situ (DCIS) and invasive tumors characteristics, respectively; together, these three cell lines allow the modeling of varying stages of breast cancer, ranging from non-malignant (MCF10A), pre-cancerous (MCF10DCIS.com), to invasive phenotype (MCF10CA1a) allowing us to investigate the correlation between cell phenotype and MCs formation. To ensure the physiological relevance of the model, the authors cultured cells in ultra-low attachment conditions with media that contained calcium, magnesium, and phosphate concentrations similar to the human body, but lacked any osteogenic agents in order to observe the real malignancy potential of spheroids just by the development of MCs. Obtained spheroids had diameters larger than 300 mm with low cell viability at the core due to limited diffusion of oxygen and nutrients (Figure 5A). Interestingly, no particles are observed in the MCF10A spheroids (Figure 5B–D). Moreover, apatite MCs were primarily detected within viable cell regions in the shells and their number and size increased with malignancy potential of the spheroids; conversely, alkaline phosphatase (ALP) decreased with malignancy potential, while osteopontin (OPN) increased. These findings support the induction of a mineralization pathway by cancer cells in a manner that is linked to their malignancy potential. This work offers an innovative exploitation of the mineralization process, which allows us to both create more reliable 3D stage-specific cancer models by inducing specific-MCs as indicators of malignancy potential and, consequently, use these platforms to deeply investigate cancer pathways.

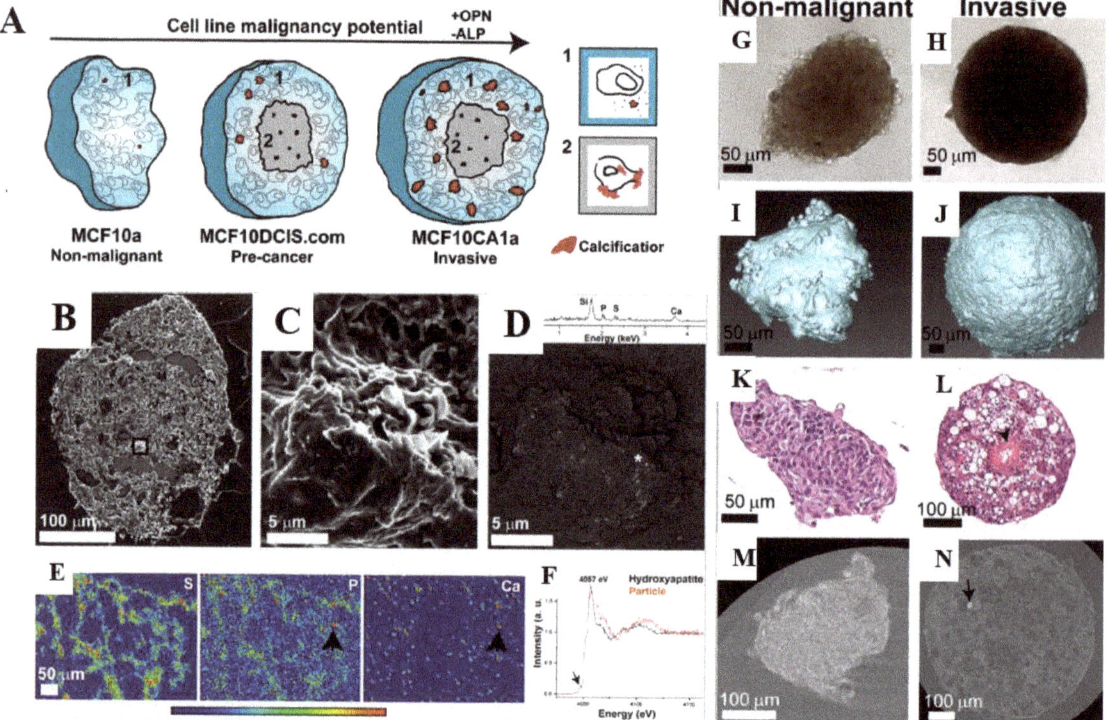

Figure 5. Panel of figures of in vitro 3D culture model of breast cancer microcalcifications from [98]. (**A**) A schematic description of the proposed mineralization pathways in the 3D in vitro breast cancer model of various tumor stages; while OPN expression levels increase, ALP expression levels decrease with an increase in malignancy potential of cell line. Viable cell region in light blue, necrotic core in gray, and calcification in red. (**B–D**) Mineralized particles in pre-cancerous MCF10DCIS.com spheroids core by SEM magnified section in images; EDS spectrum in image C of asterisk-marked area showing the presence of calcium (Ca), phosphorus (P), and sulfur (S). SRF maps of a spheroid section showing S, P, and Ca distribution in image (**E**). Ca K-edge XANES of the particle marked in (**E**) and a hydroxyapatite standard in image (**F**). (**G–N**) Characterization of non-malignant MCF10A (**G,I,K,M**) and invasive MCF10CA1a (**H,J,L,N**) spheroids at day 13 of culture; light microscope (**G,H**), 3D reconstructed volumes of spheroids (**I,J**), H&E histological staining (**K,L**), and nanoCT data stained with iodine of the spheroids cross section (**M,N**). All the figures of this panel are reproduced with permission from *Studying biomineralization pathways in a 3D culture model of breast cancer microcalcifications* by Vidavsky et al., 2018, Biomaterials, 179, 12 (doi:10.1016/j.biomaterials.2018.06.030). Copyright 2018 by Elsevier.

In conclusion, few studies exploited biomineralization-inspired process for various useful biomedical applications, from cancer modelling [73] to diagnosis markers [98], showing the need to deeply investigate the potential used of this process independently from conventional applications.

3.3. Physical Filters against Solar Radiations

Sunlight is essential for our well-being; it is responsible for regulating our internal clock, metabolism, immune systems, and for vitamin D production, essential for healthy bones. Nevertheless, it is well-known that excessive exposure to the solar radiations can cause serious damage to human health [102]. In particular, UVA (320–400 nm) and UVB (290–320 nm) radiations are the main radiations to interact with the human body, and their hazard relies in their ability to generate reactive oxygen species (ROS), which cause skin photo-aging, sunburn, dermatitis, and can also evoke long-term health effects, such as malignant tumors [103,104]. In this regard, the use of sunscreens composed of effective UV filters as protective barriers, absorbing harmful UVA and UVB radiation, has become a very important topic on which research is paying an increasing attention. UV filters can be divided in two main classes, chemical or physical filters, but nowadays physical filters are considered more attractive for sunscreens. Indeed chemical filters, despite having different advantages [105], were proven to increase environment pollution [106], and resulted to be harmful for human health [107]. For these reasons, physical filters, especially titanium dioxide (TiO_2) and zinc oxide (ZnO) [104], being able to shield the skin from both UVA and UVB radiation, are the most commonly used. However, these also have some important limitations, especially regarding their size and photocatalytic properties (Table 2). Indeed, to decrease the difficult spreading and whitening effect on the skin, the particle size is reduced to the nano-size range [108]. This entails that these particles are able to penetrate deep layers of skin, causing phototoxic reactions [109]. Furthermore, TiO_2 is known for its high photocatalytic activity, generating reactive oxygen species (ROS), which can oxidize and degrade other ingredients in the formulation, raising safety concerns [110]. Finally, the main problem, relating all common UV-filters, is that, being mainly used on the beach, the components of the cream are often released in water, causing damage to the marine environment, coral blenching, and bioaccumulation in the fauna [111]. Considering all these issues, attention is shifting towards the development of effective and safer UV-physical filters for both humans and the environment.

Recently, Battistin and co-workers [112] reported a new class of UV-physical filters through the combination of a common physical filter, TiO_2, and dihydroxyphenyl benzimidazole carboxylic acid (Oxisol) [113], an antioxidant molecule with booster effect. Boosters can be small molecules, polymers, or other particles, that act on the rheological properties of the formulation, but can also synergistically interact with the UV filters through antioxidant mechanisms or interfere with the electronic processes of UV radiations absorption. In particular, this work reported that Oxisol, functionalizing the surface of TiO_2, is able to increase the UV-protection, and also to stabilize TiO_2 nanoparticles, preventing their penetration to deeper skin layers. Furthermore, its booster activity, by means of antioxidant effects, allows a reduction in physical filter content in sunscreen formulation and a significant lowering of photocatalytic effect, typical of TiO_2. Nevertheless, Oxisol is considered as a low eco-sustainable sunscreen product. Thus, alternative, safer, and eco-sustainable sunscreen products are currently under investigation.

In the last years, the sector of sunscreens shifted attention towards formulations containing calcium phosphates (CaPs), especially hydroxyapatite (HA), the main component of animal bones, due to their excellent biocompatibility, non-toxicity, and ability to partially absorb UV radiation (Table 3) [114]. In the literature, there are some works related to hydroxyapatite as physical filter; for instance, Rehab and collaborators [115] reported the synthesis of ascorbic acid-modified, nanosized HA, stabilized with polyvinylpyrrolidone (PVP), to act as a potential biocompatible and safe constituent of sunscreens. In detail, the incorporation of the antioxidant ascorbic acid (vitamin C) [116] in HA particles maximizes photoprotection against UV damage and removes reactive oxygen species (ROS), while PVP prevent nanoparticles aggregation avoiding their skin permeation.

Another type of HA-based sunscreen has been shown by Morsy and co-authors [117], who developed a multifunctional hydroxyapatite-chitosan (HA-chitosan) gel that works as a natural antibacterial sunscreen agent for skin care. Through the simple coprecipitation method, thus avoiding use of toxic or high-cost materials, nanosized HA particles trapped within the chitosan matrix were obtained. HA acts as a physical filter against solar radiation, while chitosan acts as polymer matrix, able to avoid the agglomeration of particles and to prevent skin penetration. Additionally, chitosan acts as a natural antimicrobial agent, preventing skin wound infections caused by excessive sun exposure. Both works [115,117] focused on the intrinsic photoprotective capability of hydroxyapatite, combining it with other compounds to improve its absorption range in the UV region and to bypass the main drawbacks related to this kind of material, such as nano-size and whitening effect.

However, several studies on HA [118,119] showed its lattice has the particular ability to be modified through the doping with ions (such as Mg^{2+}, Sr^{2+}, CO_3, $Fe^{2+/3+}$, Zn^{2+}, and Ti^{4+}), thus making it a multifunctional product, adaptable according to the requests. Due to this, in the specific case of sunscreens, some works have reported that doping HA with appropriate ions can lead to an increase in the value of protection factor without necessarily having to combine other external components. In 2010, de Araujo et al. [120,121] developed, through a chemical precipitation process, four different hydroxyapatites doped with Cr^{3+}, Fe^{3+}, Zn^{2+}, and Mn^{2+} ions, having better absorption properties than pure HA in the UV region. Mostly, iron and manganese-doped HA showed the best absorption features in the UV range, necessary to be an effective sunscreen, without creating problems of toxicity or photocatalytic effect. Inspired by the previously published results [120,121], another work [122] has reported an iron-doped HA-based material containing both Fe ions (Fe^{2+}/Fe^{3+}) substituted into the hydroxyapatite lattice and iron oxide in hematite (α-Fe_2O_3) form, successfully developed from waste fish bones with a simple treatment. This was the first time an HA-based sunscreen has been synthetized, formulated in cream, and validated as proof of concept. In detail, the introduction of iron ions in the HA lattice allowed an increase in the absorption range in the UV-region, creating an effective physical filter, no photoreactivity, and a potential safe option for cream formulation, starting from waste by-products with several environmental benefits. Although iron is able to improve the photoprotective abilities of hydroxyapatite, several studies reported that titanium has a greater shielding power [123,124].

On the other hand, taking into account the photocatalytic problems associated with the use of titanium dioxide within sunscreens, some recent works shifted attention towards titanium as Ti^{4+} ions, developing titanium-doped hydroxyapatite. Yasukawa and Tamura [125] were the first to demonstrate the effective protection from solar radiations of titanium-hydroxyapatite suspensions combined with cerium ions (TiCeHA). In particular, it was revealed that the Ti^{4+} ions and Ce^{3+} ions absorbed another range of UV: UVB and UVA, respectively. Therefore, the simultaneous use of these ions further enhances the UV absorptive ability and by changing their contents in TiCeHA it is possible to create a physical filter suitable for shielding from UVA and/or UVB. Given the potential of these compounds, it would be interesting to evaluate the development of a new UV-physical filter composed of titanium-doped hydroxyapatite and biopolymers obtained by a nature inspired calcium-based biomineralization process [126]. Considering the problems associated with "classic" commercial sunscreen, having a physical filer not only able to shield solar radiations, but also safe for the human body and eco-friendly, could be the solution to overcome the main UV-filters drawbacks.

Table 2. Advantages and drawbacks of some physical filters.

Physical Filter	Advantages	Drawbacks
TiO2-Oxiol [112]	• Booster UV-shield • Stabilize nanoparticles • Antioxidant effect	• Low ecosustainable • Photocatalytic effect • No biocompatible
HA-Ascorbic Acid [115]	• High UV-shield • No toxic residual (ROS) • Biocompatible	• Low chemical stability
HA-Chitosan [117]	• Antibacterial activity • Low-cost material • Biocompatible • Eco-sustainable	• Low UV-shield
ions-doped HA (Cr^{3+}, Fe^{3+}, Zn^{2+}, Mn^{2+}, Ti^{4+}) [120,121,125]	• Biocompatible • High UV-shield • Eco-friendly • No photocatalytic effect • Stabilize nanoparticles	

Table 3. Differences between chemical, physical and hydroxyapatite (HA)-based physical filters.

Chemical Filter	Physical Filter	HA-Based Physical Filter
• Absorb UV-rays [105]; • Not degradable [106]; • Lypophilic [107]; • Partial penetration of UV-rays in the skin [107]; • Harmful for the environment [106].	• Reflect UV-rays [104]; • Not degradable [106]; • Cause whitening effect [108]; • Avoid penetration of UV-rays [104]; • Photocatalytic effect [110]; • Nanoparticles penetration [109]; • Harmful for the environment [111].	• Absorb and reflect UV-rays [114]; • Biodegradable [126]; • Avoid whitening effect [127]; • Ecosustainable [114]; • No-photocatalytic effect [127]; • Avoid particles penetration [127].

3.4. Nano and Micro Drug Delivery Systems

Bioceramics are widely used as components of implants for bone and teeth restoration. Nowadays the advanced processing techniques and the new synthesis strategies allow the incorporation of drugs, bioactive molecules, or cells within them or on their functionalized surfaces. In this regard, bioceramics and biomineralized materials can be exploited as drug delivery or controlled release in several applications, such as nanomedicine, wound healing, and bone regeneration [128,129].

Local antibiotic release is a promising and effective procedure for delivering drugs at the implantation site. With this strategy, antibiotic was loaded on a scaffold in order to both promote bone regeneration and to prevent common bacterial infections happening after surgery. In this way, scaffolds act as carriers for local antibiotic release to avoid following implant removal due to osteomyelitis (Figure 6) [130]. Different drugs can be loaded into the scaffolds, including anticancer drugs. For instance antitumoral drug-loaded scaffolds can be used to restore large bone defects after tumor extirpation, resulting in tumor inhibition with low levels of systemic toxicity [131–134].

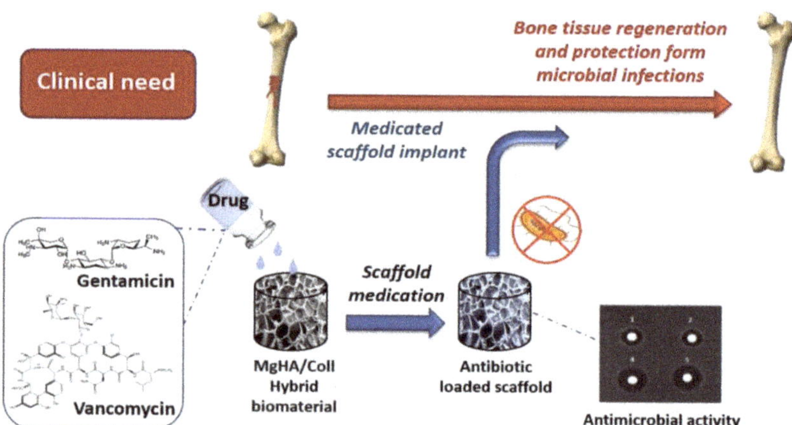

Figure 6. Schematic illustration of the loading of antibiotics within hybrid scaffolds based on collagen and magnesium-doped hydroxyapatite. Two different antibiotics, namely gentamicin and vancomycin, were introduced during biomineralization of collagen. Antibiotics proved to be tightly associated within the biomineralized scaffolds. These scaffolds were able to provide a piecemeal release of antibiotics, avoiding microbial colonization (and therefore avoiding infections) and simultaneously promoting bone tissue regeneration. The figure is reproduced from *Medicated Hydroxyapatite/Collagen Hybrid Scaffolds for Bone Regeneration and Local Antimicrobial Therapy to Prevent Bone Infections* by M. Mulazzi et al., 2021, Pharmaceutics, 13 (7), 1090 (doi:10.3390/pharmaceutics13071090) (Under a Creative Commons Attribution 4.0 International License).

Often, in orthopedic, maxillofacial, and dental surgery, whether the defect size is complex or irregular, bioceramic beads are used to induce bone tissue regeneration [135–137]. In the last decades, some research has been focused on the possibility, not only to promote tissue regeneration due to bioceramics beads, but also to modify the functionalization of them with several active molecules such as antibiotics, anticancer agents, and osteogenic agents to act themselves as drug delivery vehicles [129,138–140].

Ceramic component contributes to the mechanical stability and bioactivity of the structure; however, its adsorption of drugs often is featured by weak bonds leading to an initial burst release [140–142]. To overcome this issue, polymers can be added, forming a composite material endowed with a fine chemical and physical control of the drug adsorption and release [143,144]. These polymeric and bioceramic phases can be used as separated phases or as a single mixed phase. For example, hydrogel/bioceramic core-shell beads can be developed, by means of concentric nozzles or microfluidics exploiting both advantages of the two phases; polymers can preserve the drug, avoiding an initial excessive release, whereas ceramics contribute to the mechanical stability and bioactivity of the structure for a synergic and effective loading and sustained release of proteins [145], or drugs [146] (Figure 7), as well as cells [147].

For example, Raja and co-workers fabricated a multifunctional core-shell bead structure featured by a hydrogel shell composed of alginate including cells around a ceramic core made of α-tricalcium phosphate (α-TCP) loaded with Quercitin dihydrate, a well-known phytochemical used for the treatment of osteoporosis [129]. The core-shell beads, immersed in PBS, lead to the formation of bone-like low-crystalline apatite from α-TCP that provides structural integrity to the bead, along with a surface for the growth of cells embedded in the hydrogel shell. Researchers demonstrated a slow release of quercetin throughout the entire 120 days testing period, together with the formation of a homogenous cell layer on the ceramics structure, due to cells loaded into the hydrogel. Finally, they showed that in the region in which hydrogel and ceramics are strictly in contact, cell growth was specifically increased, highlighting the potential of the core-shell model for further material–cells interaction study [129]. This kind of composite belongs to promising class of materials able to load different types of drugs and cells to produce highly biofunctional beads, which provide an effective bone substitute for both drug delivery and bone tissue regeneration [8,147,148].

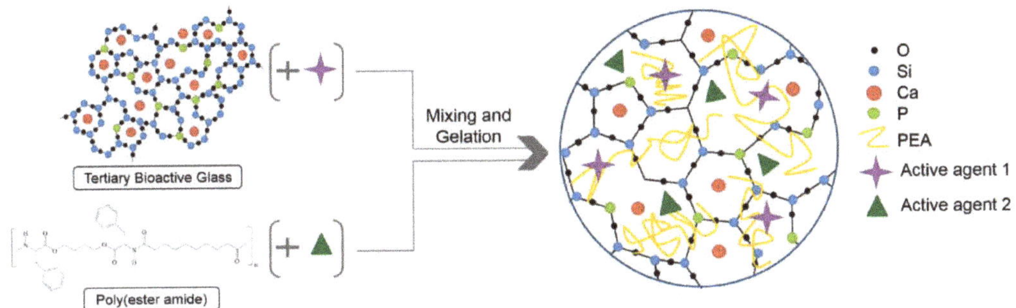

Figure 7. Schematic illustration of the two-stage synthesis approach of hybrid microparticles based on bioactive glass and poly(ester amide) (PEA). The singular components were mixed with an active agent. Subsequently the two mixtures were mixed together, and gelation was promoted, forming hybrid microparticles loaded with two different drugs. The figure is reproduced with permission from *Intrinsically fluorescent bioactive glass-poly(ester amide) hybrid microparticles for dual drug delivery and bone repair* by Aslankoohia and Mequanint et al., 2021, Materials Science and Engineering: C, 128, 112288 (doi:10.1016/j.msec.2021.112288). Copyright 2021 by Elsevier.

Moreover, calcium-based biomineralization paves the way to promising and very interesting materials where ceramics composites are nucleated onto the organic phase, as previously described, creating a single and very reactive phase that combines the advantages of the two phases. Furthermore, thanks to the possibility of introducing doping ions into apatite lattice, the resulting phase will be featured by different and new functionalities, in addition to those normally occurring in vivo, such as high bioactivity, biocompatibility, and biodegradability. Doping with magnesium or Zn ions, for example, is possible to confer antimicrobic properties essential to prevent bacterial infection or in wound healing [149,150].

The encapsulation of antibiotics in nanocarriers such as bioceramics allows the elimination of microorganisms by releasing a high antibiotic dose at a target site before the development of resistance [151]. Furthermore, many researchers demonstrated that ions present in hydroxyapatite can promote the antibacterial activity of the device. For instance, Ain and co-workers demonstrated that vancomycin-loaded HA had a slower release in comparison with pure vancomycin and also an enhanced antibacterial activity due to the presence of ions in the HA structure [140].

On the other hand, doping with Fe ions results in an interesting superparamagnetic apatite phase, able to be exploited in diagnostic field as a contrast agent or therapeutic field, due to the possibility to move it by an external magnetic field or to release drugs by means of magneto-shaking [4,139,152,153].

Concerning that point, Patricio and collaborators have developed a bio-hybrid microsphere obtained through the biomineralization of iron-doped hydroxyapatite (FeHA) within an organic matrix. In this case, the organic matrix is an animal-free recombinant peptide based on human type I collagen (RCP) enriched with RGD motif. The resulting material is bioresorbable, biocompatible, and can enhance cell adhesion. Through the fine tuning of the emulsification process, the resulting hybrid microsphere is endowed with a monomodal size dispersion, low crystallinity, and superparamagnetic properties typical of FeHA [4,23,154–156].

The resulting microspheres displayed excellent osteogenic ability with human mesenchymal stem cells, and were able to provide a slow release of recombinant human bone morphogenetic protein-2 (rhBMP-2) within 14 days. Furthermore, the release profile can be finely tuned by application of pulsed electromagnetic field, thus highlighting the potential of remote controlling the bioactivity of the new micro-devices, an interesting feature for their application in precisely designed and personalized therapies.

To conclude, the administration of therapeutic agents is still a major concern of medicine, as the systemic dose prescribed needs to be high to ensure the suitable dose in the target area, causing several collateral effects. The synergy between bioceramics and drugs therapy has paved the way to several possibilities, especially in bone pathologies, anticancer therapy, and heart diseases [157–159].

4. Conclusions and Future Perspectives

Biomimetic approaches are very promising for the design of advanced and multifunctional materials. The application of self-organization has wide potential for the tailoring structure, composition, properties, and function of materials from nano- to macroscale. Additionally, the calcium-based biomineralization process can be finely tuned by changing the environmental conditions (e.g., pH), doping ions, and organic network. Biomineralized materials can be tailored to address specific issues, including devising of materials for regenerative medicine, as well as 3D predictive models and development of drug delivery systems. Furthermore, these hybrid materials display an excellent resource to devise physical filters able to prevent UV-light-induced danger.

We believe this review will point out the future development of calcium-based biomineralization process for the creation of materials in several applications. Indeed, some issues need to be addressed, including the industrial production scale up and the sustainability— both economic and environmental—of the production.

Author Contributions: Writing—original draft preparation and editing, E.C., M.M. (Margherita Montanari), C.A., G.B. and F.F.; writing—review, M.M. (Monica Montesi), S.P., M.S. and A.T. All authors have read and agreed to the published version of the manuscript.

Funding: This research received no external funding.

Conflicts of Interest: The authors declare no conflict of interest.

References

1. Wang, W.; Liu, X.; Zheng, X.; Jin, H.J.; Li, X. Biomineralization: An Opportunity and Challenge of Nanoparticle Drug Delivery Systems for Cancer Therapy. *Adv. Heal. Mater.* **2020**, *9*, e2001117. [CrossRef]
2. Chen, Y.; Feng, Y.; Deveaux, J.G.; Masoud, M.A.; Chandra, F.S.; Chen, H.; Zhang, D.; Feng, L. Biomineralization Forming Process and Bio-inspired Nanomaterials for Biomedical Application: A Review. *Minerals* **2019**, *9*, 68. [CrossRef]
3. Tampieri, A.; Ruffini, A.; Ballardini, A.; Montesi, M.; Panseri, S.; Salamanna, F.; Fini, M.; Sprio, S. Heterogeneous chemistry in the 3-D state: An original approach to generate bioactive, mechanically-competent bone scaffolds. *Biomater. Sci.* **2018**, *7*, 307–321. [CrossRef]
4. Campodoni, E.; Patricio, T.; Montesi, M.; Tampieri, A.; Sandri, M.; Sprio, S. *Biomineralization Process Generating Hybrid Nano- and Micro-Carriers*; Woodhead Publishing: Sawston, UK, 2018; pp. 19–42. [CrossRef]
5. Tampieri, A.; D'Alessandro, T.; Sandri, M.; Sprio, S.; Landi, E.; Bertinetti, L.; Panseri, S.; Pepponi, G.; Goettlicher, J.; Bañobre-López, M.; et al. Intrinsic magnetism and hyperthermia in bioactive Fe-doped hydroxyapatite. *Acta Biomater.* **2012**, *8*, 843–851. [CrossRef]
6. Luz, G.; Mano, J. *Nanoscale Design in Biomineralization for Developing New Biomaterials for Bone Tissue Engineering (BTE)*; Woodhead Publishing: Sawston, UK, 2014; pp. 153–195. [CrossRef]
7. Luz, G.; Mano, J.F. Mineralized structures in nature: Examples and inspirations for the design of new composite materials and biomaterials. *Compos. Sci. Technol.* **2010**, *70*, 1777–1788. [CrossRef]
8. Preti, L.; Lambiase, B.; Campodoni, E.; Sandri, M.; Ruffini, A.; Pugno, N.; Tampieri, A.; Sprio, S. *Nature-Inspired Processes and Structures: New Paradigms to Develop Highly Bioactive Devices for Hard Tissue Regeneration*; IntechOpen: London, UK, 2019. [CrossRef]
9. Elisabetta, C.; Maria, D.S.; Manuela, M.; Margherita, M.; Monica, M.; Silvia, P.; Simone, S.; Anna, T.; Monica, S. *Biomimetic Approaches for the Design and Development of Multifunctional Bioresorbable Layered Scaffolds for Dental Regeneration*; CRC Press: Boca Raton, FL, USA, 2020; pp. 104–119. [CrossRef]
10. Tampieri, A.; Sandri, M.; Landi, E.; Pressato, D.; Francioli, S.; Quarto, R.; Martin, I. Design of graded biomimetic osteochondral composite scaffolds. *Biomaterials* **2008**, *29*, 3539–3546. [CrossRef]

11. Gómez-Morales, J.; Iafisco, M.; Delgado-López, J.M.; Sarda, S.; Drouet, C. Progress on the preparation of nanocrystalline apatites and surface characterization: Overview of fundamental and applied aspects. *Prog. Cryst. Growth Charact. Mater.* **2013**, *59*, 1–46. [CrossRef]

12. Pereira, D.D.M.; Habibovic, P. Biomineralization-Inspired Material Design for Bone Regeneration. *Adv. Health Mater.* **2018**, *7*, e1800700. [CrossRef] [PubMed]

13. Thula, T.T.; Rodriguez, D.E.; Lee, M.H.; Pendi, L.; Podschun, J.; Gower, L.B. In vitro mineralization of dense collagen substrates: A biomimetic approach toward the development of bone-graft materials. *Acta Biomater.* **2011**, *7*, 3158–3169. [CrossRef] [PubMed]

14. Ravindran, S.; George, A.; Tooth, B.; Genetics, D. *Engineering Mineralized and Load Bearing Tissues*; Springer International Publishing: Berlin/Heidelberg, Germany, 2015; Volume 881, pp. 129–142. [CrossRef]

15. Tampieri, A.; Sprio, S.; Sandri, M.; Valentini, F. Mimicking natural bio-mineralization processes: A new tool for osteochondral scaffold development. *Trends Biotechnol.* **2011**, *29*, 526–535. [CrossRef]

16. Nudelman, F.; Sonmezler, E.; Bomans, P.H.H.; de With, G.; Sommerdijk, N.A.J.M. Stabilization of amorphous calcium carbonate by controlling its particle size. *Nanoscale* **2010**, *2*, 2436–2439. [CrossRef]

17. Walker, J.M.; Marzec, B.; Nudelman, F. Solid-State Transformation of Amorphous Calcium Carbonate to Aragonite Captured by CryoTEM. *Angew. Chem.* **2017**, *129*, 11902–11905. [CrossRef]

18. Dellaquila, A.; Campodoni, E.; Tampieri, A.; Sandri, M. Overcoming the Design Challenge in 3D Biomimetic Hybrid Scaffolds for Bone and Osteochondral Regeneration by Factorial Design. *Front. Bioeng. Biotechnol.* **2020**, *8*, 743. [CrossRef]

19. Bharadwaz, A.; Jayasuriya, A.C. Recent trends in the application of widely used natural and synthetic polymer nanocomposites in bone tissue regeneration. *Mater. Sci. Eng. C* **2020**, *110*, 110698. [CrossRef] [PubMed]

20. Liu, S.; Hu, Y.; Zhang, J.; Bao, S.; Xian, L.; Dong, X.; Zheng, W.; Li, Y.; Gao, H.; Zhou, W. Bioactive and Biocompatible Macroporous Scaffolds with Tunable Performances Prepared Based on 3D Printing of the Pre-Crosslinked Sodium Alginate/Hydroxyapatite Hydrogel Ink. *Macromol. Mater. Eng.* **2019**, *304*, 1800698. [CrossRef]

21. Turnbull, G.; Clarke, J.; Picard, F.; Riches, P.; Jia, L.; Han, F.; Li, B.; Shu, W. 3D bioactive composite scaffolds for bone tissue engineering. *Bioact. Mater.* **2017**, *3*, 278–314. [CrossRef] [PubMed]

22. Yang, M.; He, W.; Shuai, Y.; Min, S.; Zhu, L. Nucleation of hydroxyapatite crystals by self-assembled Bombyx mori silk fibroin. *J. Polym. Sci. Part B Polym. Phys.* **2013**, *51*, 742–748. [CrossRef]

23. Patrício, T.M.F.; Panseri, S.; Sandri, M.; Tampieri, A.; Sprio, S. New bioactive bone-like microspheres with intrinsic magnetic properties obtained by bio-inspired mineralisation process. *Mater. Sci. Eng. C* **2017**, *77*, 613–623. [CrossRef] [PubMed]

24. Chatzipanagis, K.; Baumann, C.G.; Sandri, M.; Sprio, S.; Tampieri, A.; Kröger, R. In situ mechanical and molecular investigations of collagen/apatite biomimetic composites combining Raman spectroscopy and stress-strain analysis. *Acta Biomater.* **2016**, *46*, 278–285. [CrossRef] [PubMed]

25. Yan, Y.; Shi, X.; Miao, M.; He, T.; Dong, Z.H.; Zhan, K.; Yang, J.H.; Zhao, B.; Xia, B.Y. Bio-inspired design of hierarchical FeP nanostructure arrays for the hydrogen evolution reaction. *Nano Res.* **2017**, *11*, 3537–3547. [CrossRef]

26. Hou, Y.-K.; Pan, G.-L.; Sun, Y.-Y.; Gao, X.-P. LiMn$_{0.8}$Fe$_{0.2}$PO$_4$/Carbon Nanospheres@Graphene Nanoribbons Prepared by the Biomineralization Process as the Cathode for Lithium-Ion Batteries. *ACS Appl. Mater. Interfaces* **2018**, *10*, 16500–16510. [CrossRef]

27. Wu, M.; Chen, K.; Yang, S.; Wang, Z.; Huang, P.-H.; Mai, J.; Li, Z.-Y.; Huang, T.J. High-throughput cell focusing and separation via acoustofluidic tweezers. *Lab Chip* **2018**, *18*, 3003–3010. [CrossRef]

28. Zhang, W.; Zhou, S.; Sun, J.; Meng, X.; Luo, J.; Zhou, D.; Crittenden, J.C. Impact of Chloride Ions on UV/H2O2 and UV/Persulfate Advanced Oxidation Processes. *Environ. Sci. Technol.* **2018**, *52*, 7380–7389. [CrossRef]

29. Chen, W.; Wang, G.; Tang, R. Nanomodification of living organisms by biomimetic mineralization. *Nano Res.* **2014**, *7*, 1404–1428. [CrossRef]

30. Walsh, P.J.; Fee, K.; Clarke, S.A.; Julius, M.L.; Buchanan, F.J. Blueprints for the Next Generation of Bioinspired and Biomimetic Mineralised Composites for Bone Regeneration. *Mar. Drugs* **2018**, *16*, 288. [CrossRef]

31. Osorio, R.; Alfonso-Rodríguez, C.A.; Osorio, E.; Medina-Castillo, A.L.; Alaminos, M.; Toledano-Osorio, M.; Toledano, M. Novel potential scaffold for periodontal tissue engineering. *Clin. Oral Investig.* **2017**, *21*, 2695–2707. [CrossRef] [PubMed]

32. Sandri, M.; Filardo, G.; Kon, E.; Panseri, S.; Montesi, M.; Iafisco, M.; Savini, E.; Sprio, S.; Cunha, C.; Giavaresi, G.; et al. Fabrication and Pilot In Vivo Study of a Collagen-BDDGE-Elastin Core-Shell Scaffold for Tendon Regeneration. *Front. Bioeng. Biotechnol.* **2016**, *4*. [CrossRef] [PubMed]

33. Abdulghani, S.; Mitchell, G.R. Biomaterials for In Situ Tissue Regeneration: A Review. *Biomolecules* **2019**, *9*, 750. [CrossRef] [PubMed]

34. Kaczmarek, B.; Sionkowska, A.; Kozlowska, J.; Osyczka, A. New composite materials prepared by calcium phosphate precipitation in chitosan/collagen/hyaluronic acid sponge cross-linked by EDC/NHS. *Int. J. Biol. Macromol.* **2018**, *107*, 247–253. [CrossRef]

35. Weiner, S.; Dove, P.M. An Overview of Biomineralization Processes and the Problem of the Vital Effect. *Biomineralization* **2003**, *54*, 1–30. [CrossRef]

36. Zhong, L.; Qu, Y.; Shi, K.; Chu, B.; Lei, M.; Huang, K.; Gu, Y.; Qian, Z. Biomineralized polymer matrix composites for bone tissue repair: A review. *Sci. China Ser. B Chem.* **2018**, *61*, 1553–1567. [CrossRef]

37. Campodoni, E.; Dozio, S.M.; Panseri, S.; Montesi, M.; Tampieri, A.; Sandri, M. Mimicking Natural Microenvironments: Design of 3D-Aligned Hybrid Scaffold for Dentin Regeneration. *Front. Bioeng. Biotechnol.* **2020**, *8*, 836. [CrossRef]

38. Salama, A. Cellulose/calcium phosphate hybrids: New materials for biomedical and environmental applications. *Int. J. Biol. Macromol.* **2019**, *127*, 606–617. [CrossRef]
39. Wang, Q.; Tang, Y.; Ke, Q.; Yin, W.; Zhang, C.; Guo, Y.; Guan, J. Magnetic lanthanum-doped hydroxyapatite/chitosan scaffolds with endogenous stem cell-recruiting and immunomodulatory properties for bone regeneration. *J. Mater. Chem. B* **2020**, *8*, 5280–5292. [CrossRef]
40. Tang, Y.-Q.; Wang, Q.-Y.; Ke, Q.-F.; Zhang, C.-Q.; Guan, J.-J.; Guo, Y.-P. Mineralization of ytterbium-doped hydroxyapatite nanorod arrays in magnetic chitosan scaffolds improves osteogenic and angiogenic abilities for bone defect healing. *Chem. Eng. J.* **2020**, *387*, 124166. [CrossRef]
41. Furlani, F.; Marfoglia, A.; Marsich, E.; Donati, I.; Sacco, P. Strain Hardening in Highly Acetylated Chitosan Gels. *Biomacromolecules* **2021**, *22*, 2902–2909. [CrossRef]
42. Ucar, S.; Bjørnøy, S.H.; Bassett, D.C.; Strand, B.L.; Sikorski, P.; Andreassen, J.-P. Nucleation and Growth of Brushite in the Presence of Alginate. *Cryst. Growth Des.* **2015**, *15*, 5397–5405. [CrossRef]
43. Lloyd, A.W. Interfacial bioengineering to enhance surface biocompatibility. *Med. Device Technol.* **2002**, *13*, 18–21. [PubMed]
44. Finkemeier, C.G. Bone-grafting and bone-graft substitutes. *J. Bone Jt. Surg.* **2002**, *84*, 454–464. [CrossRef] [PubMed]
45. Barrère, F.; van Blitterswijk, C.A.; de Groot, K. Bone regeneration: Molecular and cellular interactions with calcium phosphate ceramics. *Int. J. Nanomed.* **2006**, *1*, 317.
46. Mountziaris, P.M.; Mikos, A.G. Modulation of the Inflammatory Response for Enhanced Bone Tissue Regeneration. *Tissue Eng. Part B Rev.* **2008**, *14*, 179–186. [CrossRef]
47. Cukierman, E.; Pankov, R.; Stevens, D.R.; Yamada, K.M. Taking Cell-Matrix Adhesions to the Third Dimension. *Science* **2001**, *294*, 1708–1712. [CrossRef] [PubMed]
48. O'Brien, F.; Harley, B.; Yannas, I.; Gibson, L. The effect of pore size on cell adhesion in collagen-GAG scaffolds. *Biomaterials* **2005**, *26*, 433–441. [CrossRef] [PubMed]
49. Tsuruga, E.; Takita, H.; Itoh, H.; Wakisaka, Y.; Kuboki, Y. Pore Size of Porous Hydroxyapatite as the Cell-Substratum Controls BMP-Induced Osteogenesis. *J. Biochem.* **1997**, *121*, 317–324. [CrossRef] [PubMed]
50. Ishida, H.; Haniu, H.; Takeuchi, A.; Ueda, K.; Sano, M.; Tanaka, M.; Takizawa, T.; Sobajima, A.; Kamanaka, T.; Saito, N. In Vitro and In Vivo Evaluation of Starfish Bone-Derived β-Tricalcium Phosphate as a Bone Substitute Material. *Materials* **2019**, *12*, 1881. [CrossRef]
51. Habibovic, P.; Yuan, H.; van der Valk, C.M.; Meijer, G.; van Blitterswijk, C.; de Groot, K. 3D microenvironment as essential element for osteoinduction by biomaterials. *Biomaterials* **2005**, *26*, 3565–3575. [CrossRef]
52. Whang, K.; Healy, K.E.; Elenz, D.R.; Nam, E.K.; Tsai, D.C.; Thomas, C.H.; Nuber, G.W.; Glorieux, F.H.; Travers, R.; Sprague, S.M. Engineering Bone Regeneration with Bioabsorbable Scaffolds with Novel Microarchitecture. *Tissue Eng.* **1999**, *5*, 35–51. [CrossRef]
53. Boccaccio, A.; Ballini, A.; Pappalettere, C.; Tullo, D.; Cantore, S.; Desiate, A. Finite Element Method (FEM), Mechanobiology and Biomimetic Scaffolds in Bone Tissue Engineering. *Int. J. Biol. Sci.* **2011**, *7*, 112–132. [CrossRef]
54. Cipitria, A.; Wagermaier, W.; Zaslansky, P.; Schell, H.; Reichert, J.; Fratzl, P.; Hutmacher, D.; Duda, G. BMP delivery complements the guiding effect of scaffold architecture without altering bone microstructure in critical-sized long bone defects: A multiscale analysis. *Acta Biomater.* **2015**, *23*, 282–294. [CrossRef]
55. Kaigler, D.; Wang, Z.; Horger, K.; Mooney, D.J.; Krebsbach, P.H. VEGF Scaffolds Enhance Angiogenesis and Bone Regeneration in Irradiated Osseous Defects. *J. Bone Miner. Res.* **2006**, *21*, 735–744. [CrossRef]
56. Dimitriou, R.; Tsiridis, E.; Giannoudis, P.V. Current concepts of molecular aspects of bone healing. *Injury* **2005**, *36*, 1392–1404. [CrossRef]
57. Tampieri, A.; Landi, E.; Valentini, F.; Sandri, M.; D'Alessandro, T.; Dediu, V.; Marcacci, M. A conceptually new type of bio-hybrid scaffold for bone regeneration. *Nanotechnology* **2010**, *22*, 015104. [CrossRef] [PubMed]
58. Parveen, S.; Misra, R.; Sahoo, S.K. Nanoparticles: A boon to drug delivery, therapeutics, diagnostics and imaging. *Nanomed. Nanotechnol. Biol. Med.* **2011**, *8*, 147–166. [CrossRef] [PubMed]
59. Cho, K.; Wang, X.; Nie, S.; Chen, Z.; Shin, D.M. Therapeutic Nanoparticles for Drug Delivery in Cancer. *Clin. Cancer Res.* **2008**, *14*, 1310–1316. [CrossRef]
60. Papakostidis, C.; Kanakaris, N.K.; Pretel, J.; Faour, O.; Morell, D.J.; Giannoudis, P.V. Prevalence of complications of open tibial shaft fractures stratified as per the Gustilo-Anderson classification. *Injury* **2011**, *42*, 1408–1415. [CrossRef]
61. Sprio, S.; Campodoni, E.; Sandri, M.; Preti, L.; Keppler, T.; Müller, F.A.; Pugno, N.M.; Tampieri, A. A Graded Multifunctional Hybrid Scaffold with Superparamagnetic Ability for Periodontal Regeneration. *Int. J. Mol. Sci.* **2018**, *19*, 3604. [CrossRef] [PubMed]
62. Hutchens, S.A.; Benson, R.S.; Evans, B.R.; O'Neill, H.; Rawn, C.J. Biomimetic synthesis of calcium-deficient hydroxyapatite in a natural hydrogel. *Biomaterials* **2006**, *27*, 4661–4670. [CrossRef]
63. Yue, K.; de Santiago, G.T.; Alvarez, M.M.; Tamayol, A.; Annabi, N.; Khademhosseini, A. Synthesis, properties, and biomedical applications of gelatin methacryloyl (GelMA) hydrogels. *Biomaterials* **2015**, *73*, 254–271. [CrossRef]
64. Alfano, M.; Nebuloni, M.; Allevi, R.; Zerbi, P.; Longhi, E.; Lucianò, R.; Locatelli, I.; Pecoraro, A.; Indrieri, M.; Speziali, C.; et al. Linearized texture of three-dimensional extracellular matrix is mandatory for bladder cancer cell invasion. *Sci. Rep.* **2016**, *6*, 36128. [CrossRef]

65. Murphy, C.A.; Costa, J.; Correia, J.S.; Oliveira, J.M.; Reis, R.L.; Collins, M. Biopolymers and polymers in the search of alternative treatments for meniscal regeneration: State of the art and future trends. *Appl. Mater. Today* **2018**, *12*, 51–71. [CrossRef]
66. Hao, Y.; Song, J.; Ravikrishnan, A.; Dicker, K.T.; Fowler, E.W.; Zerdoum, A.B.; Li, Y.; Zhang, H.; Rajasekaran, A.K.; Fox, J.M.; et al. Rapid Bioorthogonal Chemistry Enables in Situ Modulation of the Stem Cell Behavior in 3D without External Triggers. *ACS Appl. Mater. Interfaces* **2018**, *10*, 26016–26027. [CrossRef]
67. Li, W.; Xu, R.; Huang, J.; Bao, X.; Zhao, B. Treatment of rabbit growth plate injuries with oriented ECM scaffold and autologous BMSCs. *Sci. Rep.* **2017**, *7*, 44140. [CrossRef] [PubMed]
68. Florencio-Silva, R.; Sasso, G.R.D.S.; Sasso-Cerri, E.; Simões, M.D.J.; Cerri, P. Biology of Bone Tissue: Structure, Function, and Factors That Influence Bone Cells. *BioMed Res. Int.* **2015**, *2015*, 421746. [CrossRef]
69. Seeman, E. Bone Modeling and Remodeling. *Crit. Rev. Eukaryot. Gene Expr.* **2009**, *19*, 219–233. [CrossRef]
70. Bello, A.B.; Kim, D.; Kim, D.; Park, H.; Lee, S.-H. Engineering and Functionalization of Gelatin Biomaterials: From Cell Culture to Medical Applications. *Tissue Eng. Part B Rev.* **2020**, *26*, 164–180. [CrossRef]
71. Menale, C.; Campodoni, E.; Palagano, E.; Mantero, S.; Erreni, M.; Inforzato, A.; Fontana, E.; Schena, F.; Hof, R.V.; Sandri, M.; et al. Mesenchymal Stromal Cell-Seeded Biomimetic Scaffolds as a Factory of Soluble RANKL in Rankl-Deficient Osteopetrosis. *STEM CELLS Transl. Med.* **2018**, *8*, 22–34. [CrossRef] [PubMed]
72. Chocholata, P.; Kulda, V.; Dvorakova, J.; Dobra, J.K.; Babuska, V. Biological Evaluation of Polyvinyl Alcohol Hydrogels Enriched by Hyaluronic Acid and Hydroxyapatite. *Int. J. Mol. Sci.* **2020**, *21*, 5719. [CrossRef] [PubMed]
73. Bassi, G.; Panseri, S.; Dozio, S.M.; Sandri, M.; Campodoni, E.; Dapporto, M.; Sprio, S.; Tampieri, A.; Montesi, M. Scaffold-based 3D cellular models mimicking the heterogeneity of osteosarcoma stem cell niche. *Sci. Rep.* **2020**, *10*, 1–12. [CrossRef]
74. Wang, P.; Hu, J.; Ma, P.X. The engineering of patient-specific, anatomically shaped, digits. *Biomaterials* **2009**, *30*, 2735–2740. [CrossRef] [PubMed]
75. Hernandez, I.; Kumar, A.; Joddar, B. A Bioactive Hydrogel and 3D Printed Polycaprolactone System for Bone Tissue Engineering. *Gels* **2017**, *3*, 26. [CrossRef] [PubMed]
76. Hu, Y.; Wang, J.; Li, X.; Hu, X.; Zhou, W.; Dong, X.; Wang, C.; Yang, Z.; Binks, B.P. Facile preparation of bioactive nanoparticle/poly(ε-caprolactone) hierarchical porous scaffolds via 3D printing of high internal phase Pickering emulsions. *J. Colloid Interface Sci.* **2019**, *545*, 104–115. [CrossRef]
77. Huang, T.; Fan, C.; Zhu, M.; Zhu, Y.; Zhang, W.; Li, L. 3D-printed scaffolds of biomineralized hydroxyapatite nanocomposite on silk fibroin for improving bone regeneration. *Appl. Surf. Sci.* **2019**, *467–468*, 345–353. [CrossRef]
78. Romanazzo, S.; Molley, T.G.; Nemec, S.; Lin, K.; Sheikh, R.; Gooding, J.J.; Wan, B.; Li, Q.; Kilian, K.A.; Roohani, I. Synthetic Bone-Like Structures Through Omnidirectional Ceramic Bioprinting in Cell Suspensions. *Adv. Funct. Mater.* **2021**, *31*. [CrossRef]
79. Liu, X.; Miller, A.L.; Park, S.; George, M.; Waletzki, B.E.; Xu, H.; Terzic, A.; Lu, L. Two-Dimensional Black Phosphorus and Graphene Oxide Nanosheets Synergistically Enhance Cell Proliferation and Osteogenesis on 3D Printed Scaffolds. *ACS Appl. Mater. Interfaces* **2019**, *11*, 23558–23572. [CrossRef] [PubMed]
80. Yang, B.; Yin, J.; Chen, Y.; Pan, S.; Yao, H.; Gao, Y.; Shi, J. 2D-Black-Phosphorus-Reinforced 3D-Printed Scaffolds: A Stepwise Countermeasure for Osteosarcoma. *Adv. Mater.* **2018**, *30*. [CrossRef]
81. Lin, S.; Zhong, Y.; Zhao, X.; Sawada, T.; Li, X.; Lei, W.; Wang, M.; Serizawa, T.; Zhu, H. Synthetic Multifunctional Graphene Composites with Reshaping and Self-Healing Features via a Facile Biomineralization-Inspired Process. *Adv. Mater.* **2018**, *30*, e1803004. [CrossRef]
82. Unger, C.; Kramer, N.; Walzl, A.; Scherzer, M.; Hengstschläger, M.; Dolznig, H. Modeling human carcinomas: Physiologically relevant 3D models to improve anti-cancer drug development. *Adv. Drug Deliv. Rev.* **2014**, *79–80*, 50–67. [CrossRef]
83. Antoni, D.; Burckel, H.; Josset, E.; Noel, G. Three-Dimensional Cell Culture: A Breakthrough In Vivo. *Int. J. Mol. Sci.* **2015**, *16*, 5517–5527. [CrossRef]
84. Borella, G.; Da Ros, A.; Borile, G.; Porcù, E.; Tregnago, C.; Benetton, M.; Marchetti, A.; Bisio, V.; Montini, B.; Michielotto, B.; et al. Targeting mesenchymal stromal cells plasticity to reroute acute myeloid leukemia course. *Blood* **2021**. [CrossRef]
85. Bhattacharya, P.; Du, D.; Lin, Y. Bioinspired nanoscale materials for biomedical and energy applications. *J. R. Soc. Interface* **2014**, *11*, 20131067. [CrossRef]
86. Lopa, S.; Madry, H. Bioinspired Scaffolds for Osteochondral Regeneration. *Tissue Eng. Part A* **2014**, *20*, 2052–2076. [CrossRef] [PubMed]
87. Ye, B.; Luo, X.; Li, Z.; Zhuang, C.; Li, L.; Lu, L.; Ding, S.; Tian, J.; Zhou, C. Rapid biomimetic mineralization of collagen fibrils and combining with human umbilical cord mesenchymal stem cells for bone defects healing. *Mater. Sci. Eng. C* **2016**, *68*, 43–51. [CrossRef] [PubMed]
88. Rebelo, M.A.; Alves, T.F.R.; De Lima, R.; Oliveira, J.M.; Vila, M.M.D.C.; Balcão, V.; Severino, P.; Chaud, M.V.; Oliveira, J.M., Jr. Scaffolds and tissue regeneration: An overview of the functional properties of selected organic tissues. *J. Biomed. Mater. Res. Part B: Appl. Biomater.* **2015**, *104*, 1483–1494. [CrossRef]
89. Yagüe, M.F.; Abbah, S.; McNamara, L.; Zeugolis, D.I.; Pandit, A.; Biggs, M.J. Biomimetic approaches in bone tissue engineering: Integrating biological and physicomechanical strategies. *Adv. Drug Deliv. Rev.* **2014**, *84*, 1–29. [CrossRef]
90. Mravic, M.; Péault, B.; James, A.W. Current Trends in Bone Tissue Engineering. *BioMed Res. Int.* **2014**, *2014*, 865270. [CrossRef]
91. Yousefi, A.-M.; James, P.F.; Akbarzadeh, R.; Subramanian, A.; Flavin, C.; Oudadesse, H. Prospect of Stem Cells in Bone Tissue Engineering: A Review. *Stem Cells Int.* **2016**, *2016*, 6180487. [CrossRef]

92. Murphy, M.B.; Moncivais, K.; Caplan, A. Mesenchymal stem cells: Environmentally responsive therapeutics for regenerative medicine. *Exp. Mol. Med.* **2013**, *45*, e54. [CrossRef]

93. Lei, B.; Wang, L.; Chen, X.; Chae, S.-K. Biomimetic and molecular level-based silicate bioactive glass–gelatin hybrid implants for loading-bearing bone fixation and repair. *J. Mater. Chem. B* **2013**, *1*, 5153–5162. [CrossRef]

94. Gashti, M.P.; Stir, M.; Bourquin, M.; Hulliger, J. Mineralization of Calcium Phosphate Crystals in Starch Template Inducing a Brushite Kidney Stone Biomimetic Composite. *Cryst. Growth Des.* **2013**, *13*, 2166–2173. [CrossRef]

95. Zhao, R.; Wang, B.; Yang, X.; Xiao, Y.; Wang, X.; Shao, C.; Tang, R. A Drug-Free Tumor Therapy Strategy: Cancer-Cell-Targeting Calcification. *Angew. Chem. Int. Ed.* **2016**, *55*, 5225–5229. [CrossRef] [PubMed]

96. Wang, Z.; Hauser, N.; Singer, G.; Trippel, M.; Kubik-Huch, R.A.; Schneider, C.; Stampanoni, M. Non-invasive classification of microcalcifications with phase-contrast X-ray mammography. *Nat. Commun.* **2014**, *5*, 3797. [CrossRef]

97. Varsano, N.; Dadosh, T.; Kapishnikov, S.; Pereiro, E.; Shimoni, E.; Jin, X.; Kruth, H.S.; Leiserowitz, L.; Addadi, L. Development of Correlative Cryo-soft X-ray Tomography and Stochastic Reconstruction Microscopy. A Study of Cholesterol Crystal Early Formation in Cells. *J. Am. Chem. Soc.* **2016**, *138*, 14931–14940. [CrossRef]

98. Vidavsky, N.; Kunitake, J.A.; Chiou, A.E.; Northrup, P.; Porri, T.J.; Ling, L.; Fischbach, C.; Estroff, L.A. Studying biomineralization pathways in a 3D culture model of breast cancer microcalcifications. *Biomaterials* **2018**, *179*, 71–82. [CrossRef]

99. Chen, H.; Fu, S.; Fu, L.; Yang, H.; Chen, D. Simple Synthesis and Characterization of Hexagonal and Ordered Al–MCM–41 from Natural Perlite. *Minerals* **2019**, *9*, 264. [CrossRef]

100. Miller, F.R.; Santner, S.J.; Tait, L.; Dawson, P.J. MCF10DCIS.com xenograft model of human comedo ductal carcinoma in situ. *J. Natl. Cancer Inst.* **2000**, *92*, 1185–1186. [CrossRef] [PubMed]

101. Santner, S.J.; Dawson, P.J.; Tait, L.; Soule, H.D.; Eliason, J.; Mohamed, A.N.; Wolman, S.R.; Heppner, G.H.; Miller, F.R. Malignant MCF10CA1 Cell Lines Derived from Premalignant Human Breast Epithelial MCF10AT Cells. *Breast Cancer Res. Treat.* **2001**, *65*, 101–110. [CrossRef] [PubMed]

102. Palm, M.D.; O'Donoghue, M.N. Update on photoprotection. *Dermatol. Ther.* **2007**, *20*, 360–376. [CrossRef] [PubMed]

103. Smaoui, S.; Ben Hlima, H.; Ben Chobba, I.; Kadri, A. Development and stability studies of sunscreen cream formulations containing three photo-protective filters. *Arab. J. Chem.* **2017**, *10*, S1216–S1222. [CrossRef]

104. Serpone, N.; Dondi, D.; Albini, A. Inorganic and organic UV filters: Their role and efficacy in sunscreens and suncare products. *Inorganica Chim. Acta* **2007**, *360*, 794–802. [CrossRef]

105. Antoniou, C.; Kosmadaki, M.; Stratigos, A.; Katsambas, A. Sunscreens—What's important to know. *J. Eur. Acad. Dermatol. Venereol.* **2008**, *22*, 1110–1119. [CrossRef]

106. Zhong, X.; Downs, C.A.; Li, Y.; Zhang, Z.; Li, Y.; Liu, B.; Gao, H.; Li, Q. Comparison of toxicological effects of oxybenzone, avobenzone, octocrylene, and octinoxate sunscreen ingredients on cucumber plants (*Cucumis sativus* L.). *Sci. Total Environ.* **2020**, *714*, 136879. [CrossRef]

107. Hiller, J.; Klotz, K.; Meyer, S.; Uter, W.; Hof, K.; Greiner, A.; Göen, T.; Drexler, H. Systemic availability of lipophilic organic UV filters through dermal sunscreen exposure. *Environ. Int.* **2019**, *132*, 105068. [CrossRef]

108. Monteiro-Riviere, N.A.; Wiench, K.; Landsiedel, R.; Schulte, S.; Inman, A.O.; Riviere, J.E. Safety Evaluation of Sunscreen Formulations Containing Titanium Dioxide and Zinc Oxide Nanoparticles in UVB Sunburned Skin: An In Vitro and In Vivo Study. *Toxicol. Sci.* **2011**, *123*, 264–280. [CrossRef]

109. Schulz, J.; Hohenberg, H.; Pflücker, F.; Gärtner, E.; Will, T.; Pfeiffer, S.; Wepf, R.; Wendel, V.; Gers-Barlag, H.; Wittern, K.-P. Distribution of sunscreens on skin. *Adv. Drug Deliv. Rev.* **2002**, *54*, S157–S163. [CrossRef]

110. Shao, Y.; Schlossman, D. Effect of particle size on performance of physical sunscreen formulas. In Proceedings of the PCIA Conference, Shanghai, China, March 1999; pp. 1–9.

111. Levine, A. Sunscreen use and awareness of chemical toxicity among beach goers in Hawaii prior to a ban on the sale of sunscreens containing ingredients found to be toxic to coral reef ecosystems. *Mar. Policy* **2020**, *117*, 103875. [CrossRef]

112. Battistin, M.; Dissette, V.; Bonetto, A.; Durini, E.; Manfredini, S.; Marcomini, A.; Casagrande, E.; Brunetta, A.; Ziosi, P.; Molesini, S.; et al. A New Approach to UV Protection by Direct Surface Functionalization of TiO_2 with the Antioxidant Polyphenol Dihydroxyphenyl Benzimidazole Carboxylic Acid. *Nanomaterials* **2020**, *10*, 231. [CrossRef] [PubMed]

113. Bino, A.; Baldisserotto, A.; Scalambra, E.; Dissette, V.; Vedaldi, D.E.; Salvador, A.; Durini, E.; Manfredini, S.; Vertuani, S. Design, synthesis and biological evaluation of novel hydroxy-phenyl-1H-benzimidazoles as radical scavengers and UV-protective agents. *J. Enzym. Inhib. Med. Chem.* **2017**, *32*, 527–537. [CrossRef]

114. Hossan, J.; Gafur, M.A.; Kadir, M.R.; Mainul, M. Preparation and Characterization of Gelatin- Hydroxyapatite Composite for Bone Tissue Engineering. *Int. J. Eng. Technol.* **2014**, *57*, 113–122.

115. Amin, R.M.; Elfeky, S.; Verwanger, T.; Krammer, B. A new biocompatible nanocomposite as a promising constituent of sunscreens. *Mater. Sci. Eng. C* **2016**, *63*, 46–51. [CrossRef] [PubMed]

116. Lin, J.-Y.; Selim, M.; Shea, C.R.; Grichnik, J.M.; Omar, M.M.; Monteiro-Riviere, N.; Pinnell, S.R. UV photoprotection by combination topical antioxidants vitamin C and vitamin E. *J. Am. Acad. Dermatol.* **2003**, *48*, 866–874. [CrossRef] [PubMed]

117. Morsy, R.; Ali, S.S.; El-Shetehy, M. Development of hydroxyapatite-chitosan gel sunscreen combating clinical multidrug-resistant bacteria. *J. Mol. Struct.* **2017**, *1143*, 251–258. [CrossRef]

118. Landi, E.; Tampieri, A.; Mattioli-Belmonte, M.; Celotti, G.; Sandri, M.; Gigante, A.; Fava, P.; Biagini, G. Biomimetic Mg- and Mg,CO3-substituted hydroxyapatites: Synthesis characterization and in vitro behaviour. *J. Eur. Ceram. Soc.* **2006**, *26*, 2593–2601. [CrossRef]

119. Landi, E.; Tampieri, A.; Celotti, G.; Sprio, S.; Sandri, M.; Logroscino, G. Sr-substituted hydroxyapatites for osteoporotic bone replacement. *Acta Biomater.* **2007**, *3*, 961–969. [CrossRef]

120. de Araujo, T.; de Souza, S.; Miyakawa, W.; de Sousa, E. Phosphates nanoparticles doped with zinc and manganese for sunscreens. *Mater. Chem. Phys.* **2010**, *124*, 1071–1076. [CrossRef]

121. De Araujo, T.S.; De Souza, S.O.; De Sousa, E.M.B. Effect of Zn^{2+}, Fe^{3+} and Cr^{3+} addition to hydroxyapatite for its application as an active constituent of sunscreens. *J. Physics: Conf. Ser.* **2010**, *249*. [CrossRef]

122. Piccirillo, C.; Rocha, C.; Tobaldi, D.M.; Pullar, R.C.; Labrincha, J.A.; Ferreira, M.O.; Castro, P.M.L.; Pintado, M.M.E. A hydroxyapatite–Fe2O3 based material of natural origin as an active sunscreen filter. *J. Mater. Chem. B* **2014**, *2*, 5999–6009. [CrossRef]

123. Popov, A.; Priezzhev, A.V.; Lademann, J.; Myllylä, R. Alteration of skin light-scattering and absorption properties by application of sunscreen nanoparticles: A Monte Carlo study. *J. Quant. Spectrosc. Radiat. Transf.* **2011**, *112*, 1891–1897. [CrossRef]

124. Morlando, A.; Cardillo, D.; Devers, T.; Konstantinov, K. Titanium doped tin dioxide as potential UV filter with low photocatalytic activity for sunscreen products. *Mater. Lett.* **2016**, *171*, 289–292. [CrossRef]

125. Yasukawa, A.; Tamura, J. Preparation and structure of titanium-cerium-calcium hydroxyapatite particles and their ultraviolet protecting ability. *Colloids Surf. A Physicochem. Eng. Asp.* **2020**, *609*, 125705. [CrossRef]

126. Tampieri, A.; Sandri, M.; Sprio, S. Physical Solar Filter Consisting of Substituted Hydroxyapatite in An Organic Matrix. US 10,813,856 B2, 27 October 2020.

127. Burnett, M.E.; Wang, S.Q. Current sunscreen controversies: A critical review. *Photodermatol. Photoimmunol. Photomed.* **2011**, *27*, 58–67. [CrossRef] [PubMed]

128. Manzano, M.; Lamberti, G.; Galdi, I.; Vallet-Regí, M. Anti-Osteoporotic Drug Release from Ordered Mesoporous Bioceramics: Experiments and Modeling. *AAPS PharmSciTech* **2011**, *12*, 1193–1199. [CrossRef]

129. Raja, N.; Park, H.; Choi, Y.-J.; Yun, H.-S. Multifunctional Calcium-Deficient Hydroxyl Apatite–Alginate Core–Shell-Structured Bone Substitutes as Cell and Drug Delivery Vehicles for Bone Tissue Regeneration. *ACS Biomater. Sci. Eng.* **2021**, *7*, 1123–1133. [CrossRef]

130. Mulazzi, M.; Campodoni, E.; Bassi, G.; Montesi, M.; Panseri, S.; Bonvicini, F.; Gentilomi, G.A.; Tampieri, A.; Sandri, M. Medicated Hydroxyapatite/Collagen Hybrid Scaffolds for Bone Regeneration and Local Antimicrobial Therapy to Prevent Bone Infections. *Pharmaceutics* **2021**, *13*, 1090. [CrossRef] [PubMed]

131. Barroug, A.; Glimcher, M.J. Hydroxyapatite crystals as a local delivery system for cisplatin: Adsorption and release of cisplatin in vitro. *J. Orthop. Res.* **2002**, *20*, 274–280. [CrossRef]

132. Barroug, A.; Kuhn, L.T.; Gerstenfeld, L.C.; Glimcher, M.J. Interactions of cisplatin with calcium phosphate nanoparticles: In vitro controlled adsorption and release. *J. Orthop. Res.* **2004**, *22*, 703–708. [CrossRef]

133. Chen, Q.; Liang, J.; Wang, S.; Wang, D.; Wang, R. An asymmetric approach toward chiral multicyclic spirooxindoles from isothiocyanato oxindoles and unsaturated pyrazolones by a chiral tertiary amine thiourea catalyst. *Chem. Commun.* **2013**, *49*, 1657–1659. [CrossRef]

134. Srivastava, P.; Hira, S.K.; Srivastava, D.N.N.; Singh, V.K.; Gupta, U.; Singh, R.; Singh, R.A.; Manna, P.P. ATP-Decorated Mesoporous Silica for Biomineralization of Calcium Carbonate and P2 Purinergic Receptor-Mediated Antitumor Activity against Aggressive Lymphoma. *ACS Appl. Mater. Interfaces* **2018**, *10*, 6917–6929. [CrossRef]

135. Descamps, M.; Duhoo, T.; Monchau, F.; Lu, J.; Hardouin, P.; Hornez, J.; Leriche, A. Manufacture of macroporous β-tricalcium phosphate bioceramics. *J. Eur. Ceram. Soc.* **2008**, *28*, 149–157. [CrossRef]

136. Okazaki, Y.; Abe, Y.; Yasuda, K.; Hiasa, K.; Hirata, I. Osteoclast Response to Bioactive Surface Modification of Hydroxyapatite. *Open J. Stomatol.* **2014**, *04*, 340–344. [CrossRef]

137. Pokhrel, S. Hydroxyapatite: Preparation, Properties and Its Biomedical Applications. *Adv. Chem. Eng. Sci.* **2018**, *08*, 225–240. [CrossRef]

138. Iafisco, M.; Sandri, M.; Panseri, S.; Delgado-López, J.M.; Gómez-Morales, J.; Tampieri, A. Magnetic Bioactive and Biodegradable Hollow Fe-Doped Hydroxyapatite Coated Poly(l-lactic) Acid Micro-nanospheres. *Chem. Mater.* **2013**, *25*, 2610–2617. [CrossRef]

139. Iafisco, M.; Drouet, C.; Adamiano, A.; Pascaud, P.; Montesi, M.; Panseri, S.; Sarda, S.; Tampieri, A. Superparamagnetic iron-doped nanocrystalline apatite as a delivery system for doxorubicin. *J. Mater. Chem. B* **2015**, *4*, 57–70. [CrossRef] [PubMed]

140. Ain, Q.-U.; Munir, H.; Jelani, F.; Anjum, F.; Bilal, M. Antibacterial potential of biomaterial derived nanoparticles for drug delivery application. *Mater. Res. Express* **2019**, *6*, 125426. [CrossRef]

141. Caplin, J.D.; García, A.J. Implantable antimicrobial biomaterials for local drug delivery in bone infection models. *Acta Biomater.* **2019**, *93*, 2–11. [CrossRef] [PubMed]

142. Mondal, S.; Pal, U. 3D hydroxyapatite scaffold for bone regeneration and local drug delivery applications. *J. Drug Deliv. Sci. Technol.* **2019**, *53*, 101131. [CrossRef]

143. Dorati, R.; De Trizio, A.; Genta, I.; Merelli, A.; Modena, T.; Conti, B. Formulation and in vitro characterization of a composite biodegradable scaffold as antibiotic delivery system and regenerative device for bone. *J. Drug Deliv. Sci. Technol.* **2016**, *35*, 124–133. [CrossRef]

144. El-Husseiny, M.; Patel, S.; Macfarlane, R.J.; Haddad, F.S. Biodegradable antibiotic delivery systems. *J. Bone Jt. Surg.* **2011**, *93*, 151–157. [CrossRef]
145. Wu, C.; Fan, W.; Gelinsky, M.; Xiao, Y.; Chang, J.; Friis, T.; Cuniberti, G. In situ preparation and protein delivery of silicate–alginate composite microspheres with core-shell structure. *J. R. Soc. Interface* **2011**, *8*, 1804–1814. [CrossRef]
146. Aslankoohi, N.; Mequanint, K. Intrinsically fluorescent bioactive glass-poly(ester amide) hybrid microparticles for dual drug delivery and bone repair. *Mater. Sci. Eng. C* **2021**, *128*, 112288. [CrossRef]
147. Etter, J.N.; Karasinski, M.; Ware, J.; Oldinski, R.A. Dual-crosslinked homogeneous alginate microspheres for mesenchymal stem cell encapsulation. *J. Mater. Sci. Mater. Med.* **2018**, *29*, 143. [CrossRef]
148. Radwan, N.H.; Nasr, M.; Ishak, R.A.; Awad, G.A. Moxifloxacin-loaded in situ synthesized Bioceramic/Poly(L-lactide-co-ε-caprolactone) composite scaffolds for treatment of osteomyelitis and orthopedic regeneration. *Int. J. Pharm.* **2021**, *602*, 120662. [CrossRef]
149. Alshemary, A.Z.; Akram, M.; Goh, Y.-F.; Tariq, U.; Butt, F.K.; Abdolahi, A.; Hussain, R. Synthesis, characterization, in vitro bioactivity and antimicrobial activity of magnesium and nickel doped silicate hydroxyapatite. *Ceram. Int.* **2015**, *41*, 11886–11898. [CrossRef]
150. Ballardini, A.; Montesi, M.; Panseri, S.; Vandini, A.; Balboni, P.G.; Tampieri, A.; Sprio, S. New hydroxyapatite nanophases with enhanced osteogenic and anti-bacterial activity. *J. Biomed. Mater. Res. Part A* **2017**, *106*, 521–530. [CrossRef]
151. Huh, A.J.; Kwon, Y.J. "Nanoantibiotics": A new paradigm for treating infectious diseases using nanomaterials in the antibiotics resistant era. *J. Control. Release* **2011**, *156*, 128–145. [CrossRef]
152. Sprio, S.; Sandri, M.; Iafisco, M.; Panseri, S.; Montesi, M.; Ruffini, A.; Adamiano, A.; Ballardini, A.; Tampieri, A.B.A.A. *Nature-Inspired Nanotechnology and Smart Magnetic Activation: Two Groundbreaking Approaches toward a New Generation of Biomaterials for Hard Tissue Regeneration*; IntechOpen: London, UK, 2016. [CrossRef]
153. Campodoni, E.; Adamiano, A.; Dozio, S.M.; Panseri, S.; Montesi, M.; Sprio, S.; Tampieri, A.; Sandri, M. Development of innovative hybrid and intrinsically magnetic nanobeads as a drug delivery system. *Nanomedicine* **2016**, *11*, 2119–2130. [CrossRef] [PubMed]
154. Patrício, T.M.F.; Panseri, S.; Montesi, M.; Iafisco, M.; Sandri, M.; Tampieri, A.; Sprio, S. Superparamagnetic hybrid microspheres affecting osteoblasts behaviour. *Mater. Sci. Eng. C* **2018**, *96*, 234–247. [CrossRef] [PubMed]
155. Patrício, T.M.F.; Mumcuoglu, D.; Montesi, M.; Panseri, S.; Witte-Bouma, J.; Garcia, S.F.; Sandri, M.; Tampieri, A.; Farrell, E.; Sprio, S. Bio-inspired polymeric iron-doped hydroxyapatite microspheres as a tunable carrier of rhBMP-2. *Mater. Sci. Eng. C* **2020**, *119*, 111410. [CrossRef] [PubMed]
156. Tatiana, F.P.; Simone, S.; Monica, S.; Natalia, G.; Monica, M.; Silvia, P.; Bas, K.; Anna, T. Bio-inspired superparamagnetic microspheres for bone tissue engineering applications. *Front. Bioeng. Biotechnol.* **2016**, *4*. [CrossRef]
157. Miragoli, M.; Ceriotti, P.; Iafisco, M.; Vacchiano, M.; Salvarani, N.; Alogna, A.; Carullo, P.; Ramirez-Rodríguez, G.B.; Patrício, T.; Degli Esposti, L.; et al. Inhalation of peptide-loaded nanoparticles improves heart failure. *Sci. Transl. Med.* **2018**, *10*, eaan6205. [CrossRef]
158. Degli Esposti, L.; Carella, F.; Adamiano, A.; Tampieri, A.; Iafisco, M. Calcium phosphate-based nanosystems for advanced targeted nanomedicine. *Drug Dev. Ind. Pharm.* **2018**, *44*, 1223–1238. [CrossRef]
159. Barbanente, A.; Nadar, R.A.; Degli Esposti, L.; Palazzo, B.; Iafisco, M.; Beucken, J.J.J.P.V.D.; Leeuwenburgh, S.C.G.; Margiotta, N. Platinum-loaded, selenium-doped hydroxyapatite nanoparticles selectively reduce proliferation of prostate and breast cancer cells co-cultured in the presence of stem cells. *J. Mater. Chem. B* **2020**, *8*, 2792–2804. [CrossRef]

Journal of
Composites Science

MDPI

Article

Materials and Manufacturing Techniques for Polymeric and Ceramic Scaffolds Used in Implant Dentistry

Mutlu Özcan [1,*], Dachamir Hotza [2], Márcio Celso Fredel [3], Ariadne Cruz [4] and Claudia Angela Maziero Volpato [4]

1   Center of Dental Medicine, Division of Dental Biomaterials, Clinic for Reconstructive Dentistry, University of Zürich, Plattenstrasse 11, CH-8032 Zürich, Switzerland
2   Technological Center, Department of Chemical Engineering, Federal University of Santa Catarina, 88040-900 Florianópolis, Santa Catarina, Brazil; dhotza@gmail.com
3   Technological Center, Department of Mechanical Engineering, Federal University of Santa Catarina, 88040-900 Florianópolis, Santa Catarina, Brazil; m.fredel@ufsc.br
4   Health Sciences Center, Department of Dentistry, Federal University of Santa Catarina, 88040-900 Florianópolis, Santa Catarina, Brazil; ariadne.cruz@ufsc.br (A.C.); claudia.m.volpato@outlook.com (C.A.M.V.)
*   Correspondence: mutluozcan@hotmail.com; Tel.: +41-44-634-5600

Abstract: Preventive and regenerative techniques have been suggested to minimize the aesthetic and functional effects caused by intraoral bone defects, enabling the installation of dental implants. Among them, porous three-dimensional structures (scaffolds) composed mainly of bioabsorbable ceramics, such as hydroxyapatite (HAp) and β-tricalcium phosphate (β-TCP) stand out for reducing the use of autogenous, homogeneous, and xenogenous bone grafts and their unwanted effects. In order to stimulate bone formation, biodegradable polymers such as cellulose, collagen, glycosaminoglycans, polylactic acid (PLA), polyvinyl alcohol (PVA), poly-ε-caprolactone (PCL), polyglycolic acid (PGA), polyhydroxylbutyrate (PHB), polypropylenofumarate (PPF), polylactic-co-glycolic acid (PLGA), and poly L-co-D, L lactic acid (PLDLA) have also been studied. More recently, hybrid scaffolds can combine the tunable macro/microporosity and osteoinductive properties of ceramic materials with the chemical/physical properties of biodegradable polymers. Various methods are suggested for the manufacture of scaffolds with adequate porosity, such as conventional and additive manufacturing techniques and, more recently, 3D and 4D printing. The purpose of this manuscript is to review features concerning biomaterials, scaffolds macro and microstructure, fabrication techniques, as well as the potential interaction of the scaffolds with the human body.

Keywords: biomaterials; bone grafts; bone repair; dental implants; scaffolds

Citation: Özcan, M.; Hotza, D.; Fredel, M.C.; Cruz, A.; Volpato, C.A.M. Materials and Manufacturing Techniques for Polymeric and Ceramic Scaffolds Used in Implant Dentistry. *J. Compos. Sci.* **2021**, *5*, 78. https://doi.org/10.3390/jcs5030078

Academic Editor: Francesco Tornabene

Received: 19 February 2021
Accepted: 8 March 2021
Published: 11 March 2021

Publisher's Note: MDPI stays neutral with regard to jurisdictional claims in published maps and institutional affiliations.

1. Introduction

Osseointegration and dental implants were introduced in dentistry more than 40 years ago, thanks to the pioneering studies of Per-Ingvar Brånemark and collaborators [1–3]. Since then, unitary, partial, and total dental losses have been rehabilitated by implant-supported prosthesis successfully and predictably [4]. Dental implants are devices usually made of pure grade IV titanium and are surgically installed in healthy bone areas. After the osseointegration period, which is about 3 to 6 months, they can be restored by the dental prostheses, thus collaborating with the restoration of the masticatory function and the return of oral comfort and aesthetics to the patient [5].

However, the loss of alveolar bone that occurs before or after a tooth extraction is responsible for altering the original volume of the alveolar ridge, as well as for the formation of bone defects (Figures 1–3). After tooth extraction, an average alveolar bone loss of about 30% (in the vertical direction) and 40–50% (in the horizontal direction) occurs for up to 6 months [6]. If no treatment is made, bone loss advances, reaching 40–60% reductions in bone crest volume within 3 years [7].

197

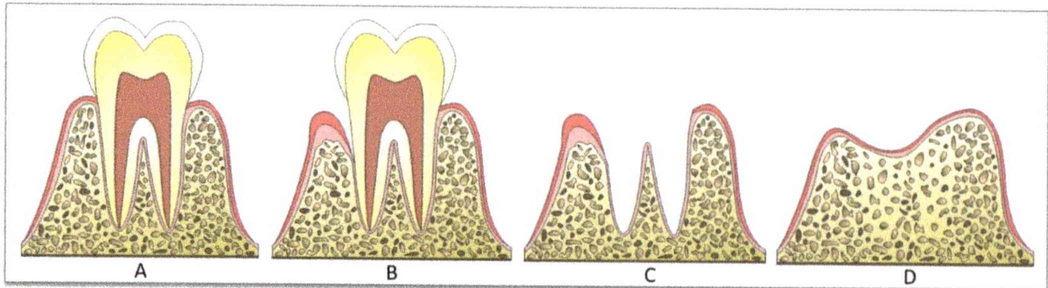

Figure 1. Diagram showing the change in the volume of the alveolar bone ridge due to bone loss before tooth extraction. (**A**) Proper dental implantation. (**B**) Alveolar bone loss due to periodontal disease. (**C**) Bone condition after tooth extraction. (**D**) Alveolar bone healing.

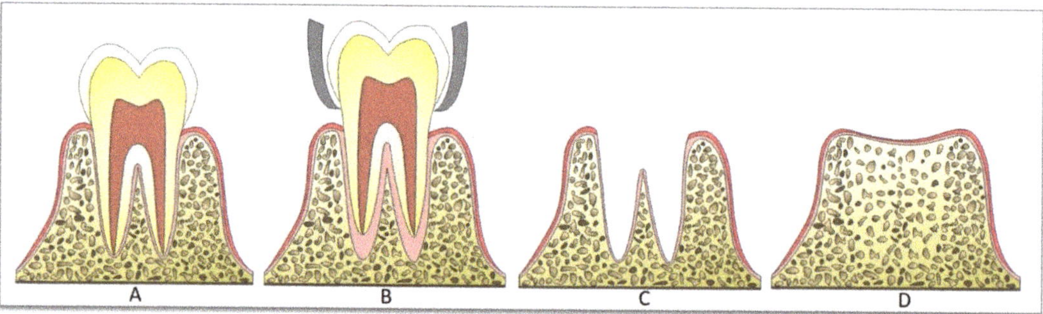

Figure 2. Diagram showing the change in the volume of the alveolar bone ridge resulting from the atraumatic tooth extraction. (**A**) Proper dental implantation. (**B**) Dental extraction with no fracture of the alveolar bone wall. (**C**) Bone condition after tooth extraction. (**D**) Alveolar bone healing.

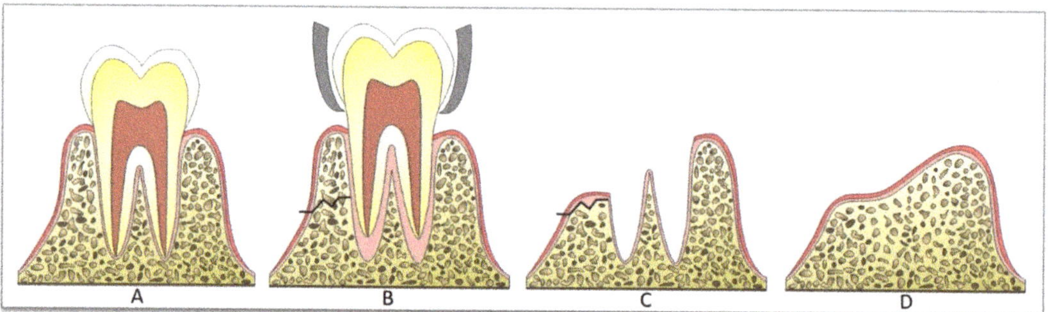

Figure 3. Diagram showing the change in the volume of the alveolar bone ridge resulting from the traumatic tooth extraction. (**A**) Proper dental implantation. (**B**) Dental extraction with fracture of the alveolar bone wall. (**C**) Bone condition after tooth extraction. (**D**) Alveolar bone healing.

Bone loss before tooth extraction may be related to periodontal diseases, periapical pathologies, and trauma to the dentition and/or bone [6]. It is important to mention that bone loss after a tooth extraction is also related to the type of the surgical procedure, being aggravated mainly by invasive and traumatic surgeries [8]. In addition, bone loss resulting from tumors or genetic disorders has also been reported [9,10].

Bone remodeling that occurs after tooth loss certainly results in the formation of a bone defect that makes dental implant placement difficult or even unfeasible depending on the size and location [8]. In cases of implants, positioned in deficient bone or extraction cavities with compromised bone walls, horizontal and/or vertical defects can be formed, exposing the implant body and compromising the short- and long-term functional and aesthetic results [11]. Therefore, the reestablishment and maintenance of the dimensions of the alveolar ridge after tooth loss are essential to ensure a favorable and predictable result with osseointegrated implants [8,12].

Additionally, in order to minimize the aesthetic and functional effects caused by intraoral bone defects, the clinical use of scaffolds as preventive and regenerative techniques have spread widely, enabling the installation of dental implants and, consequently, implant-supported prosthesis rehabilitation. This review describes the features of biomaterials used as scaffolds to promote bone formation, macro and microstructures of scaffolds, fabrication techniques, as well as the potential interaction of the scaffolds with the human body. Scaffolds composed of mainly bioabsorbable ceramics, such as hydroxyapatite (HAp) and β-tricalcium phosphate (β-TCP), and biodegradable polymers like cellulose, collagen, glycosaminoglycans, polylactic acid (PLA), polyvinyl alcohol (PVA), poly-ε-caprolactone (PCL), polyglycolic acid (PGA), polyhydroxylbutyrate (PHB), polypropylenofumarate (PPF), polylactic-co-glycolic acid (PLGA), and poly L-co-D, L lactic acid (PLDLA) are described. This present review also examines hybrid scaffolds that can combine the tunable macro/microporosity and osteoinductive properties of ceramic materials with the chemical/physical properties of biodegradable polymers. Various methods for the manufacture of scaffolds with adequate porosity, such as conventional and additive manufacturing techniques and, more recently, 3D and 4D printing are discussed. Finally, this review briefly discusses the new trends and future directions in developing scaffolds for bone formation and presents relevant information regarding the main materials and manufacturing techniques for scaffolds used in implant dentistry, including the trends in material composition and manufacturing techniques.

2. Bone-Grafting Techniques

Preventive (such as atraumatic tooth extraction and filling the socket soon after extraction) or regenerative techniques (such as grafting to gain bone volume after healing the ridge) have been suggested to minimize the esthetic and phonetics effects caused by the bone defects and enable the placement of dental implants [11,13] (Figures 4 and 5). Both techniques employ bone grafts to promote bone repair and the reduction of bone defects. In preventive techniques, the bone grafts help to maintain the volume for cell infiltration and proliferation, as well as assist in closing the surgical wound [14]. In regenerative techniques, the bone grafts have been used to increase the vertical and/or horizontal volume of the alveolar ridge, being the guided bone regeneration (GBR) indicated as the best technique, with satisfactory results over time [15–17].

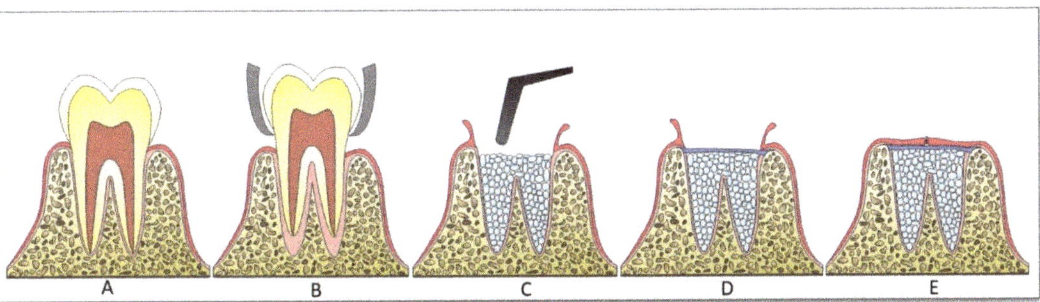

Figure 4. Diagram showing a preventive technique after tooth extraction. (**A**) Dental implantation. (**B**) Dental extraction. (**C**) Filling the dental socket with biomaterial. (**D**) Closure with a membrane. (**E**) Suturing the grafted area.

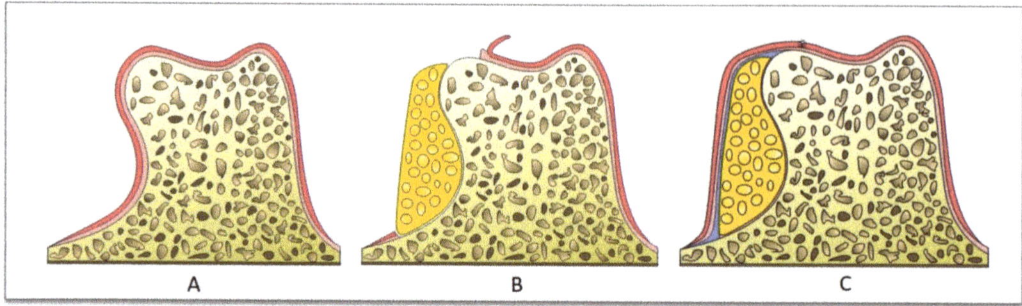

Figure 5. Diagram showing a regenerative technique in an area with the bone defect. (**A**) Alveolar ridge with a bone defect in thickness. (**B**) Bone graft adapted to the area of the defect and membrane positioned. (**C**) Suturing the grafted area.

Since bone repair depends on mechanisms of osteoconduction, osteoinduction, and osteogenesis, the ideal bone graft should guide the bone growth three-dimensionally, establishing cell recruitment, inducing differentiation of resident bone cells, and providing cells at the implantation site [18] (Figure 6). For many years, autogenous bone grafts had been considered the reference standard for the treatment of bone defects. In addition to having imunocompatible cells, they are osteogenic, osteoconductive, and osteoinductive presenting characteristics of bioabsorption and angiogenesis, which guarantees high clinical predictability [19,20]. While osteoinductors are biomaterials that stimulate undifferentiated cells to differentiate into osteoblasts, osteoconductors act as a framework for the proliferation of blood vessels, perivascular tissue, and osteoprogenitor cells of the patient. Osteogenitors biomaterials are capable of forming bone tissue by themselves since they have viable precursor cells and/or osteoblasts [21].

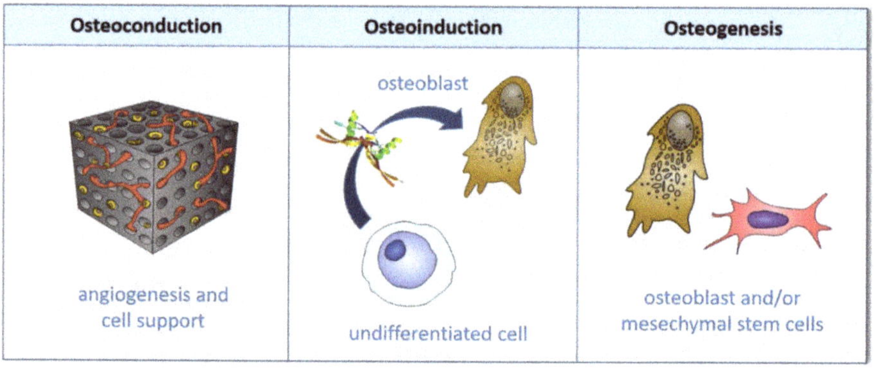

Osteoconduction	**Osteoinduction**	**Osteogenesis**
angiogenesis and cell support	osteoblast undifferentiated cell	osteoblast and/or mesechymal stem cells

Figure 6. Diagram showing mechanisms of osteoconduction, osteoinduction, and osteogenesis.

In autogenous grafts, the patient is both the donor and the graft receptor. When used to correct intra-oral bone defects, autogenous bone grafts have been obtained from the oral cavity (such as the mental area, mandible body, or maxillary tuberosity) or extra-oral donor areas (such as the iliac crest, tibia, or skullcap) [19]. However, limitations such as the restricted availability of bone for removal, increased surgical morbidity, high rates of graft bone remodeling, and difficulty in predicting the rate of degradation over time, have been associated with autogenous bone grafts [22–24]. Therefore, autogenous bone grafts have declined in use over time, especially from the extra-oral area.

In order to minimize the inherent limitations of autogenous bone grafts, bone substitutes such as homologous, xenogenous, and alloplastic grafts have been suggested in the literature [19,25–27]. Homologous bone grafts are originated from another individual of the same species from cadavers (e.g., DFDBA: demineralized freeze-dried bone allograft; FDBA: freeze-dried bone allograft), while xenogenous grafts are obtained from another species (ex: CB-BB: chemically deproteinized bovine bone; TD-BB: thermally deproteinized bovine bone). Despite being available in large amounts, the main disadvantages of homologous bone grafts are high costs, the requirement for complex sterilization and storage techniques, difficulty in predicting the rate of degradation over time, the risks of disease transmission, variable osteoinductive and osteoconductive properties, and lower osteogenic potential compared to autografts [28,29]; while risks of zoonotic diseases transmission (e.g., bovine spongiform encephalopathy), prion infections, and immunological activation of diseases have been related to xenogenous grafts [30,31].

Conversely, alloplastic bone grafts are fabricated from inorganic or synthetic biomaterials, which, despite not having the osteogenesis and osteoinduction capacity, considerably reduce surgical morbidity rates and the risk of disease transmission [32,33]. Therefore, alloplastic bone grafts have been developed to overcome the limitations of autogenous, homologous, and xenogenous grafts. Indeed, these biomaterials have demonstrated advantages such as a reduction in surgical time, an abundance of materials with no amount limitation, ease of handling, no risk of disease transmission, and very low antigenicity potential [34–36].

Alloplastic bone grafts are fabricated from absorbable or non-absorbable synthetic materials, with different sizes and shapes, and variations in their physical and chemical properties, such as granule morphology, crystalline or amorphous phase, and pore size and interconnectivity [37]. Composed of osteoconductive biomaterials, these grafts provide a framework, which will be populated by cells originated from osteoprogenitor cells (from the defect margins), thus promoting bone neoformation until the biomaterial is completely replaced by the new bone [33,35,36]. Alloplastic grafts can be used alone, or in association with autogenous bone, biomaterials, or bioactive substances [35,37,38].

3. Scaffolds

With the advancement and diversity of alloplastic bone grafts, the concept of "bone tissue engineering (BTE)" stands out, which aims to combine biological knowledge concerning the histology and morphology of bone tissues with the development of appropriate biomaterials and techniques for the three-dimensional (3D) structure's construction, capable of simulating the bone environment on a micro and nanoscale [39]. These structures, better known as scaffolds, are carriers for cellular interactions (migration, adhesion, and cell proliferation), allowing the deposition of a new bone extracellular matrix on its porous surface [40–42]. Additionally, they also provide temporary support for newly formed bone tissue and vessels [43,44].

Metals (tantalum, magnesium, titanium and titanium alloys, nickel-titanium alloy [nitinol]); natural polymers (collagen, gelatin, silk fibroin, chitosan, alginate, hyaluronic acid); synthetic polymers (polylactic acid [PLA], polyglycolic acid [PGA], polylactic-co-glycolic acid [PLGA], polycaprolactone [PCL], polyvinyl alcohol [PVA], polypropylene fumarate [PPF], polyurethane [PU]); bioactive ceramics (hydroxyapatite [HAp], tricalcium phosphate [TCP], beta-tricalcium phosphate [β-TCP], calcium sulfate [$CaSO_4$], akermanite [Ca, Si, Mg], diopside [$MgCaSi_2O_6$], bioactive glass [BGs]); and bioinert ceramics (aluminum oxide, zirconia) have been proposed for the manufacture of scaffolds-based bone tissue engineering [41,45,46]. Additionally, the materials most used in clinical practices to repair intra-oral defects are HAp, dicalcium, tricalcium phosphates, and bioactive glasses [47–50]. The materials used in scaffolds for implant dentistry are shown in Table 1.

Table 1. Materials, advantages, disadvantages of scaffolds for implant dentistry.

Materials		Advantages	Disadvantages
Poly-lactic acid (PLA) [46,51]		- Water-soluble - Crystallinity tunable by changing hydroxylation degree	- Non-hydrophobic - Shortage of cell adhesion
Poly-glycolic acid (PGA) [46]		- Water-soluble - Crystallinity tunable by changing hydroxylation degree	- Non-hydrophobic - Shortage of cell adhesion
Polylactic-co-glycolic acid (PLGA) [52,53]		- Water-soluble - Crystallinity tunable by changing hydroxylation degree - Easily synthesized - Biodegradable in non-toxic by-products - Controlled degradation time	- Non-hydrophobic - Shortage of cell adhesion
Polycaprolactone (PCL) [54]		- Crosslink in situ and print by injection - Elastic behavior	- Degradation rate in years
Polyvynil alcohol (PVA) [55,56]		- Ability to manufacture scaffolds with various characteristics such as shape, porosity, and degradation rate - Flexibility - Mechanically strong - Water-soluble - Compatible with several polymers	- Non-soluble in organic solvents - Cross-linking of polymers to maintain integrity
Polypropylene fumarate (PPF) [57]		- Adjustable mechanically strong - Adjustable rates of degradation	- Cross-linking of polymers to maintain integrity

Table 1. *Cont.*

Materials		Advantages	Disadvantages
Hydroxyapatite (HAp) [44,46,58]		- Highly biocompatibility - Nontoxic - Hydrophilicity - Provides calcium and phosphorus for new tissue	- Poor mechanical strength - Lack of organic phase
Tricalcium phosphates (TCP) [59,60]		- Provides calcium and phosphorus for new tissue	- Poor mechanical strength - Lack of organic phase
Bioactive glass (BG) [46,61]		- Bioactive - Bond-bonding affinity	- High solubility - Limitation of shaping
Zirconia (ZrO$_2$) [62,63]		- Mechanically strong - High fracture toughness - Osseointegration potential - Radiopacity	- No biological activity

More recently, hybrid scaffolds, which combine polymers and ceramics have been proposed to associate the advantages of polymeric materials with the favorable properties of ceramic such as bioactivity and mechanical resistance [41]. It is relevant to mention that there is still no single synthetic biomaterial that offers all the desirable properties for a scaffold; thus, the association of biomaterials combines the best properties of each one, in order to meet the needs of the bone-grafted area [48]. These 3D structures can also be associated with growth factors (such as recombinant human bone morphogenetic protein-2 (rhBMP-2) and platelet-derived factors-BB), bioactive substances (such as simvastatin), and specific cells which stimulate bone tissue regeneration (such as mesenchymal stem cells or osteoblasts) [64–66].

4. Expected Properties for a Scaffold

Ideally, a scaffold should be composed of biocompatible materials that are easily adaptable to the bone defect; must present controlled bioabsorption and in line with bone formation; promote high surface wettability inducing cell adsorption and proliferation; having appropriate surface chemical properties to enable cell adhesion; have an interconnected pore architecture; present satisfactory mechanical properties to support intra-oral loads in the defect area; allow sterilization, and present industrial viability to be manufactured in different sizes and shapes [65–69].

Therefore, during the manufacture of a scaffold, the biomaterial design must be taken into account, the desired morphology, pore size and interconnectivity, and the mechanical properties [49,69]. If the scaffold conformation is similar to the defect shape, its adaptation will be more effective, quickly establishing a solid interface and the complete integration between the surface of the biomaterial and the bone tissue [70]. This intimate adaptation, associated with the chemical properties of the surface and a porous structure interconnected by the association of macro and micropores, facilitates cell dynamics, enabling the adhesion, proliferation, and migration of bone-related cells with subsequent deposition of osteoid tissue inside the scaffold [46,70,71].

However, although the importance of the presence of porosities to create a microenvironment for cell proliferation is clear, there is still no consensus on the ideal pore size for a scaffold in bone repair. The literature presents different values, ranging from 40 to 500 μm in diameter [41,71,72]. These variations are probably related to the nature and variability of the bone (cortical and/or spongy) where the bone graft will be used and the biomaterial used to manufacture the scaffold [72,73].

Studies suggest that smaller pores (with around 40 μm) favor cell agglomeration, while larger pores (with approximately 100 μm) accelerate cell migration [74,75]. It is recommended to use pores with at least 100 μm in diameter to ensure successful diffusion of nutrients and oxygen, which enables cell survival and stimulates bone growth [76,77]. Other studies indicate that diameters of 5 μm are suitable for neovascularization, 5–15 μm for fibroblast growth, 20 μm for hepatocyte growth, 20–125 μm for skin regeneration in adult mammals, 40–100 μm for osteoid tissue growth, and diameters between 100–350 μm favor bone regeneration [78]. Additionally, the use of nanoscale pores can increase the surface area, which is advantageous for the apatite formation and proteins and/or osteoblasts fixation [79,80]. Interconnectivity among the pores must also be obtained, since a network of interconnected pores increases the diffusion rates to the center of the scaffold, allowing vascularization, improving the transport of nutrients and oxygen, and facilitating the removal of metabolic waste [76,77].

It is also reported that porous scaffolds with nano, micro, and macroporosities, can perform better than macroporous scaffolds [81]. However, the reproduction of varying degrees of porosity increases the complexity and the challenge in making reproducible scaffolds, especially when composed of a single material [68]. Additionally, high porosity scaffolds demonstrated a reduction in the mechanical properties, directly impacting its structural integrity [68,82]. The mechanical characteristics of a scaffold must be similar to those of native tissue, especially concerning the resistance to stresses suffered in vivo, until the new tissue formed occupies the scaffold matrix [82].

The scaffold degradation capacity and speed are also important parameters to consider. Ideally, the scaffold degradation should be concomitant with the bone formation process, since its mechanical properties decrease as a result of its degradation [83]. In physiological and artificial aqueous environments, biomaterials can degrade via some mechanisms, including (i) physicochemical degradation (chain scission and dissolution in an aqueous environment), (ii) enzymatic activity, (iii) cellular degradation (e.g., inflammation, foreign body response), and (iv) mechanical fragmentation due to a loss of structural integrity resulting from the former mechanisms [84,85]. The rate of scaffold degradation is determined by factors such as the configurational structure, crystallinity, molecular weight, morphology, stresses, porosity, and implantation site [86]. Polymeric and resorbable ceramic biomaterials can be degraded; however, polymeric scaffolds have higher degradation rates [87].

5. Polymeric Biomaterials Applied to Fabricate Scaffolds Used in Implant Dentistry

Polymeric materials are composed of a long repeating chain of monomers formed by covalent bonds. Natural polymers such as proteins (fibroin, collagen, gelatin, fibrinogen, elastin, keratin, actin, myosin), polysaccharides (cellulose, amylose, dextran, chitosan, glycosaminoglycans), and polynucleotides (DNA, RNA) have a better interaction with biological systems due to its bioactive, bioadhesive, and hydrophilic properties [87], while synthetic polymers (synthetic aliphatic polyesters: poly-glycolic acid (PGA), poly-lactic acid (PLA), poly-lactic acid-co-glycolic acid (PLGA), polycaprolactone (PCL) demonstrated more possibilities for chemical and mechanical modifications [51].

The polymer's biodegradation occurs by hydrolytic scission in its main chain, resulting in soluble, non-toxic oligomers or monomers. The main biodegradation processes occur by two mechanisms: (i) hydrolysis or enzymatic digestion of the main chain, promoting gradual erosion of the polymer, and (ii) rupture of the crosslinking links, generating water-soluble fragments, which are transported away from the site deployment [84,85].

5.1. Poly-Lactic Acid (PLA)

PLA is a polyester obtained by condensing hydroxyl and carboxyl groups of the lactic acid monomer or by opening the lactide ring. In addition to its biocompatibility and biodegradability, it has low rigidity, good processability, and is thermally stable [88]. Chemically, PLA is considered an organic acid with an asymmetric carbon that has two enantiomers (L+ and D−) and a racemic DL [51], being used as a precursor in the manufacture of polymers. Approved for use in the biomedical area since the 1970s due to its biocompatibility, when PLA comes into contact with the human body, it undergoes hydrolytic degradation via a mass erosion mechanism by a random splitting of the ester bonds, decomposing into lactic acid and producing water and carbon dioxide via the Krebs cycle. Its degradation depends on characteristics such as degree of crystallinity, molar mass, type of isomerism, and changes in pH [89]. In the treatment of peri-implant, periodontal, and bone defects, PLA is used in the form of membranes.

5.2. Poly-Glycolic Acid (PGA)

PGA is a biodegradable, thermoplastic polymer, and the simplest linear aliphatic polyester. It can be prepared to start from glycolic acid using polycondensation or ring-opening polymerization. However, high-molecular-weight PGA could not be obtained because it was unstable and easily degradable compared with other synthetic polymers [90]. Polyglycolide fiber is a clinically well-known non-woven fabric, which has rapid absorption as an advantage [91]. Conversely, the polyglycolide mesh has low integrity mechanics in vitro. Therefore, its application in bone in an isolated form is inadequate. PGA combining with materials that promote a greater reinforcement to bone tissue can obtain a stable combination. The association of the PGA mesh with a PLLA solution allowed a substantial increase in resistance to compression than PGA alone [92]. Currently, PGA and its copolymers poly(lactic-co-glycolic acid) with lactic acid, poly(glycolide-co-caprolactone) with ε-caprolactone, and poly (glycolide-co-trimethylene carbonate) with trimethylene carbonate have been widely used as biomaterials in the biomedical field [93].

5.3. Polylactic-Co-Glycolic Acid (PLGA)

PLGA is a biocompatible biomaterial, easily synthesized, and biodegradable in non-toxic by-products [52,53,94]. This copolymer has been used in a variety of therapeutic devices approved by the Food and Drug Administration due to its high rate of biodegradability and biocompatibility. PLGA is synthesized by the ring-opening copolymerization of two different monomers, PGA and PLA. Therefore, modifications of physical-chemical characteristics can be performed by the composition of the original monomers (PLA and PGA). Depending on the ratio of PLA to PGA used in the polymerization, different forms of PLGA can be obtained. These forms are usually identified concerning the molar proportion of the monomers used (e.g., PLGA 75:25 identifies a copolymer whose composition is 75% PLA and 25% PGA). The crystallinity of PLGAs will vary from 100% amorphous to 100% crystalline depending on the block structure and molar ratio [95].

Since PLGA is highly biocompatible and non-toxic, in addition to being easily processed in different devices, the clinical applications of PLGA have increased in recent years, especially in the field of orthopedics as devices for fixing fractures in the craniomaxillofacial region, support for cell growth, and a device for controlled drug release [96]. Additionally, the PLGA membrane is also indicated for periodontal, peri-implant, and bone regeneration. It is important to mention that although PLGA is not considered osteoinductive, it allows the incorporation and release of biomolecules with substantivity [96].

PLGA is degraded faster than PLA because of glycolic acid incorporation in the polymer chain through de-esterification. PLGA scaffolds are often used as bone reconstruction materials. They can be synthesized in personalized shape and to satisfy the required absorption time. There are various methods for processing these porous synthetic scaffolds. Nevertheless, PLGA has demonstrated reduced cell adhesion and proliferation in response to its hydrophobicity [54,97].

5.4. Polycaprolactone (PCL)

PCL consists of hexanoate units and represents an important biodegradable aliphatic polymer. It is synthesized by poly-condensation of 6-hydroxyhexanoic acid and ring-opening polymerization of ε-caprolactone [54]. Due to the interconnected pores, high porosity, and elastic behavior, the 3D DPCL electrospun nanofibrous has a similar structure to the extracellular matrix and has demonstrated unique features for tissue formation [98,99]. PCL has been used to fabricate several types of hybrid scaffold [93,99].

5.5. Polyvinyl Alcohol (PVA)

PVA is a biodegradable synthetic polymer which is synthesized by hydrolysis of poly(vinyl acetate [55,56]. Some unique features of PVA (e.g., solubility, flexibility, biocompatibility, biodegradability, mechanical strength) make it an important choice as a polymeric scaffold for bone tissue engineering.

This polymer is interesting for electrospinning due to the presence of a hydroxyl group in its repeating unit, which makes it cross-linkable using its interconnected hydrogen bonding [100,101]. PVA is the most commonly used water-soluble synthetic polymer for biomedical applications [100]. PVA is not soluble in organic solvents and only sparsely soluble in ethanol. Due to PVA compatibility with several polymers, it can be easily mixed up with several biomaterials, extending its applicability. Different studies demonstrated that the mechanical property of PVA can be improved without compromising the degradability through the inclusion of reinforced agents [102,103].

The physicochemical property of PVA is determined by the degree of hydrolysis during the synthesis procedure. Because PVA is a water-soluble polymer, before any biological application, cross-linking of polymers is important to maintain integrity. Therefore, the degree of cross-linking plays an important role in deciding the stability in the biological environment, fluid uptake, degradation property, among others. For biomedical applications, physical cross-linking is more useful as it does not leave any residual toxic crosslinking agents [56,100].

5.6. Polypropylene Fumarate (PPF)

Since its introduction by Yaszemski et al. [57], PPF has been used preclinically for bone regeneration. PPF demonstrates several medical requirements including biocompatibility, mechanical properties, osteoconductivity, and capacity to be sterilized [57,104,105]. This synthetic polymer degrades via hydrolysis of its ester bonds. Additionally, the degradation time depends on the molecular mass of the backbone chain, the types of crosslinker used, and the crosslinking density [104,105]. PPF is degraded in non-toxic fumaric acid and propylene glycol, equal favorable for in vivo applications [106]. In PPF cross-linked, the strength is adequate to guide and allow cell attachment and tissue formation in vivo. Moreover, the PPF degradation occurs in a timeframe adequate to bone healing and remodeling [107].

6. Ceramic Biomaterials Used in Scaffolds Applied in Implant Dentistry

Ceramics are inorganic, non-metallic, and crystalline materials, which can be classified as bioinert and bioactive. Bioinert ceramics have no interaction with living tissue, while bioactive ceramics are capable of promoting adherence to living bone tissue [108]. The ceramics most used in bone tissue engineering are bioactive, also known as bioceramics, with emphasis on hydroxyapatite and β-tricalcium phosphate [109,110].

These bioceramics contain calcium salts that stimulate the formation and precipitation of calcium phosphates in bone tissue [111]. However, due to their low structural rigidity, they cannot be used in areas of great mechanical stress, because of the risk of fracture [112]. To address these mechanical limitations, bioinert ceramics, such as zirconia, have been suggested for use alone or associated with bioactive ceramics [113].

6.1. Hydroxyapatite (HAp)

Hydroxyapatite, a hydrated calcium phosphate $(Ca_{10}(PO_4)_6(OH)_2)$, is a mineral present in vertebrates (about 55% of the bone composition, 96% of dental enamel composition, and 70% of dentin), which acts as a reserve of calcium and phosphorus [58,110]. For use as a graft material, it can be obtained by deproteinizing bone tissue (natural HAp, usually from bovine tissue) or by precipitating aqueous solutions from phosphates (synthetic HAp) [114]. Natural and synthetic HAp are thermodynamically stable at physiological pH and actively participate in bone bonds, forming a strong chemical bond with bone tissue [58]. The HAp surface allows the interaction of dipole-type bonds, causing water molecules and also proteins and collagen to be adsorbed on the surface, thus inducing tissue regeneration [59].

Synthetic HAp has been the most widely used clinically, characterized by being a biocompatible and osteoconductive material that presents high stability in aqueous media [115]. It is commercialized in the form of dense or porous ceramics, in blocks, granules, or coatings, being used in the repair of bone defects, an increase of alveolar ridge, guided regeneration of bone tissues, and buccomaxillofacial reconstructions [116,117]. Compared with natural HAp, synthetic HAp has a higher crystallinity, which results in slower degradation that can last 4 to 5 years [116]. Therefore, scaffolds manufactured in HAp maintain their geometric shape for a longer time during the regeneration of bone tissue [117].

However, in some clinical situations, the rate of HAp degradation may be out of step with bone formation [118]. When compared to other calcium phosphates (amorphous tricalcium phosphate: 25.7 to 32.7 g/L; calcium monophosphate monohydrate: about 18 g/L; anhydrous calcium monophosphate: about 17 g/L), the rate of HAp reabsorption is considered to be quite low (about 0.0094 g/L) [119,120]. Thus, studies have suggested replacing phosphate groups (PO_4^{3-}) with carbonate groups (CO_3^{2-}) (carbonated or carboapatite HAp), which modifies the crystalline structure of HAp, increasing its solubility and, consequently, its clinical application [116,120].

6.2. Tricalcium Phosphates (TCP)

When subjected to high-temperature treatments, HAp can give rise to other phases such as tricalcium phosphates (α and β-TCP) that are also frequently used as bioceramic materials. α-TCP and β-TCP have the same chemical composition ($Ca_3(PO_4)_2$); however, the crystallographic structures are different, and the α phase is more soluble. Additionally, α- and β-TCP have different densities: α-TCP (2.86 g/cm^3) and β-TCP (3.07 g/cm^3); the last being closer to that of HAp (3.16 g/cm^3) [59].

Biomaterials composed of calcium phosphate (CaP) can be manufactured in both porous and dense forms as bulk, granules, and powders, besides the de-coating form. These biomaterials demonstrated biocompatibility, safety, availability, low morbidity, and are affordable. CaP bioceramics are now in common use for different medical and dental applications such as treatment of bone defects and fractures, total joint replacement, spinal surgery, dental implants, peri-implants and periodontal therapy, and craniomaxillofacial reconstruction [121].

CaP-based biomaterials are bioactive and have a composition and structure similar to the mineral phase of bone. Despite the osteoconductive property [60], CaP-based biomaterials have a high affinity for protein adsorption and growth factors [122]. The osteoinductive property can be achieved by: (i) structural or chemical optimization of the biomaterials themselves; and/or (ii) incorporation of osteoinductive substances, such as rhBMP [123,124].

Notwithstanding the several advantages of CaP bioceramics, these biomaterials demonstrated poor mechanical strength, lack of organic phase, presence of impurities, micro-scale grain size, non-homogenous particle size and shape, prolonged fabrication time, and difficult porosity control [125]. However, several modifications of fabrication parameters have been performed and the physicochemical properties of these biomaterials are thereby improved [126].

6.3. Bioactive Glass (BG)

Bioactive glass (BG) was first developed by Hench et al. in 1971, with the $45S_5$ composition through the use of Na_2O-CaO-SiO_2-P_2O_5 phase diagram [61], which demonstrated biocompatibility and bone-bonding ability. These synthetic materials based on silica are highly bioactive, due to the calcium and phosphate ions in their composition [127]. When BGs are exposed to bone or biologic fluids, their structure fully reacts to form internal silica gel cores with calcium phosphate-rich surface. Therefore, the internal silica gel core degrades, leaving an external calcium phosphate bulk, which is structured as a hydroxy-carbonated apatite layer that improves protein adsorption to BGs' surface and integration with surrounding tissue [128,129]. The Ca:P ratio, composition, and microstructure of BGs determines the rate of ion release from the BGs' surface.

Inside the degraded BGs, osteoprogenitor cells differentiate and form new bone. BGs are particularly attractive for bone repair due to their controllable degradation, osteogenic potential, and bone-bonding affinity [130]. It is relevant to mention that BGs degradation rate is highly tunable due to changing their chemical compositions or material processing methods. Therefore, BGs can be designed with a specific degradation rate to respond to the precise requirement of a certain bone repair.

Many variations of BGs are currently being used in periodontics and implantology. They are generally composed of silica (45%), calcium oxide (24.5%), sodium oxide (24.5%), and pyrophosphate (6%), named $45S_5$. Clinically, this composition of BGs has been used in restorative dentistry, periodontics, implantology, and maxillofacial area for periodontal, peri-implant, and bone defects [131,132]. Recently, mesoporous BGs have been developed, which enables greater degradation control [133].

6.4. Zirconia (ZrO$_2$)

Zirconia is a structural ceramic that has been used for biomedical applications due to its biocompatibility, osseointegration potential, radiopacity, favorable mechanical properties, and in particular, its toughness [134–136]. When a crack occurs in zirconia, an internal tension is generated due to its propagation, transforming some grains from tetragonal to monoclinic (t→m), which increases the volume by about 5% [137]. As a result, compressive stress is generated, acting on the crack tip and hindering its propagation [137,138]. This phenomenon of "containment" of the crack is known as "transformation toughening", and since the discovery by Garvie et al. [139], it has been the focus of research for the biomedical application of zirconia.

Due to this favorable behavior, zirconia can supply the mechanical needs of a scaffold, so that it does not deform when submitted to loading and can be used to increase atrophic alveolar ridges or to replace the bone loss in the maxillofacial area [62]. Additionally, zirconia scaffolds can be manufactured by various techniques, resulting in different degrees of porosity, control of the geometric structure, and micro-roughness, which allows a good interconnection structure between the pores to support the growth of osteoblasts, vessels, and new bone [63,140].

However, despite offering superior properties, such as corrosion resistance, low friction coefficient, great wear resistance, hardness, and resistance to fracture propagation, zirconia scaffolds do not have the same efficiency in integration with bone tissue as phosphate-based ceramics [141]. Thus, nanocrystalline calcium phosphate powders, tricalcium phosphates, and/or bioactive glass have been associated with zirconia scaffolds, in the form of coatings or infiltrations, to increase biological activity, healing capacity, and osteogenesis within the adjacent tissue [138,140,142–144]. The current trend of using hybrid scaffolds, through the association of different materials, has been the path that tissue bone engineering has been seeking to obtain artificial structures more similar to bone biology.

7. Techniques for Manufacturing Scaffolds

Due to the several biomedical areas that benefit from tissue bone engineering, the rapid advance in the manufacture of 3D structures has been accompanied by the development and improvement of methods that aim to achieve the desired criteria for a scaffold. Scaffolds can be manufactured by conventional or additive manufacturing techniques and more recently, by 3D and 4D printing techniques [145,146]. Conventional techniques include methods such as solvent casting and particle leaching, freeze-drying, thermally induced phase separation, gas foaming, powder-forming, polymeric sponge replica method, and electrospinning [145,147–150], while among additive manufacturing techniques stereolithography, fused-deposition modeling, selective stand out laser sintering and electron beam melting stand out [145,151].

7.1. Conventional Techniques

Conventional techniques for manufacturing scaffolds use subtraction methods, in which part of the materials is removed so that the desired properties are achieved [152]. Generally, these techniques are easy to made and present low cost; however, these techniques may have limitations, such as the difficulty of obtaining structures with complex geometries [73]. The chemicals in the solvents used may not be completely removed from the scaffolding, being toxic to the newly formed tissue and the surrounding tissue of the host [153]. Table 2 describes the most commonly used conventional techniques and the scaffolds that can be obtained from them.

Table 2. Conventional manufacturing techniques: description and typical scaffold materials.

Technique	Description	Scaffold Materials
Solvent casting and particle leaching [154]	A polymer solution is dissolved in a solvent rich in crystals of soluble salts or organic particles. After removing the solvent by evaporation, these particles come together to form a matrix. The system is immersed in water, allowing the dissolution of the salt matrix and the removal of the produced polymeric structure, which is highly porous. The structures produced are simple but may contain some solvent residue. The centrifugation and layer technique can be combined to minimize these limitations.	- PLGA
Freeze drying [155,156]	The polymeric material is dissolved in a solvent and the solution obtained is cooled below its freezing point taking the solvent to solidification. This system is taken to a freeze dryer, previously adjusted with a temperature below the freezing point of the solvent and a pressure below atmospheric pressure to promote the sublimation of the solvent. The result is the formation of a porous structure, with multiple empty spaces and channels connected.	- Gelatine - HAp - PLA - PCL - Chitosan
Thermally-induced phase separation [148,157]	A polymer is dissolved in a solvent at high temperature, followed by rapid cooling. The solvent is separated from the polymeric structure due to the change in the solubility coefficient caused by the temperature reduction, forming one phase rich in polymer and another poor. The polymeric phase solidifies, while the other phase is removed, resulting in a highly porous polymeric structure. This technique can be used in association with other techniques to manufacture 3D structures with controlled pore morphology, such as leaching.	- PPLA - Chitosan
Gas foaming [158]	Blowing agents are used to pressurize molded polymers. These agents generate gas bubbles that act as porosity builders, causing expansion in volume and reduction in the density of polymers. When associated with the replica technique, the polymeric foam is impregnated with a ceramic suspension. The structure sintered at high temperature, degrades the polymer, resulting in a porous ceramic structure.	- HAp - β-TCP
Powder-forming [116]	A suspension of ceramic particles is prepared in an appropriate liquid to form a paste. From this paste, green bodies are produced in different ways. Subsequent sintering results in porous scaffolds.	- PLGA - HAp
Electrospinning [159,160]	An electric field is used to form fibers with diameters ranging from micrometer to nanometer scale. A typical apparatus consists of an infusion pump, syringe set, and metallic needle for the formation of the spinning droplet, a collector, and the electrical system. The potential difference applied by the electrical system generates high electric fields and its strength exceeds the surface tension of the droplet, elongating it. After evaporation of the solvent, the fibers are collected.	- PLA - β-TCP

7.2. Additive Manufacturing Techniques

Additive manufacturing (AM), commonly known as 3D printing (3DP), includes techniques based on the traditional principles of rapid prototyping, which are used to manufacture a physical object, using three-dimensional computer-aided data (CAD) [161]. They employ additive processes, where the manufacture of three-dimensional physical models is undertaken layer by layer. The production of parts with low volume and with the high complexity of format is facilitated, due to better control of properties such as compressive strength, elastic modulus, dissolution, and mass transport [160,162]. Also, a specific geometry, with a particular shape, size, and porosity (uniform or functionally graded) can be achieved.

AM techniques have the advantage of manufacturing patient-specific designs, which can be obtained from the computed tomography scan of the bone defect. This is particularly important when repairing more complex injuries [163]. These techniques do not use toxic organic solvents and allow better control of pore architecture, pore volume, and percentage porosity, in addition to the mechanical properties of the scaffold. Thus, AM techniques are superior to conventional methods, where it is difficult to control the pore size, shape, and pore interconnectivity [152]. Moreover, AM techniques increasingly allow the manufacture of hybrid scaffolds, combining the advantages of the selected materials [30,48,113,142].

AM or 3DP techniques—such as stereolithography (SL), fused-deposition modeling (FDM), and selective laser sintering (SLS)—combine computer-aided design (CAD), computer-aided manufacturing (CAM), and computer numerical control (CNC) [164]. More recently, additively manufactured structures using smart (intelligent) materials that can modify in a pre-defined form or perform a pre-defined function according to the stimuli are characterized in "4D printing" processes [146]. The techniques used are the same as those mentioned for 3DP; however, the nature of the materials used are different, which must present "shape memory" or "self-performance" [165]. Table 3 describes the most widely used AM techniques and which scaffold materials can be obtained from them.

Table 3. Additive manufacturing techniques: description and typical scaffold materials.

Techniques	Description	Scaffold Materials
Stereolithography (SL) [166,167]	Solid objects are manufactured, layer by layer, by curing a photoreticulable liquid resin of ultraviolet or visible light beams, directed by a dynamic mirror system. A mobile platform moves the cured part. Therefore, another layer can solidify producing a three-dimensional structure.	- PEG - PPF
Fused Deposition Modeling (FDM) [168]	Thermoplastic filaments, consisting of an extruded material or composite, are melted and deposited layerwise on a build platform until the object is formed.	- ABS - PLA - nylon
Selective Laser Sintering (SLS) [169,170]	In this technique, also known as selective laser melting (SLM), the poorly compacted powder is sintered with a high-power laser (e.g., CO_2), particle by particle, uniting them in a controlled manner, forming thin layers. The layers are joined to each other according to predefined computer-aided data (CAD) parameters. The interaction of the laser beam with the powder increases the temperature of the powder above the glass transition temperature and below the melting temperature, causing the melting and bonding of the particles to form a solid mass. The process results in solid or porous structures with superior mechanical properties, custom density, and elastic modulus, and a post-processing phase is required to remove the remaining power.	- PPLA - HPa
Bioprinting [171]	Cells and biomaterials are printed using inkjet, extrusion, or laser-assisted bioprinting techniques with micrometric precision. Jet-based bioprinting produces 2D and 3D structures by applying layers of bio-ink on a substrate. In extrusion-based bioprinting, a mixture of hydrogels is injected by pressure. Afterward, the hydrogels are solidified physically or chemically, and the 3D structures are manufactured by stacking. In laser bioprinting, a receptor material made of glass covered with a layer of gold absorbs the laser, and in this way, a drop is created at high pressure, which in turn transfers materials to the substrate.	- alginate - chitosan - collagen - fibrin

8. Future Studies

The progress of scaffolds for bone formation during the last few decades has been remarkable. As described in this review, scaffolds can be composed with different materials and combinations, as well as, using several manufacturing techniques. Due to the notable developments in biotechnology and manufacturing technologies in the last few years, emergent smart scaffolds have been arising. However, the clinical application of some of such scaffolds needs time. It is necessary to further clarify the interaction between the surface of the scaffolds and tissues and study the degradation process of such materials in different kinds of human bone (trabecular/cortical, different densities in different age groups). Moreover, it is arduous to understand all these biological events in depth, especially taking into account that in some situations the scaffolds will be grafted simultaneously to the dental implant or that, after the grafting procedure, the dental implant will be installed.

9. Conclusions

In summary, conventional and 3D printing manufacturing techniques and associated materials are revolutionizing the development of biomaterials for scaffolds in implant dentistry. Clinical applications include patient-specific implants and prostheses; engineering scaffolding for the regeneration of tissues, and customization of drug-delivering systems. Currently, there are only a limited number of biodegradable materials available for the manufacture of materials and composites, particularly by 3D printing techniques. Therefore, there is a great need for research to manufacture new biomaterials and biocomposites with adjustable properties that can restore functionality at the application site. Low-cost and readily available lactic acid-based polymers (such as PLA and PCL) are focused, mainly due to their ability to work well in most types of 3DP technologies. Also, they have excellent mechanical and biodegradable properties. These polymers can be mixed with ceramic biomaterials (such as HAp, TCP, bioglass) and used as composites to provide greater printability, mechanical stability, and better integration of tissues for dentistry applications.

Author Contributions: M.Ö.: Critically revised and edited the manuscript for important intellectual content. D.H.: Edited and revised the manuscript for important intellectual content. M.C.F.: Drafted the manuscript and revised the manuscript for important intellectual content. A.C.: Contributed to the acquisition and analysis of articles; drafted the manuscript. C.A.M.V.: Contributed to the acquisition and analysis of articles; drafted the manuscript. All authors have read and agreed to the published version of the manuscript.

Funding: This research received no external funding.

Conflicts of Interest: The authors declare no conflict of interest.

References

1. Brånemark, P.I.; Hansson, B.O.; Adell, R.; Breine, U.; Lindström, J.; Hallén, O.; Ohman, A. Osseointegrated implants in the treatment of the edentulous jaw. Experience from a 10-year period. *Scan. J. Plast. Rec. Surg.* **1977**, *16*, 1–132.
2. Adell, R.; Lekholm, U.; Rockler, B.; Brånemark, P.I. A 15-year study of osseointegrated implants in the treatment of the edentulous jaw. *Int. J. Oral Surg.* **1981**, *10*, 387–416.
3. Adell, R.; Eriksson, B.; Lekholm, U.; I Brånemark, P.; Jemt, T. Long-term follow-up study of osseointegrated implants in the treatment of totally edentulous jaws. *Int. J. Oral Maxillofac. Implant.* **1990**, *5*, 347–359.
4. Howe, M.-S.; Keys, W.; Richards, D. Long-term (10-year) dental implant survival: A systematic review and sensitivity meta-analysis. *J. Dent.* **2019**, *84*, 9–21. [CrossRef]
5. Srinivasan, M.; Meyer, S.; Mombelli, A.; Müller, F. Dental implants in the elderly population: A systematic review and meta-analysis. *Clin. Oral Implant. Res.* **2017**, *28*, 920–930. [CrossRef]
6. Van Der Weijden, F.; Dell'Acqua, F.; Slot, D.E. Alveolar bone dimensional changes of post-extraction sockets in humans: A systematic review. *J. Clin. Periodontol.* **2009**, *36*, 1048–1058. [CrossRef]
7. Tallgren, A. The continuing reduction of the residual alveolar ridges in complete denture wearers: A mixed-longitudinal study covering 25 years. *J. Prosthet. Dent.* **2003**, *89*, 427–435. [CrossRef]
8. Araujo, M.G.; Lindhe, J. Dimensional ridge alterations following tooth extraction. An experimental study in the dog. *J. Clin. Periodontol.* **2005**, *32*, 212–218. [CrossRef]
9. Shokri, T.; Wang, W.; Vincent, A.; Cohn, J.E.; Kadakia, S.; Ducic, Y. Osteoradionecrosis of the Maxilla: Conservative Management and Reconstructive Considerations. *Semin. Plast. Surg.* **2020**, *34*, 106–113. [CrossRef]
10. Wu, C.; Pan, W.; Feng, C.; Su, Z.; Duan, Z.; Zheng, Q.; Hua, C.; Li, C. Grafting materials for alveolar cleft reconstruction: A systematic review and best-evidence synthesis. *Int. J. Oral. Maxillofac. Surg.* **2018**, *47*, 345–356. [CrossRef]
11. Chiapasco, M.; Casentini, P.; Zaniboni, M. Bone augmentation procedures in implant dentistry. *Int. J. Oral Maxillofac. Implant.* **2009**, *24*, 237–259.
12. Cardaropoli, D.; Cardaropolli, G. Preservation of the postextraction alveolar ridge: A clinical and histologic study. *Int. J. Periodontics Restor. Dent.* **2008**, *5*, 469–477.
13. Perelman-Karmon, M.; Kozlovsky, A.; Lilov, R.; Artzi, Z. Socket site preservation using bovine bone mineral with and without a bioabsorbable collagen membrane. *Int. J. Periodontics Restor. Dent.* **2012**, *32*, 459–465.
14. Esposito, M.; Grusovin, M.G.; Kwan, S.; Worthington, H.V.; Coulthard, P. Interventions for replacing missing teeth: Bone augmentation techniques for dental implant treatment. *Cochrane Database Syst. Rev.* **2003**, *16*, CD003607. [CrossRef]
15. Simion, M.; Jovanovic, A.S.; Trisi, P.; Scarano, A.; Piattelli, A. Vertical ridge augmentation around dental implants using a mebrane technique and autogenous bone or allografts in humans. *Int. J. Periodontics Restor. Dent.* **1998**, *18*, 8–23.
16. Buser, D.; Dula, K.; Hirt, H.P.; Schenk, R.K. Lateral ridge augmentation using autografts and barrier membranes: A clinical study with 40 partially edentulous patients. *J. Oral Maxillofac. Surg.* **1996**, *54*, 420–432. [CrossRef]
17. Simion, M.; Fontana, F.; Rasperini, G.; Maiorana, C. Long-term evaluation of osseointegrated implants placed in sites augmented with sinus floor elevation associated with vertical ridge augmentation: A retrospective study of 38 consecutive implants with 1- to 7-year follow-up. *Int. J. Periodontics Restor. Dent.* **2004**, *24*, 208–221.
18. Elsalanty, M.E.; Genecov, D.G. Bone Grafts in Craniofacial Surgery. *Craniomaxillofacial Trauma Reconstr.* **2009**, *2*, 125–134. [CrossRef]
19. Sheikh, Z.; Sima, C.; Glogauer, M. Bone Replacement Materials and Techniques Used for Achieving Vertical Alveolar Bone Augmentation. *Materials* **2015**, *8*, 2953–2993. [CrossRef]
20. Tessier, P.; Kawamoto, H.; Matthews, D.; Posnick, J.; Raulo, Y.; Tulasne, J.F.; Wolfe, S.A. Autogenous Bone Grafts and Bone Substitutes—Tools and Techniques: I. A 20,000-Case Experience in Maxillofacial and Craniofacial Surgery. *Plast. Reconstr. Surg.* **2005**, *116*, 6S–24S. [CrossRef]
21. Al Ruhaimi, K.A. Bone graft substitutes: A comparative qualitative histologic review of current osteoconductive grafting materials. *Int. J. Oral Maxillofac. Implant.* **2001**, *16*, 105–114.
22. Acocella, A.; Bertolai, R.; Colafranceschi, M.; Sacco, R. Clinical, histological and histomorphometric evaluation of the healing of mandibular ramus bone block grafts for alveolar ridge augmentation before implant placement. *J. Cranio-Maxillofac. Surg.* **2010**, *38*, 222–230. [CrossRef]

23. Scarano, A.; Degidi, M.; Iezzi, G.; Pecora, G.; Piattelli, M.; Orsini, G.; Caputi, S.; Perrotti, V.; Mangano, C.; Piattelli, A. Maxillary Sinus Augmentation with Different Biomaterials: A Comparative Histologic and Histomorphometric Study in Man. *Implant. Dent.* **2006**, *15*, 197–207. [CrossRef]
24. Keles, G.C.; Sumer, M.; Cetinkaya, B.O.; Tutkun, F.; Simsek, S.B. Effect of Autogenous Cortical Bone Grafting in Conjunction with Guided Tissue Regeneration in the Treatment of Intraosseous Periodontal Defects. *Eur. J. Dent.* **2010**, *4*, 403–411. [CrossRef]
25. Iasella, J.M.; Greenwell, H.; Miller, R.L.; Hill, M.; Drisko, C.; Bohra, A.A.; Scheetz, J.P. Ridge Preservation with Freeze-Dried Bone Allograft and a Collagen Membrane Compared to Extraction Alone for Implant Site Development: A Clinical and Histologic Study in Humans. *J. Periodontol.* **2003**, *74*, 990–999. [CrossRef]
26. Araújo, M.G.; Lindhe, J. Socket grafting with the use of autologous bone: An experimental study in the dog. *Clin. Oral Implant. Res.* **2011**, *22*, 9–13. [CrossRef]
27. Carvalho, P.H.D.A.; Trento, G.D.S.; Moura, L.B.; Cunha, G.; Gabrielli, M.A.C.; Pereira-Filho, V.A. Horizontal ridge augmentation using xenogenous bone graft—Systematic review. *Oral Maxillofac. Surg.* **2019**, *23*, 271–279. [CrossRef]
28. Goulet, J.A.; Senunas, L.E.; De Silva, G.L.; Greenfield, M.L.V.H. Autogenous Iliac Crest Bone Graft: Complications and Functional Assessment. *Clin. Orthop. Relat. Res.* **1997**, *339*, 76–81. [CrossRef]
29. Coquelin, L.; Fialaire-Legendre, A.; Roux, S.; Poignard, A.; Bierling, P.; Hernigou, P.; Chevallier, N.; Rouard, H. In Vivo and In Vitro Comparison of Three Different Allografts Vitalized with Human Mesenchymal Stromal Cells. *Tissue Eng. Part A* **2012**, *18*, 1921–1931. [CrossRef]
30. Wenz, B.; Oesch, B.; Horst, M. Analysis of the risk of transmitting bovine spongiform encephalopathy through bone grafts derived from bovine bone. *Biomaterials* **2001**, *22*, 1599–1606. [CrossRef]
31. Kim, Y.; Nowzari, H.; Rich, S.K. Risk of Prion Disease Transmission through Bovine-Derived Bone Substitutes: A Systematic Review. *Clin. Implant. Dent. Relat. Res.* **2013**, *15*, 645–653. [CrossRef]
32. Chavda, S.; Levin, L. Human Studies of Vertical and Horizontal Alveolar Ridge Augmentation Comparing Different Types of Bone Graft Materials: A Systematic Review. *J. Oral Implant.* **2018**, *44*, 74–84. [CrossRef]
33. Papageorgiou, S.N.; Papageorgiou, P.N.; Deschner, J.; Götz, W. Comparative effectiveness of natural and synthetic bone grafts in oral and maxillofacial surgery prior to insertion of dental implants: Systematic review and network meta-analysis of parallel and cluster randomized controlled trials. *J. Dent.* **2016**, *48*, 1–8. [CrossRef]
34. Shetty, V.; Han, T.J. Alloplastic materials in reconstructive periodontal surgery. *Dent. Clin. N. Am.* **1991**, *35*, 521–530.
35. Samavedi, S.; Whittington, A.R.; Goldstein, A.S. Calcium phosphate ceramics in bone tissue engineering: A review of properties and their influence on cell behavior. *Acta Biomater.* **2013**, *9*, 8037–8045. [CrossRef]
36. Wang, E.; Han, J.; Zhang, X.; Wu, Y.; Deng, X.-L. Efficacy of a mineralized collagen bone-grafting material for peri-implant bone defect reconstruction in mini pigs. *Regen. Biomater.* **2019**, *6*, 107–111. [CrossRef]
37. Teng, F.; Zhang, Q.; Wu, M.; Rachana, S.; Ou, G. Clinical use of ridge-splitting combined with ridge expansion osteotomy, sandwich bone augmentation, and simultaneous implantation. *Br. J. Oral Maxillofac. Surg.* **2014**, *52*, 703–708. [CrossRef]
38. Galarraga-Vinueza, M.; Magini, R.; Henriques, B.; Teughels, W.; Fredel, M.; Hotza, D.; Souza, J.; Boccaccini, A.R. Nanostructured biomaterials embedding bioactive molecules. In *Nanostructured Biomaterials for Cranio-Maxillofacial and Oral Applications*; Elsevier: Amsterdam, The Netherlands, 2018; Volume 1, pp. 143–158.
39. Braddock, M.; Houston, P.; Campbell, C.; Ashcroft, P. Born again bone: Tissue engineering for bone repair. *News Physiol. Sci.* **2001**, *16*, 208–213. [CrossRef]
40. Ikada, Y. Challenges in tissue engineering. *J. R. Soc. Interface* **2006**, *3*, 589–601. [CrossRef]
41. Qu, H.; Fu, H.; Han, Z.; Sun, Y. Biomaterials for bone tissue engineering scaffolds: A review. *RSC Adv.* **2019**, *9*, 26252–26262. [CrossRef]
42. Tutak, W.; Sarkar, S.; Lin-Gibson, S.; Farooque, T.M.; Jyotsnendu, G.; Wang, D.; Kohn, J.; Bolikal, D.; Simon, C.G. The support of bone marrow stromal cell differentiation by airbrushed nanofiber scaffolds. *Biomaterials* **2013**, *34*, 2389–2398. [CrossRef]
43. Ramay, H.R.; Zhang, M. Preparation of porous hydroxyapatite scaffolds by combination of the gel-casting and polymer sponge methods. *Biomaterials* **2003**, *24*, 3293–3302. [CrossRef]
44. Zhao, J.; Xiao, S.; Lu, X.; Wang, J.; Weng, J. A study on improving mechanical properties of porous HA tissue engineering scaffolds by hot isostatic pressing. *Biomed. Mater.* **2006**, *1*, 188–192. [CrossRef]
45. Roseti, L.; Parisi, V.; Petretta, M.; Cavallo, C.; Desando, G.; Bartolotti, I.; Grigolo, B. Scaffolds for bone tissue engineering: State of the art and new perspectives. *Mater. Sci. Eng. C Mater. Biol. Appl.* **2017**, *78*, 1246–1262. [CrossRef]
46. Shuai, C.; Li, Y.; Feng, P.; Guo, W.; Yang, W.; Peng, S. Positive feedback effects of Mg on the hydrolysis of poly-l-lactic acid (PLLA): Promoted degradation of PLLA scaffolds. *Polym. Test.* **2018**, *68*, 27–33. [CrossRef]
47. Saffar, J.-L.; Colombier, M.; Detienville, R. Bone Formation in Tricalcium Phosphate-Filled Periodontal Intrabony Lesions. Histological Observations in Humans. *J. Periodontol.* **1990**, *61*, 209–216. [CrossRef]
48. Schepers, E.; De Clercq, M.; Ducheyne, P.; Kempeneers, R. Bioactive glass particulate material as a filler for bone lesions. *J. Oral Rehabil.* **1991**, *18*, 439–452. [CrossRef]
49. Pereira, H.F.; Cengiz, I.F.; Silva, F.S.; Reis, R.L.; Oliveira, J.M. Scaffolds and coatings for bone regeneration. *J. Mater. Sci. Mater. Med.* **2020**, *31*, 27. [CrossRef]
50. Kumar, P.; Kumar, V.; Kumar, R.; Kumar, R.; Pruncu, C.I. Fabrication and characterization of ZrO_2 incorporated SiO_2–CaO–P_2O_5 bioactive glass scaffolds. *J. Mech. Behav. Biomed. Mater.* **2020**, *109*, 103854. [CrossRef]

51. Middleton, J.C.; Tipton, A.J. Synthetic biodegradable polymers as orthopedic devices. *Biomaterials* **2001**, *21*, 2335–2346. [CrossRef]
52. Erbetta, C.D.C.; Alves, R.J.; Resende, J.M.; Freitas, R.; de Sousa, R.G. Synthesis and Characterization of Poly(D,L-Lactide-co-Glycolide) Copolymer. *J. Biomater. Nanobiotechnol.* **2012**, *3*, 208–225. [CrossRef]
53. Miao, X.; Tan, D.M.; Li, J.; Xiao, Y.; Crawford, R.W. Mechanical and biological properties of hydroxyapatite/tricalcium phosphate scaffolds coated with poly(lactic-co-glycolic acid). *Acta Biomater.* **2008**, *4*, 638–645. [CrossRef]
54. Liu, X.; Okada, M.; Maeda, H.; Fujii, S.; Furuzono, T. Hydroxyapatite/biodegradable poly(l-lactide–co-ε-caprolactone) composite microparticles as injectable scaffolds by a Pickering emulsion route. *Acta Biomater.* **2011**, *7*, 821–828. [CrossRef]
55. Chandra, R. Biodegradable polymers. *Prog. Polym. Sci.* **1998**, *23*, 1273–1335. [CrossRef]
56. Dibbern-Brunelli, D.; Atvars, T.D.Z.; Joekes, I.; Barbosa, V.C. Mapping phases of poly(vinyl alcohol) and poly(vinyl acetate) blends by FTIR microspectroscopy and optical fluorescence microscopy. *J. Appl. Polym. Sci.* **1998**, *69*, 645–655. [CrossRef]
57. Yaszemski, M.J.; Payne, R.G.; Hayes, W.C.; Langer, R.; Mikos, A.G. In vitro degradation of a poly(propylene fumarate)-based composite material. *Biomaterials* **1996**, *17*, 2127–2130. [CrossRef]
58. Jeong, J.; Kim, J.H.; Shim, J.H.; Hwang, N.S.; Heo, C.Y. Bioactive calcium phosphate materials and applications in bone regeneration. *Biomater. Res.* **2019**, *23*, 4. [CrossRef]
59. De Lima, I.R.; Alves, G.G.; Fernandes, G.V.D.O.; Dias, E.P.; Soares, G.D.A.; Granjeiro, J.M. Evaluation of the in vivo biocompatibility of hydroxyapatite granules incorporated with zinc ions. *Mater. Res.* **2010**, *13*, 563–568. [CrossRef]
60. Giannoudis, P.V.; Dinopoulos, H.; Tsiridis, E. Bone substitutes: An update. *Injury* **2005**, *36*, S20–S27. [CrossRef]
61. Rahaman, M.N.; Day, D.E.; Bal, B.S.; Fu, Q.; Jung, S.B.; Bonewald, L.F.; Tomsia, A.P. Bioactive glass in tissue engineering. *Acta Biomater.* **2011**, *7*, 2355–2373. [CrossRef]
62. Aboushelib, M.N.; Shawky, R. Osteogenesis ability of CAD/CAM porous zirconia scaffolds enriched with nano-hydroxyapatite particles. *Int. J. Implant. Dent.* **2017**, *3*, 21. [CrossRef]
63. Alizadeh, A.; Moztarzadeh, F.; Ostad, S.N.; Azami, M.; Geramizadeh, B.; Hatam, G.; Bizari, D.; Tavangar, S.M.; Vasei, M.; Ai, J. Synthesis of calcium phosphate-zirconia scaffold and human endometrial adult stem cells for bone tissue engineering. *Artif. Cells Nanomed. Biotechnol.* **2014**, *44*, 66–73. [CrossRef]
64. El-Rashidy, A.A.; Roether, J.A.; Harhaus, L.; Kneser, U.; Boccaccini, A.R. Regenerating bone with bioactive glass scaffolds: A review of in vivo studies in bone defect models. *Acta Biomater.* **2017**, *62*, 1–28. [CrossRef]
65. Que, R.A.; Chan, S.W.P.; Jabaiah, A.M.; Lathrop, R.H.; Da Silva, N.A.; Wang, S.-W. Tuning cellular response by modular design of bioactive domains in collagen. *Biomaterials* **2015**, *53*, 309–317. [CrossRef]
66. Cicciù, M.; Scott, A.; Cicciù, D.; Tandon, R.; Maiorana, C. Recombinant Human Bone Morphogenetic Protein-2 Promote and Stabilize Hard and Soft Tissue Healing for Large Mandibular New Bone Reconstruction Defects. *J. Craniofacial Surg.* **2014**, *25*, 860–862. [CrossRef]
67. Okamoto, M.; John, B. Synthetic biopolymer nanocomposites for tissue engineering scaffolds. *Prog. Polym. Sci.* **2013**, *38*, 1487–1503. [CrossRef]
68. Liu, X.; Ma, P.X. Polymeric Scaffolds for Bone Tissue Engineering. *Ann. Biomed. Eng.* **2004**, *32*, 477–486. [CrossRef]
69. Hollister, S.J. Scaffold engineering: A bridge to where? *Biofabrication* **2009**, *1*, 012001. [CrossRef]
70. Anselme, K. Osteoblast adhesion on biomaterials. *Biomaterials* **2000**, *21*, 667–681. [CrossRef]
71. Costa-Pinto, A.R.; Reis, R.L.; Neves, N.M. Scaffolds Based Bone Tissue Engineering: The Role of Chitosan. *Tissue Eng. Part B Rev.* **2011**, *17*, 331–347. [CrossRef]
72. Murphy, C.M.; Haugh, M.G.; O'Brien, F.J. The effect of mean pore size on cell attachment, proliferation and migration in collagen–glycosaminoglycan scaffolds for bone tissue engineering. *Biomaterials* **2010**, *31*, 461–466. [CrossRef]
73. Karageorgiou, V.; Kaplan, D.L. Porosity of 3D biomaterial scaffolds and osteogenesis. *Biomaterials* **2005**, *26*, 5474–5491. [CrossRef]
74. Thein-Han, W.; Misra, R. Biomimetic chitosan–nanohydroxyapatite composite scaffolds for bone tissue engineering. *Acta Biomater.* **2009**, *5*, 1182–1197. [CrossRef]
75. Akay, G.; Birch, M.; Bokhari, M. Microcellular polyHIPE polymer supports osteoblast growth and bone formation in vitro. *Biomaterials* **2004**, *25*, 3991–4000. [CrossRef]
76. Rouwkema, J.; Rivron, N.C.; van Blitterswijk, C.A. Vascularization in tissue engineering. *Trends Biotechnol.* **2008**, *26*, 434–441. [CrossRef]
77. Prananingrum, W.; Naito, Y.; Galli, S.; Bae, J.; Sekine, K.; Hamada, K.; Tomotake, Y.; Wennerberg, A.; Jimbo, R.; Ichikawa, T. Bone ingrowth of various porous titanium scaffolds produced by a moldless and space holder technique: An in vivo study in rabbits. *Biomed. Mater.* **2016**, *11*, 015012. [CrossRef]
78. Yang, S.; Leong, K.-F.; Du, Z.; Chua, C.-K. The Design of Scaffolds for Use in Tissue Engineering. Part I. Traditional Factors. *Tissue Eng.* **2001**, *7*, 679–689. [CrossRef]
79. Xia, Y.; Sun, J.; Zhao, L.; Zhang, F.; Liang, X.-J.; Guo, Y.; Weir, M.D.; Reynolds, M.A.; Gu, N.; Xu, H.H. Magnetic field and nano-scaffolds with stem cells to enhance bone regeneration. *Biomaterials* **2018**, *183*, 151–170. [CrossRef]
80. Funda, G.; Taschieri, S.; Bruno, G.A.; Grecchi, E.; Paolo, S.; Girolamo, D.; Del Fabbro, M. Nanotechnology Scaffolds for Alveolar Bone Regeneration. *Materials* **2020**, *13*, 201. [CrossRef]
81. Woodard, J.R.; Hilldore, A.J.; Lan, S.K.; Park, C.; Morgan, A.W.; Eurell, J.A.C.; Clark, S.G.; Wheeler, M.B.; Jamison, R.D.; Johnson, A.J.W. The mechanical properties and osteoconductivity of hydroxyapatite bone scaffolds with multi-scale porosity. *Biomaterials* **2007**, *28*, 45–54. [CrossRef]

82. Bose, S.; Roy, M.; Bandyopadhyay, A. Recent advances in bone tissue engineering scaffolds. *Trends Biotechnol.* **2012**, *30*, 546–554. [CrossRef]

83. Thomson, R.C.; Yaszemski, M.J.; Powers, J.M.; Mikos, A.G. Fabrication of biodegradable polymer scaffolds to engineer trabecular bone. *J. Biomater. Sci. Polym. Ed.* **1995**, *7*, 23–38. [CrossRef]

84. Davison, N.L.; Groot, F.B.-D.; Grijpma, D.W. Degradation of Biomaterials. *Tissue Eng.* **2015**, 177–215. [CrossRef]

85. Wang, L.; Wu, S.; Cao, G.; Fan, Y.; Dunne, N.; Li, X. Biomechanical studies on biomaterial degradation and co-cultured cells: Mechanisms, potential applications, challenges and prospects. *J. Mater. Chem. B* **2019**, *7*, 7439–7459. [CrossRef]

86. Gunatillake, P.A.; Adhikari, R. Biodegradable synthetic polymers for tissue engineering. *Eur. Cells Mater.* **2003**, *5*, 1–16. [CrossRef]

87. Dhandayuthapani, B.; Yoshida, Y.; Maekawa, T.; Kumar, D.S. Polymeric Scaffolds in Tissue Engineering Application: A Review. *Int. J. Polym. Sci.* **2011**, *2011*, 1–19. [CrossRef]

88. Carrasco, F.; Pagès, P.; Gámez-Pérez, J.; Santana, O.; Maspoch, M. Processing of poly(lactic acid): Characterization of chemical structure, thermal stability and mechanical properties. *Polym. Degrad. Stab.* **2010**, *95*, 116–125. [CrossRef]

89. Park, K.; Xanthos, M. A study on the degradation of polylactic acid in the presence of phosphonium ionic liquids. *Polym. Degrad. Stab.* **2009**, *94*, 834–844. [CrossRef]

90. Carothers, W.H.; Dorough, G.L.; Van Natta, F.J. Studies of polymerization and ring formation. X. the reversible polymerization of six-membered cyclic esters. *J. Am. Chem. Soc.* **1932**, *54*, 761–772. [CrossRef]

91. Puelacher, W.; Vacanti, J.; Ferraro, N.; Schloo, B.; Vacanti, C. Femoral shaft reconstruction using tissue-engineered growth of bone. *Int. J. Oral Maxillofac. Surg.* **1996**, *25*, 223–228. [CrossRef]

92. Mooney, D.J.; Mazzoni, C.L.; Breuer, C.; McNamara, K.; Hern, D.; Vacanti, J.P.; Langer, R. Stabilized polyglycolic acid fibre-based tubes for tissue engineering. *Biomaterials* **1996**, *17*, 115–124. [CrossRef]

93. Amiryaghoubi, N.; Fathi, M.; Pesyan, N.N.; Samiei, M.; Barar, J.; Omidi, Y. Bioactive polymeric scaffolds for osteogenic repair and bone regenerative medicine. *Med. Res. Rev.* **2020**, *40*, 1833–1870. [CrossRef]

94. Mano, J.F.; A Sousa, R.; Boesel, L.F.; Neves, N.M.; Reis, R.L. Bioinert, biodegradable and injectable polymeric matrix composites for hard tissue replacement: State of the art and recent developments. *Compos. Sci. Technol.* **2004**, *64*, 789–817. [CrossRef]

95. Astete, C.E.; Sabliov, C.M. Synthesis and characterization of PLGA nanoparticles. *J. Biomater. Sci.* **2006**, *17*, 247–289. [CrossRef]

96. Sordi, M.B.; Da Cruz, A.C.C.; Aragones, Á.; Cordeiro, M.M.R.; Magini, R.D.S. PLGA+HA/βTCP scaffold incorporating simvastatin: A promising biomaterial for bone tissue engineering. *J. Oral Implant.* **2020**. [CrossRef]

97. Shen, H.; Hu, X.; Yang, F.; Bei, J.; Wang, S. Cell affinity for bFGF immobilized heparin-containing poly(lactide-co-glycolide) scaffolds. *Biomaterials* **2011**, *32*, 3404–3412. [CrossRef]

98. Labet, M.; Thielemans, W. Synthesis of polycaprolactone: A review. *Chem. Soc. Rev.* **2009**, *38*, 3484–3504. [CrossRef]

99. Fu, S.; Ni, P.; Wang, B.; Chu, B.; Peng, J.; Zheng, L.; Zhao, X.; Luo, F.; Wei, Y.; Qian, Z. In vivo biocompatibility and osteogenesis of electrospun poly(ε-caprolactone)–poly(ethylene glycol)–poly(ε-caprolactone)/nano-hydroxyapatite composite scaffold. *Biomaterials* **2012**, *33*, 8363–8371. [CrossRef]

100. Supaphol, P.; Chuangchote, S. On the electrospinning of poly(vinyl alcohol) nanofiber mats: A revisit. *J. Appl. Polym. Sci.* **2008**, *108*, 969–978. [CrossRef]

101. Ding, B.; Kim, H.-Y.; Lee, S.-C.; Shao, C.-L.; Lee, D.-R.; Park, S.-J.; Kwag, G.-B.; Choi, K.-J. Preparation and characterization of a nanoscale PVA fibers aggregate produced by electro spinning method. *J. Polym. Sci. Part B Polym. Phys.* **2002**, *40*, 1261–1268. [CrossRef]

102. Jeong, B.-H.; Hoek, E.M.; Yan, Y.; Subramani, A.; Huang, X.; Hurwitz, G.; Ghosh, A.K.; Jawor, A. Interfacial polymerization of thin film nanocomposites: A new concept for reverse osmosis membranes. *Compos. Sci. Technol.* **2007**, *294*, 1–7. [CrossRef]

103. Wong, K.K.H.; Hutter, J.L.; Zinke-Allmang, M.; Wan, W. Physical properties of ion beam treated electrospun poly(vinyl alcohol) nanofibers. *Eur. Polym. J.* **2009**, *45*, 1349–1358. [CrossRef]

104. Dom, A.J.; Manor, N.; Elmalak, O. Biodegradable bone cement compositions based on acrylate and epoxide terminated poly(propylene fumarate) oligomers and calcium salt compositions. *Biomaterials* **1996**, *17*, 411–417. [CrossRef]

105. He, S.; Timmer, M.; Yaszemski, M.; Yasko, A.; Engel, P.; Mikos, A. Synthesis of biodegradable poly(propylene fumarate) networks with poly(propylene fumarate)–diacrylate macromers as crosslinking agents and characterization of their degradation products. *Polymer* **2001**, *42*, 1251–1260. [CrossRef]

106. Timmer, M.D.; Shin, H.; Horch, R.A.; Ambrose, C.G.; Mikos, A.G. In Vitro Cytotoxicity of Injectable and Biodegradable Poly(propylene fumarate)-Based Networks: Unreacted Macromers, Cross-Linked Networks, and Degradation Products. *Biomacromolecules* **2003**, *4*, 1026–1033. [CrossRef]

107. Walker, J.M.; Bodamer, E.; Krebs, O.; Luo, Y.; Kleinfehn, A.; Becker, M.L.; Dean, D. Effect of Chemical and Physical Properties on the In Vitro Degradation of 3D Printed High Resolution Poly(propylene fumarate) Scaffolds. *Biomacromolecules* **2017**, *18*, 1419–1425. [CrossRef]

108. Vallet-Regi, M.; González-Calbet, J.M. Calcium phosphates as substitution of bone tissues. *Prog. Solid State Chem.* **2004**, *32*, 1–31. [CrossRef]

109. Tuukkanen, J.; Nakamura, M. Hydroxyapatite as a Nanomaterial for Advanced Tissue Engineering and Drug Therapy. *Curr. Pharm. Des.* **2017**, *23*, 3786–3793. [CrossRef]

110. Daculsi, G.; LeGeros, R. Biphasic calcium phosphate (BCP) bioceramics: Chemical, physical and biological properties. In *Encyclopedia of Biomaterials and Biomedical Engineering*; CRC Press: Boca Raton, FL, USA, 2006; pp. 1–9.

111. Matassi, F.; Nistri, L.; Paez, D.C.; Innocenti, M. New biomaterials for bone regeneration. *Clin. Cases Miner. Bone Metab.* **2011**, *8*, 21–24.

112. Leong, K.; Cheah, C.; Chua, C. Solid freeform fabrication of three-dimensional scaffolds for engineering replacement tissues and organs. *Biomaterial* **2003**, *24*, 2363–2378. [CrossRef]

113. Gaihre, B.; Jayasuriya, A.C. Comparative investigation of porous nano-hydroxyapaptite/chitosan, nano-zirconia/chitosan and novel nano-calcium zirconate/chitosan composite scaffolds for their potential applications in bone regeneration. *Mater. Sci. Eng. C Mater. Biol. Appl.* **2018**, *91*, 330–339. [CrossRef]

114. Best, S.M.; Porter, A.E.; Thian, E.S.; Huang, J. Bioceramics: Past, present and for the future. *J. Eur. Ceram. Soc.* **2008**, *28*, 1319–1327. [CrossRef]

115. LeGeros, R.Z. Properties of Osteoconductive Biomaterials: Calcium Phosphates. *Clin. Orthop. Relat. Res.* **2002**, *395*, 81–98. [CrossRef]

116. Rezwan, K.; Chen, Q.Z.; Blaker, J.J.; Boccaccini, A.R. Biodegradable and bioactive porous polymer/inorganic composite scaffolds for bone tissue engineering. *Biomaterials* **2006**, *27*, 3413–3431. [CrossRef]

117. Abukawa, H.; Papadaki, M.; Abulikemu, M.; Leaf, J.; Vacanti, J.P.; Kaban, L.B.; Troulis, M.J. The Engineering of Craniofacial Tissues in the Laboratory: A Review of Biomaterials for Scaffolds and Implant Coatings. *Dent. Clin. N. Am.* **2006**, *50*, 205–216. [CrossRef]

118. Bohner, M. Calcium orthophosphates in medicine: From ceramics to calcium phosphate cements. *Injury* **2000**, *31*, D37–D47. [CrossRef]

119. Dorozhkin, S.V. Biphasic, triphasic and multiphasic calcium orthophosphates. *Acta Biomater.* **2012**, *8*, 963–977. [CrossRef]

120. Montazeri, L.; Javadpour, J.; Shokrgozar, M.A.; Bonakdar, S.; Javadian, S. Hydrothermal synthesis and characterization of hydroxyapatite and fluorhydroxyapatite nano-size powders. *Biomed. Mater.* **2010**, *5*, 045004. [CrossRef]

121. Vallet-Regí, M. Ceramics for medical applications. *J. Chem. Soc. Dalton Trans.* **2001**, 97–108. [CrossRef]

122. Kroese-Deutman, H.C.; Ruhé, P.; Spauwen, P.H.; Jansen, J.A. Bone inductive properties of rhBMP-2 loaded porous calcium phosphate cement implants inserted at an ectopic site in rabbits. *Biomaterials* **2005**, *26*, 1131–1138. [CrossRef]

123. Boden, S.D. Bioactive Factors for Bone Tissue Engineering. *Clin. Orthop. Relat. Res.* **1999**, *367*, S84–S94. [CrossRef]

124. Yuan, H.; Zou, P.; Yang, Z.; Zhang, X.; De Bruijn, J.D.; De Groot, K. Bone morphogenetic protein and ceramic-induced osteogenesis. *J. Mater. Sci. Mater. Med.* **1998**, *9*, 717–721. [CrossRef]

125. Dorozhkin, S.V. Bioceramics of calcium orthophosphates. *Biomaterials* **2010**, *31*, 1465–1485. [CrossRef]

126. Pawar, A.M.; Sawant, K. Bioactive glass in dentistry: A systematic review. *Saudi J. Oral Sci.* **2020**, *7*, 3. [CrossRef]

127. Da Cruz, A.C.C.; Pochapski, M.T.; Tramonti, R.; da Silva, J.C.Z.; Antunes, A.C.; Pilatti, G.L.; Santos, F.A. Evaluation of physical–chemical properties and biocompatibility of a microrough and smooth bioactive glass particles. *J. Mater. Sci. Mater. Med.* **2008**, *19*, 2809–2817. [CrossRef]

128. Vogel, M.; Voigt, C.; Gross, U.M.; Müller-Mai, C.M. In vivo comparison of bioactive glass particles in rabbits. *Biomaterials* **2001**, *22*, 357–362. [CrossRef]

129. Bosetti, M.; Hench, L.; Cannas, M. Interaction of bioactive glasses with peritoneal macrophages and monocytesin vitro. *J. Biomed. Mater. Res.* **2002**, *60*, 79–85. [CrossRef]

130. Wilson, T.; Parikka, V.; Holmbom, J.; Ylänen, H.; Penttinen, R. Intact surface of bioactive glass S53P4 is resistant to osteoclastic activity. *J. Biomed. Mater. Res. Part A* **2006**, *77*, 67–74. [CrossRef]

131. Skallevold, H.E.; Rokaya, D.; Khurshid, Z.; Zafar, M.S. Bioactive Glass Applications in Dentistry. *Int. J. Mol. Sci.* **2019**, *20*, 5960. [CrossRef]

132. Mesquita-Guimarães, J.; Leite, M.A.; Souza, J.C.M.; Henriques, B.; Silva, F.S.; Hotza, D.; Boccaccini, A.R.; Fredel, M.C. Processing and strengthening of 58S bioactive glass-infiltrated titania scaffolds. *J. Biomed. Mater. Res. Part A* **2017**, *105*, 590–600. [CrossRef]

133. Galarraga-Vinueza, M.E.; Mesquita-Guimarães, J.; Magini, R.S.; Souza, J.C.M.; Fredel, M.C.; Boccaccini, A.R. Mesoporous bioactive glass embedding propolis and cranberry antibiofilm compounds. *J. Biomed. Mater. Res. Part A* **2018**, *106*, 1614–1625. [CrossRef]

134. Chevalier, J.; Gremillard, L. Ceramics for medical applications: A picture for the next 20 years. *J. Eur. Ceram. Soc.* **2009**, *29*, 1245–1255. [CrossRef]

135. Kelly, J.R.; Denry, I. Stabilized zirconia as a structural ceramic: An overview. *Dent. Mater.* **2008**, *24*, 289–298. [CrossRef]

136. De Bortoli, L.S.; Schabbach, L.M.; Fredel, M.C.; Hotza, D.; Henriques, B. Ecological footprint of biomaterials for implant dentistry: Is the metal-free practice an eco-friendly shift? *J. Clean. Prod.* **2018**, *213*, 723–732. [CrossRef]

137. Kelly, P.M.; Rose, L.F. The martensitic transformation in ceramics—Its role in transformation toughening. *Prog. Mater. Sci.* **2002**, *47*, 463–557. [CrossRef]

138. Piconi, C.; Maccauro, G. Zirconia as a ceramic biomaterial. *Biomaterials* **1999**, *20*, 1–25. [CrossRef]

139. Garvie, R.C.; Hannink, R.H.; Pascoe, R.T. Ceramic steel? *Nature* **1975**, *258*, 703–704.

140. Mondal, D.; So-Ra, S.; Sarkar, S.K.; Min, Y.K.; Yang, H.M.; Lee, B.T. Fabrication of multilayer ZrO_2–biphasic calcium phosphate–poly-caprolactone unidirectional channeled scaffold for bone tissue formation. *J. Biomater. Appl.* **2012**, *28*, 462–472. [CrossRef]

141. Malmström, J.; Slotte, C.; Adolfsson, E.; Norderyd, O.; Thomsen, P. Bone response to free form-fabricated hydroxyapatite and zirconia scaffolds: A histological study in the human maxilla. *Clin. Oral Implant. Res.* **2009**, *20*, 379–385. [CrossRef]

142. An, S.-H.; Matsumoto, T.; Miyajima, H.; Nakahira, A.; Kim, K.-H.; Imazato, S. Porous zirconia/hydroxyapatite scaffolds for bone reconstruction. *Dent. Mater.* **2012**, *28*, 1221–1231. [CrossRef]

143. Pattnaik, S.; Nethala, S.; Tripathi, A.; Saravanan, S.; Moorthi, A.; Selvamurugan, N. Chitosan scaffolds containing silicon dioxide and zirconia nano particles for bone tissue engineering. *Int. J. Biol. Macromol.* **2011**, *49*, 1167–1172. [CrossRef]

144. Mesquita-Guimarães, J.; Ramos, L.; Detsch, R.; Henriques, B.; Fredel, M.; Silva, F.; Boccaccini, A. Evaluation of in vitro properties of 3D micro-macro porous zirconia scaffolds coated with 58S bioactive glass using MG-63 osteoblast-like cells. *J. Eur. Ceram. Soc.* **2019**, *39*, 2545–2558. [CrossRef]

145. Kumar, A.; Nune, K.C.; Murr, L.E.; Misra, R.D.K. Biocompatibility and mechanical behaviour of three-dimensional scaffolds for biomedical devices: Process–structure–property paradigm. *Int. Mater. Rev.* **2016**, *61*, 20–45. [CrossRef]

146. Tamay, D.G.; Usal, T.D.; Alagoz, A.S.; Yucel, D.; Hasirci, N.; Hasirci, V. 3D and 4D Printing of Polymers for Tissue Engineering Applications. *Front. Bioeng. Biotechnol.* **2009**, *7*, 164. [CrossRef]

147. Castro, V.O.; Fredel, M.C.; Aragones, Á.; Barra, G.M.D.O.; Cesca, K.; Merlini, C. Electrospun fibrous membranes of poly (lactic-co-glycolic acid) with β-tricalcium phosphate for guided bone regeneration application. *Polym. Test.* **2020**, *86*, 106489. [CrossRef]

148. Dos Santos, V.I.; Merlini, C.; Aragones, Á.; Cesca, K.; Fredel, M.C. In vitro evaluation of bilayer membranes of PLGA/hydroxyapatite/β-tricalcium phosphate for guided bone regeneration. *Mater. Sci. Eng. C* **2020**, *112*, 110849. [CrossRef]

149. Kumar, P.; Dehiya, B.S.; Sindhu, A.; Kumar, R.; Pruncu, C.I.; Yadav, A. Fabrication and characterization of silver nanorods incorporated calcium silicate scaffold using polymeric sponge replica technique. *Mater. Des.* **2020**, *195*, 109026. [CrossRef]

150. Baino, F.; Fiume, E.; Barberi, J.; Kargozar, S.; Marchi, J.; Massera, J.; Verné, E. Processing methods for making porous bioactive glass-based scaffolds—A state-of-the-art review. *Int. J. Appl. Ceram. Technol.* **2019**, *16*, 1762–1796. [CrossRef]

151. Carlier, A.; Skvortsov, G.A.; Hafezi, F.; Ferraris, E.; Patterson, J.; Koc, B.; Van Oosterwyck, H. Computational model-informed design and bioprinting of cell-patterned constructs for bone tissue engineering. *Biofabrication* **2016**, *8*, 025009. [CrossRef]

152. Cheung, H.-Y.; Lau, K.-T.; Lu, T.-P.; Hui, D. A critical review on polymer-based bio-engineered materials for scaffold development. *Compos. Part B Eng.* **2007**, *38*, 291–300. [CrossRef]

153. Coelho, P.G.; Hollister, S.J.; Flanagan, C.L.; Fernandes, P.R. Bioresorbable scaffolds for bone tissue engineering: Optimal design, fabrication, mechanical testing and scale-size effects analysis. *Med. Eng. Phys.* **2015**, *37*, 287–296. [CrossRef]

154. Prasad, A.; Sankar, M.; Katiyar, V. State of Art on Solvent Casting Particulate Leaching Method for Orthopedic ScaffoldsFabrication. *Mater. Today Proc.* **2017**, *4*, 898–907. [CrossRef]

155. Brougham, C.M.; Levingstone, T.J.; Shen, N.; Cooney, G.M.; Jockenhoevel, S.; Flanagan, T.C.; O'Brien, F.J. Freeze-Drying as a Novel Biofabrication Method for Achieving a Controlled Microarchitecture within Large, Complex Natural Biomaterial Scaffolds. *Adv. Healthc. Mater.* **2017**, *6*. [CrossRef]

156. Kumar, P.; Saini, M.; Dehiya, B.S.; Umar, A.; Sindhu, A.; Mohammed, H.; Al-Hadeethi, Y.; Guo, Z. Fabrication and in-vitro biocompatibility of freeze-dried CTS-nHA and CTS-nBG scaffolds for bone regeneration applications. *Int. J. Biol. Macromol.* **2020**, *149*, 1–10. [CrossRef]

157. Conoscenti, G.; Schneider, T.; Stoelzel, K.; Pavia, F.C.; Brucato, V.; Goegele, C.; La Carrubba, V.; Schulze-Tanzil, G. PLLA scaffolds produced by thermally induced phase separation (TIPS) allow human chondrocyte growth and extracellular matrix formation dependent on pore size. *Mater. Sci. Eng. C Mater. Biol. Appl.* **2017**, *80*, 449–459. [CrossRef]

158. Moghadam, M.Z.; Hassanajili, S.; Esmaeilzadeh, F.; Ayatollahi, M.; Ahmadi, M. Formation of porous HPCL/LPCL/HA scaffolds with supercritical CO_2 gas foaming method. *J. Mech. Behav. Biomed. Mater.* **2017**, *69*, 115–127. [CrossRef]

159. Jiang, L.; Wang, L.; Wang, N.; Gong, S.; Wang, L.; Li, Q.; Shen, C.; Turng, L.-S. Fabrication of polycaprolactone electrospun fibers with different hierarchical structures mimicking collagen fibrils for tissue engineering scaffolds. *Appl. Surf. Sci.* **2018**, *427*, 311–325. [CrossRef]

160. Lin, W.; Chen, M.; Qu, T.; Li, J.; Man, Y. Three-dimensional electrospun nanofibrous scaffolds for bone tissue engineering. *J. Biomed. Mater. Res. Part B Appl. Biomater.* **2020**, *108*, 1311–1321. [CrossRef]

161. Bhushan, B.; Caspers, M. An overview of additive manufacturing (3D printing) for microfabrication. *Microsyst. Technol.* **2017**, *23*, 1117–1124. [CrossRef]

162. Bendtsen, S.T.; Quinnell, S.P.; Wei, M. Development of a novel alginate-polyvinyl alcohol-hydroxyapatite hydrogel for 3D bioprinting bone tissue engineered scaffolds. *J. Biomed. Mater. Res. Part A* **2017**, *105*, 1457–1468. [CrossRef]

163. Sachlos, E.; Czernuszka, J. Making tissue engineering scaffolds work. Review on the application of solid freeform fabrication technology to the production of tissue engineering scaffolds. *Eur. Cells Mater.* **2003**, *5*, 29–39.

164. Bose, S.; Ke, D.; Sahasrabudhe, H.; Bandyopadhyay, A. Additive manufacturing of biomaterials. *Prog. Mater. Sci.* **2018**, *93*, 45–111. [CrossRef]

165. Rychter, P.; Pamula, E.; Orchel, A.; Posadowska, U.; Krok-Borkowicz, M.; Kaps, A.; Smigiel-Gac, N.; Smola, A.; Kasperczyk, J.; Prochwicz, W.; et al. Scaffolds with shape memory behavior for the treatment of large bone defects. *J. Biomed. Mater. Res. Part A* **2015**, *103*, 3503–3515. [CrossRef]

166. Schmidleithner, C.; Malferrari, S.; Palgrave, R.G.; Bomze, D.; Schwentenwein, M.; Kalaskar, D.M. Application of high resolution DLP stereolithography for fabrication of tricalcium phosphate scaffolds for bone regeneration. *Biomed. Mater.* **2019**, *14*, 045018. [CrossRef]

167. Madrid, A.P.M.; Vrech, S.M.; Sanchez, M.A.; Rodriguez, A.P. Advances in additive manufacturing for bone tissue engineering scaffolds. *Mater. Sci. Eng. C Mater. Biol. Appl.* **2019**, *100*, 631–644. [CrossRef]
168. Distler, T.; Fournier, N.; Grünewald, A.; Polley, C.; Seitz, H.; Detsch, R.; Boccaccini, A.R. Polymer-Bioactive Glass Composite Filaments for 3D Scaffold Manufacturing by Fused Deposition Modeling: Fabrication and Characterization. *Front. Bioeng. Biotechnol.* **2020**, *8*, 552. [CrossRef]
169. Mazzoli, A. Selective laser sintering in biomedical engineering. *Med. Biol. Eng. Comput.* **2013**, *51*, 245–256. [CrossRef]
170. Qin, T.; Li, X.; Long, H.; Bin, S.; Xu, Y. Bioactive Tetracalcium Phosphate Scaffolds Fabricated by Selective Laser Sintering for Bone Regeneration Applications. *Materials* **2020**, *13*, 2268. [CrossRef]
171. Ashammakhi, N.; Hasan, A.; Kaarela, O.; Byambaa, B.; Sheikhi, A.; Gaharwar, A.K.; Khademhosseini, A. Advancing Frontiers in Bone Bioprinting. *Adv. Healthc. Mater.* **2019**, *8*, e1801048. [CrossRef]

MDPI

St. Alban-Anlage 66

4052 Basel

Switzerland

Tel. +41 61 683 77 34

Fax +41 61 302 89 18

www.mdpi.com

Journal of Composites Science Editorial Office

E-mail: jcs@mdpi.com

www.mdpi.com/journal/jcs

www.ingramcontent.com/pod-product-compliance
Lightning Source LLC
LaVergne TN
LVHW071258170726
843515LV00006BA/1429